Interdisciplinary Evolution Research

Volume 1

Series editors

Nathalie Gontier, Lisbon, Portugal
Olga Pombo, Lisbon, Portugal

For further volumes:
http://www.springer.com/series/13109

About the Series

The time when only biologists studied evolution has long since passed. Accepting evolution requires us to come to terms with the fact that everything that exists must be the outcome of evolutionary processes. Today, a wide variety of academic disciplines are therefore confronted with evolutionary problems, ranging from physics and medicine, to linguistics, anthropology and sociology. Solving evolutionary problems also necessitates an inter- and transdisciplinary approach, which is why the Modern Synthesis is currently extended to include drift theory, symbiogenesis, lateral gene transfer, hybridization, epigenetics and punctuated equilibria theory.

The series Interdisciplinary Evolution Research aims to provide a scholarly platform for the growing demand to examine specific evolutionary problems from the perspectives of multiple disciplines. It does not adhere to one specific academic field, one specific school of thought, or one specific evolutionary theory. Rather, books in the series thematically analyze how a variety of evolutionary fields and evolutionary theories provide insights into specific, well-defined evolutionary problems of life and the socio-cultural domain.

Editors-in-chief of the series are Nathalie Gontier and Olga Pombo. The Series is edited from within the Applied Evolutionary Epistemology Lab, more information on the lab is available at http://appeel.fc.ul.pt.

Marco Pina · Nathalie Gontier
Editors

The Evolution of Social Communication in Primates

A Multidisciplinary Approach

Springer

Editors
Marco Pina
Nathalie Gontier
AppEEL, Centre for Philosophy of Science
University of Lisbon
Lisbon
Portugal

ISSN 2199-3068 ISSN 2199-3076 (electronic)
ISBN 978-3-319-02668-8 ISBN 978-3-319-02669-5 (eBook)
DOI 10.1007/978-3-319-02669-5
Springer Cham Heidelberg New York Dordrecht London

Library of Congress Control Number: 2014940710

Printed on acid-free paper

Springer is part of Springer Science+Business Media (www.springer.com)

Preface and Acknowledgments

Writing a preface is always accompanied by a celebration of seeing many hours of hard labor coming to a successful end. The volume that lies before you, *The Evolution of Social Communication in Primates: A Multidisciplinary Approach*, is both an expansion as well as an outgrowth of the 2012 International Conference *From Grooming to Speaking: Recent Trends in Social Primatology and Human Ethology*, which was organized by the Center for Philosophy of Science at the Faculty of Science of the University of Lisbon (http://cfcul.fc.ul.pt), in collaboration with Minding Animals International (http://www.mindinganimals.com); as well as several 2013 events that were organized by the Center's Applied Evolutionary Epistemology Lab (http://appeel.fc.ul.pt).

We are very grateful towards the authors who contributed a chapter to this volume, as well as to Kristin Andrews, Louise Barrett, Barend Beekhuizen, Márcia Belchior, Rod Bennison, Rudie Botha, Michael Bradie, Mindaugas Briedis, Jeremy Burman, Massimiliano L. Cappuccio, Raymond Corbey, Paul Davies, Merlin Donald, Daniel Dor, Luc Faucher, José Miguel Fernández-Dols, Tecumseh Fitch, Hajime Fukuchi, Robert Gordon, Robert Lickliter, Jorge Marques da Silva, Larissa Mendoza Straffon, Dominic Mitchell, Mark Nelissen, Elisabetta Palagi, Alexandra Pereira, Olga Pombo, Eugenia Ramirez Goicoechea, Ricardo Santos, Emanuele Serrelli, Chris Sinha, James Steele, Natalie Uomini, Max van Duijn, Arie Verhagen, Luis Vicente, Slawomir Wacewicz, and Wendy Woodward for the advice and support they gave in what concerns both the organization of the events and the preparations for this volume.

We are also much obliged towards the whole Springer team, in particular Anette Lindqvist, Annette Schneider, and Sabine Schwarz for their guidance, advice, and patience.

Finally, Nathalie Gontier wants to kindly thank Marco Pina for taking the lead in the editing process and handling most of the administration involved.

Marco Pina
Nathalie Gontier

Contents

Contributors

Patrícia Arriaga Centro de Investigação e Intervenção Social, ISCTE—Instituto Universitário de Lisboa (ISCTE-IUL), Lisboa, Portugal

Antonio Benítez-Burraco Department of Spanish Philology and Teaching, University of Huelva, Huelva, Spain

Stefaan Blancke Department of Philosophy and Moral Sciences, Ghent University, Ghent, Belgium

Cedric Boeckx ICREA—Catalan Institute for Advanced Studies, Barcelona, Spain; Department of General Linguistics, Universitat de Barcelona, Barcelona, Spain

Maria Botero Department of Psychology and Philosophy, Sam Houston State University, Huntsville, TX, USA

Alexandre Castro-Caldas Institute of Health Sciences, Portuguese Catholic University, Lisbon, Portugal

Francisco Esteves Centro de Investigação e Intervenção Social, ISCTE—Instituto Universitário de Lisboa (ISCTE-IUL), Lisboa, Portugal; Department of Social Sciences, Mid Sweden University, Sundsvall, Sweden

Augusta Gaspar Centro de Investigação e Intervenção Social, ISCTE—Instituto Universitário de Lisboa (ISCTE-IUL), Lisboa, Portugal

Nathalie Gontier Applied Evolutionary Epistemology Lab, Faculty of Science, Centre for Philosophy of Science, University of Lisbon, Lisbon, Portugal

William D. Hopkins Neuroscience Institute and Language Research Center, Georgia State University, Atlanta, GA, USA; Yerkes National Primate Research Center, Emory University, Atlanta, GA, USA

Mary Lee Jensvold Department of Anthropology and Museum Studies, Central Washington University, Ellensburg, WA, USA

David A. Leavens School of Psychology, University of Sussex, Brighton, UK

Ana Mineiro Institute of Health Sciences, Portuguese Catholic University, Lisbon, Portugal

Shoji Nagataki Department of International Liberal Studies, Chukyo University, Nagoya, Japan

Rita Nolan Department of Philosophy, Stony Brook University, Stony Brook, NY, USA

Marco Pina Applied Evolutionary Epistemology Lab, Faculty of Science, Centre for Philosophy of Science, University of Lisbon, Lisbon, Portugal

Timothy P. Racine Department of Psychology, Simon Fraser University, Burnaby, BC, Canada

Adam See Department of Philosophy, Graduate Center of the City University of New York, New York, NY, USA

Sandra Swart History Department, Stellenbosch University, Stellenbosch, Republic of South Africa

Donna Tafreshi Department of Psychology, Simon Fraser University, Burnaby, BC, Canada

Jared P. Taglialatela Department of Biology and Physics, Kennesaw State University, Kennesaw, GA, USA

Mónica Tamariz Language Evolution and Computation Research Unit, Department of Linguistics and English Language, School of Philosophy, Psychology and Language Sciences, The University of Edinburgh, Edinburgh, UK

Ian Tattersall Division of Anthropology, American Museum of Natural History, New York, NY, USA

Joseph J. Thompson Department of Psychology, Simon Fraser University, Burnaby, BC, Canada

Olga Vasileva Department of Psychology, Simon Fraser University, Burnaby, BC, Canada

Tyler J. Wereha Department of Psychology, Simon Fraser University, Burnaby, BC, Canada

Jordan Zlatev Department of Cognitive Semiotics, Centre for Languages and Literature, Lund University, Lund, Sweden

Studying Social Communication in Primates: From Ethology and Comparative Zoology to Social Primatology, Evolutionary Psychology, and Evolutionary Linguistics

Nathalie Gontier and Marco Pina

Abstract Research fields adhere to particular epistemic frameworks that outline the methodological rules of conduct on how to study and interpret primate behavior as both social and communicative. Since the onset of social communication studies, epistemic focus has shifted from behaviorist observations to an examination of the cognitive and neurological capacities that underlie the observed communicative behavior and subsequently, toward an investigation of the evolutionary units, levels, and mechanisms whereby social communication evolved. This volume brings together scholars from within these diverse fields who (1) investigate the historical and epistemic roots of the primate communication/human language divide; (2) identify and analyze the building blocks of social communication; (3) examine how primate social communication strategies are evolutionary precursors of human language; and (4) analyze how social communication differs from human language. In their chapters, the contributors explain the merits and pitfalls of their field-specific epistemic approaches. They compare them to other theoretical frameworks and they give guidelines on how theory formation on the origin and evolution of social communication in primates can be enhanced by allowing for epistemic plurality.

Keywords Social communication • Language • Epistemology • Philosophy of science

Emotions, expressions, vocal signaling, and manual and bodily gestures are evolved means whereby primates, including humans, communicate socially. Additionally, humans have invented signed and vocal languages that not only enable social communication but also abstract, symbolic, and creative thought on the past, present, future, and the inexistent. The development and evolution of

N. Gontier (✉) · M. Pina
Applied Evolutionary Epistemology Lab, Centre for Philosophy of Science,
Faculty of Science, University of Lisbon, Lisbon, Portugal
e-mail: nlgontier@fc.ul.pt

M. Pina and N. Gontier (eds.), *The Evolution of Social Communication in Primates*, Interdisciplinary Evolution Research 1, DOI: 10.1007/978-3-319-02669-5_1,

social communication in humans and other primates has been studied from within multiple disciplines, ranging from ethology and comparative zoology, over primatology and comparative psychology, to evolutionary psychology and evolutionary linguistics. In this volume, contributors examine the epistemic frameworks of these various fields and they give directions for future research.

1 Introduction to the Theme and Outline of the Book

The study of human language is preceded by a rich history reaching back as far as the early Greek philosophers' works. Early philosophers understood language primarily as a knowledge device that enables the expression of abstract thought. The study of social, non-verbal, and verbal communicative behavior in other primate species, and the recognition that human language too can be investigated as a communicative behavior, originated much later in time. From the seventeenth century onward, utopian thinkers and social contract philosophers considered human languages a communicative means to bond societies both socially and politically (see Gontier 2009 for an overview). These ideas were foundational for nineteenth century natural history scholars, and rising fields such as historical linguistics and biology provided the first historical and evolutionary accounts on the origins of verbal and non-verbal communication in our and other species. This history results in the fact that scholars today continue to distinguish human language from other forms of animal and primate communication.

This volume brings together scholars from within diverse fields who:

(1) Investigate the historical and epistemic roots of the primate communication/human language divide;
(2) Identify and analyze the building blocks of social communication in primates;
(3) Examine how primate social communication strategies can be understood as evolutionary precursors of human language; and
(4) Evaluate how social communication differs from human language.

We make no attempt to provide a complete account on the various data and theories that exist on the origin and evolution of social communication in primates and the origin of language in humans. Indeed, such cannot be the subject of one book alone. With this anthology, we do hope to provide an introductory review of some of the main methodological and theoretical frameworks that are currently available to investigate the origin of both social communication and human language.

Today, due to the nature of academia and how disciplines are structured, beginning scholars often have to make crucial and limited choices on the types of methodologies they will apply and the kind of theoretical frameworks they work from. This has resulted in numerous high-standard and specialized volumes that investigate the subject of this book from within one specific school or theoretical framework. Results of this research are then presented at excellent high-profile conferences such as Evolang; Language, Culture, and Mind; Ways to Protolanguage; or at annual meetings of the International Primatological Society; the International Society of Zoological

Studies; the American Association of Physical Anthropologists; the Animal Behavior Society; the Human Behavior and Evolution Society; the International Cognitive Linguistics Society; the European Society for Philosophy and Psychology; and so on. Nonetheless, these conferences and their proceedings or journals are targeted at field-specific audiences that work within certain but not other disciplines. The unfortunate result is that the scientific outputs often remain juxtaposed.

In this anthology, we have invited our writers to provide reviews of how the research programs that underlie their specific fields define studies on primate communication and human language. The contributors give an overview of the gathered data, they explain the methodologies used to collect them, and they demonstrate how such data contributes to overall theory formation on the subjects at hand. Rather than present new data, the authors thus highlight the numerous methodologies and epistemic frameworks that are currently at a researcher's disposal. This book does not select a "winning methodology" or research school. The aim of this volume is to provide the reader with ways to break into the research, by showing how rich and informed research on the origin of social communication and human language can be when we allow for epistemic plurality. How the results of these various lines of research can be combined into broader, more encompassing theories on the origin of social communication and language goes beyond the scope of this volume.

2 Brief Sketch of the Various Epistemic Frameworks Available for Researching Social Communication and Language in Primates

What are the epistemic frameworks that guide researchers in their studies on primate communication and human language? Current research methodologies and theoretical frameworks on communication and language originated around the turn of the last century, when they emancipated from classic philosophical traditions.

The field of ethology arose in the 1930s, mostly in Europe, as an outgrowth of both naturalized epistemology and comparative zoology. Inspired by early scholars such as von Uexküll (1909), Heinroth (1910), Haldane and Huxley (1927), Lorenz (1937, 1941, 1958) instigated the scientific study of animal "instincts" and developed theories on imprinting and fixed action patterns, and Tinbergen (1963) defined what became known as the 4 questions of ethology. In America, the field of comparative zoology was defined by both Louis and Alexander Agassiz (for a historical account, see Winsor 1991). These scholars contributed by defining how we can observe animal behavior and how we can conduct both comparative developmental and evolutionary research on animal behavior, including communication.

At around the same time, especially in America, modern comparative psychology turned behaviorism into a school. With their focus on learning and conditioning in humans and other primates, scholars such as Thorndike (1911), Watson (1913), and Skinner (1957, 1986) introduced the empirical and experimental study of behavioral development. Skinner developed a learning theory of human language, averring that language is a behavior and that much of verbal behavior can be learned through operant conditioning.

From within the field of linguistics, Chomsky (1959) fiercely argued against Skinner's behaviorist theories on language learning. Chomsky critiqued behaviorism based upon, what was later called, the poverty of stimulus argument: during development, human children are competent to learn language even though the language performances they are exposed to are imperfect. Chomsky (1965, 1972) therefore postulated an innate language faculty or a brain-based cognitive "language organ."

This I(nternal) language faculty differs from the various E(xternal) languages that exist in the world today, and I-language, Chomsky professed, requires cognitive rather than behaviorist research.

Both the competence/performance and I-language/E-language distinctions divided the field of linguistics into synchronic and diachronic (historical and comparative socio- and anthropological) linguistics. Synchronic linguistics investigated what was structurally universal to all languages, with the assumption that what was universally shared lends insight into the core of I-language. The answer given was that semantically, all languages allow for displacement, or the ability to use symbols that refer to non-observable events in the world, and syntactically, all languages are recursive and compositional.

This characterization of human language held consequences for how animal communication was defined and contended to differ from human language. Chomsky reasoned that animal communication lacks displacement, compositionality, and recursion and that language therefore qualitatively differs from any other type of communication we find in primates. Chomsky never denied that primates have social communicative skills. What he did do was annihilate evolutionary continuity between primate communication and human language, because for Chomsky, the two were incomparable. The defining characteristics of language are not that it enables social communication, but that it enables one to express abstract and creative thought. It is for these reasons that human language is considered qualitatively distinct from primate communication.

The I/E language distinction also held consequences for how communicative and social aspects of human languages became defined and studied. Diachronic studies on the historical origin, dispersal, and death of languages; comparative research on everyday language use; and psychological and behavioral studies on the various types of non-verbal behavior that accompanies human language performance can be investigated from within ethology, comparative psychology, zoology, sociology, and anthropology but, for Chomsky, they contribute little to an understanding of I-language and language competence.

In the meantime, comparative psychology had indeed developed a tradition of investigating non-verbal communicative behavior such as emotions, expressions, and gestures, in both humans and other primates. Such research had proven that there is quite some overlap in the types of behaviors humans and primates use to communicate socially, which again raised questions about evolutionary continuity. Cross-fostering experiments were introduced where humans taught non-human primates to sign human languages (Fouts and Mills 1997; Gardner and Gardner 1969; Gardner et al. 1989; Patterson 1978; Terrace 1979); and to learn artificial languages such as Yerkish (Rumbaugh 1977, Savage-Rumbaugh 1986). Many of the original cross-fostering experiments were conducted with the following two goals in mind. One was to flesh out the limits and possibilities of reinforcement or operant learning to gain insight into the boundaries of the nature/nurture, continuity/discontinuity, and quantitative/qualitative debate (for a discussion, see Gontier 2006); another was to test Chomsky's predicaments on human language. The results of these cross-fostering experiments are that primates are competent in learning a limited amount of symbols, and they can compose rudimentary recursive sentences. Another outcome of these experiments was that our cousins learn human language more easily when they are socially motivated to use it as a means to communicate, rather than when they are conditioned to associate ASL constructs with events in the world.

The modern field of primatology emancipated from comparative psychological and ethological schools in the late 1960s, when pioneers such as Fossey (1983) and Goodall (1986) started collecting data on how primates behave and communicate in natural settings. The result of these careful observations was that scholars identified the various ways in which primates interact socially. Maternal care, food sharing, fighting, and sex require interactions with conspecifics. These interactions are accompanied by behavior such as grooming, emotional displays, expressions, vocal calls, manual gestures, eye gazing, and joint attention, and these behaviors can be characterized as socially communicative.

With the rise of sociobiology, the ontogenetic observations and cross-species comparisons were placed in an evolutionary context. Mayr's distinction between ultimate and proximate causes of evolution and the deciphering of the genetic code in the 1950s provided early scholars hope that soon, the genetic basis of primate behavior, cognition, and anatomy would be discovered and that scholars would be able to deduce from that how these traits evolved. In order to understand nurture, we need to understand nature, and early sociobiologists synthesized selection theory with the data obtained from fieldwork and behaviorist experiments, and they developed the first theories on the evolution of human and non-human primate social behavior (Axelrod 1981; Hamilton 1964; Wilson 1975; Morris et al. 1979).

The epistemic tenets of behaviorism were also criticized by cognitive developmental psychologists (Piaget 1972). Overall, advances in the cognitive and neurological sciences allowed scholars to associate postulated mental capacities to specific structures and areas of the brain and to initiate comparative brain research. One important outcome of this cognitive revolution (for discussions, see Baars 1986; Piattelli-Palmarini 1980) was the rise of the field of biolinguistics (Bickerton 1984; Puppel 1995; Jenkins 2000), which investigates how mental capacities and brain structures underlie language. Other outcomes were cognitive research on theory of mind in human and other primates (Byrne and Whiten 1988; Tomasello and Call 1997; Whiten and Byrne 1997). Unanswered questions of these fields today include whether the mind operates in a modular or domain-general fashion.

By the beginning of the 1990s, also the cognitive turn became partly criticized and partly expanded by the "social turn" and "adaptationist turn." By expanding on the early works in sociobiology, evolutionary psychologists such as Cosmides and Tooby (1994) conjectured that human behavior primarily needs to be understood by making use of natural selection theory and by studying our hominin past, much more than by studying behavior or cognition as it unfolds in modern human infants or non-human primates. They question the possibility to straightforwardly draw inferences on phylogeny from ontogeny, and they underline that chimpanzees are our cousins and thus evolved separately from our hominin ancestors. Such conjectures of course do not invalidate the results of species-specific behavioral research, but it does shift epistemic focus when human behavior in particular is the topic of investigation.

Although a great deal of progress has been made in acquiring knowledge on the physiological, developmental, and sociocultural mechanisms that, respectively, underlie primate anatomy, cognition, and behavior, scholars have so far neither been able to ground these traits either in genes or in specific brain structures (as the early sociobiologists hoped would be the case) nor have they been able to demonstrate how exactly these traits evolved. Because many of the proximate causes of behavior remain poorly identified, evolutionary psychologists place epistemic focus on the ultimate causes of behavior and investigate how natural selection in particular might be the underlying evolutionary mechanism by which these traits evolve (for a discussion, see Gontier 2012a, b). Epistemic questions they ask are: Why did social behavior and cognition evolve? And what are the adaptive benefits?

Pinker and Bloom (1990) followed this shift in epistemic focus and advanced that of primary importance is an understanding of what language is for, i.e., what are the functional adaptive benefits of human language? This differs from Chomsky's epistemic outlook, because he investigated what is unique to human language, and what is unique to human language does not necessarily provide a reason for why it evolved (Gontier 2010). Pinker and Bloom conjectured that language is an adaptation that evolved by means of natural selection. The evolutionary adaptive benefit that language is argued to provide humans is enhanced social communication. Insofar as other primates evolved means to communicate socially, there is epistemic ground to examine how these forms of communication are evolutionary precursors of human language. The field of evolutionary linguistics (Hurford et al. 1998; Knight et al. 2000) and to some extent evolutionary anthropology (Boyle et al. 2007; Mellars and Stringer 1989; McBrearty and Brooks 2000) are direct outgrowths of evolutionary psychology (Barrett et al. 2002). Both fields examine the adaptive functions and selection pressures that enabled verbal and non-verbal communication as well as human sociocultural behavior and how communication is a form of sociocultural behavior (see Gontier 2012b for a discussion). Innovative to their approach is that many of the developed adaptationist theories are currently tested by a varied set of computational and experimental modeling and simulation techniques.

As this brief sketch demonstrates, new disciplines often get established by their adherents overruling the epistemic programs and methodological toolkits of previous disciplines and by reinterpreting the acquired data according to new epistemic rules, or by shifting epistemic focus. The above-described paradigm shifts have often been characterized as transitions from instructionism or behaviorism over cognitivism to selectionism. But the fact of the matter is that today, scholars remain active in all these fields.

Scholars have developed many methodologies and theoretical frameworks on how we can study and understand primate communication. Currently, researchers remain divided on whether human language is merely one type of social communication that has evolved within the primate lineage, or whether language has non-communicative properties. This divide also continues to impact non-human primate social communication studies. Their social communicative skills are either understood as evolutionary precursors to human communication or, in attempts to demonstrate what is specific and unique to non-human primate communication, their species-specific abilities to communicate socially nonetheless become defined by differentiating their traits from our own. And eventually, both communication and language are studied from within the same theoretical assumptions and by using the same methodologies.

For this book, we have invited representatives of all disciplines who demonstrate how both communication and language can be studied, what scientific data has been gathered, and how theories are build.

3 Introduction to the Contributors and Their Chapters

The volume is divided into four parts. In the first part, the academic origins of the epistemic outlooks on the nature of social communication and human language are analyzed. In the second part, scholars identify the various behaviors that have been understood as socially communicative, in both humans and other primates. In the third part, scholars horn in on how primate social communication evolved and how it can be understood as an evolutionary precursor to human language. And in the final part, scholars highlight how human language differs from other forms of primate social communication.

3.1 Philosophical and Historical Roots of Social Communication Studies

The study of language and social communication has deep historical and philosophical roots in academic culture. Philosophers of science and historians can therefore provide valuable insights into the origin of conceptual frameworks as well as the rhetoric and politics of science that justify or bias how we study the origin of social communication and human language in primates.

In his "Lord Monboddo's *Ourang Outang* and the Origin and Progress of Language", the philosopher **Stefaan Blancke** gives a historical account on the origin and evolution of language debates as they emerged in the Enlightenment literature. In the eighteenth century, trade and colonization confronted Western civilization with other cultures. Incoming travel accounts reported on the observations of "ape-men" in the woods, or "*Homo Sylvestris*" (Tyson 1699), today designated as orangutans, that demonstrate anatomical form and communicative skills comparable to our own. Blancke first reviews how, from within the Romantic Movement, debates on human nature set forth the groundwork of dichotomies that still plague current theorizing on non-verbal communication and human language. These dichotomies include the innate/acquired or nature/nurture debate (is language the result of learning and enculturation in civilization or is it an innate biological capacity); the qualitative/quantitative distinction (does human rational, abstract linguistic thought differ qualitatively from non-verbal social communication in animals and primates; and can one find grades of complexity in the various languages man uses to communicate); and the continuity/discontinuity debate (does man, with his capacity for language stand apart from nature, or is there continuity with other types of communication). Blancke relates these dichotomies to ideas of nature being created and creation being orderly structured into a Great Chain of Being that steadily progresses from the simple to the complex. Accordingly, the first languages were understood to be more "passionate" and construed of iconic gestures and vocalizations stringed together with little syntactic rules, while younger languages were considered to be more "rational" and containing more abstract symbolizations and complex grammar. In the second part of his chapter, Blancke zooms in on James Burnett aka Lord Monboddo, a Scottish intellectual of the eighteenth century who published one of the first Romantic accounts on the natural (meaning historical, not evolutionary) origins of language in his work *The Origin and Progress of Language*. As Blancke demonstrates, Lord Monboddo entertained a somewhat idiosyncratic view on the nature of human language, for he ascertained that it is not a defining feature of our species. He further conjectured that humans demonstrate affinity with non-human primates, especially orangutans. Orangutans, he contended, are "lower" humans that can provide insight into our emotional, rational, and linguistic capacities; and to make his point, he hypothesized that if orangutans were enculturated properly, they would be able to acquire language and other traits associated with civilization.

This challenge of enculturating non-human primates with the hope to finally settle on the innate/acquired, continuity/discontinuity, and qualitative/quantitative dichotomies was put to test during the numerous cross-fostering experiments conducted in the beginning of the twentieth century. The early cross-fostering experiments where the chimpanzee Gua was raised together with a human child by the Kellog family (Kellog and Kellog 1933), as well as the speech experiments conducted by the Heyes family with the chimpanzee Vicky are reviewed by **Sandra Swart**, in her chapter on "Ferality and Morality: The Politics of the "Forbidden Experiment" in the Twentieth Century". Swart is a historian, and she sketches the broader historical and scientific contexts in which these reinforcement-based behavioristic learning experiments were conducted; how incoming results were received by the larger scientific community; and she investigates the scientific motivations the researchers entertained to justify comparison of human ontogeny with non-human primate behavior. Why conduct such experiments; what results were the researchers aiming for; and which ideologies on human nature were they supporting or fighting? In fact, why were such experiments "allowed"? Swart demonstrates that rather than providing insight into the origin and evolution of communication, these experiments aimed to understand the beginnings of human nature and thus the limits of non-human primates. Language skills were a mere example of the latter boundary. In the second part of her paper, she turns to debates on the "forbidden experiment," i.e., raising human children by non-human primates. Although raising a series of ethical issues, Swart demonstrates that scholars such as Winthrop Kellog took the idea of such experiments seriously. In attempts to find leverage, Kellog even referred to intellectuals such as Montesquieu who similarly had speculated on the potential of such experiments. Experiments where humans were reared by other primates were never conducted in practice, which made scholars resort to the quest for alternatives. Inspired by Linnaeus, Darwin, and Galton, who had referred to "beast-children" in their works, such alternatives were found by reports on "feral" children. Reports on real "Tarzans" and "Mowglis" in the wild, such as Lucas, the "Baboon Boy" of South Africa, were often dubious and fraudulent, but nonetheless taken quite seriously by these early intellectuals. Swart reviews how pioneering and leading scholars, including Raymond Dart and John Foley, authenticated several of the cases reported, and she analyzes the polemics of the scholars involved. She ends with a critical assessment of the usefulness of the results obtained from such studies and weighs them against the ethical concerns they raised.

3.2 *The Elements of Social Communication in Primates and Humans*

Many scholars agree that social communication can make use of gestures, vocalizations, and expressions. But what do such behaviors communicate? Do they convey emotions, shared intention, or symbolic abstract thought? Is joint attention sufficient to communicate, or does it require intentionality and theory of mind? In fact, when do behavioral acts become understood as communicative and as social? And how does one study all of the above? In the second part of the volume, and from within ethology, primatology, and psychology, scholars introduce the building blocks of social communication; they highlight the biological, cognitive, and cultural requirements to communicate socially; and they theorize when it originates in primates.

From the mid-1960s onward, interspecies cross-fostering and language learning experiments took on a more scientific approach. Because vocal language learning experiments had proven unsuccessful, several experimental projects were launched where psychologists, primatologists, and linguists investigated non-human primates' capacity to learn signed and artificial languages. The famous Project Washoe, launched by Beatrix and Allen Gardner at the University of Nevada in Reno, was the first experiment whereby a chimpanzee was taught lexical and grammatical structures of American Sign Language. Later, the project was expanded to include other chimpanzees. This expansion enabled the study of chimpanzee–chimpanzee ASL conversations as well as transmission modes among chimpanzees across multiple generations. Many of the language-trained chimpanzees eventually found their home at Central Washington University, in the Chimpanzee and Human Communication Institute. Originally run by Roger Fouts, this Institute then became directed by **Mary Lee Jensvold**. In her chapter "Experimental Conversations: Sign Language Studies with Chimpanzees", Jensvold reviews the historico-theoretical context, methodological preliminaries, and various experimental setups used over more than 40 years of inter-specific and intra-generational research on ASL comprehension, conversation, and transmission. One of the major outcomes of this research is that reinforcement learning had but limited success. Rather, chimpanzees more successfully learned ASL constructions when social immersion techniques were used, i.e., when a sociocultural environment was created that made communicating in ASL constructs meaningful. This research therefore demonstrates, as is the case with human children, that language learning not merely depends upon having a cognitive capacity to learn language, but that language learning is a culturally embedded practice that requires social motivation. A second major outcome of these experiments was that chimpanzees were able to learn not only a lexicon, but also basic grammar. Finally, she demonstrates that once learned, ASL becomes a permanent and meaningful way to communicate with caregivers as well as with other chimpanzees. Jensvold discusses cases of displacement (that demonstrate successful use of arbitrary signs); wh-question experiments (that indicate syntactic skills such as compositionality); and conversation analyses that focused on turn-taking, on topic/off topic conversations, and reactions to negative-response statements (that prove correct comprehension, usage and additivity); and she compares the chimpanzee's skills with that of human ASL learners. Numerous cases of the expansions of the message to facilitate comprehension of the communicative partner, private signing in contexts of imaginative play, and successful transmission of signing to offspring (as was the case with Washoe's adopted son Loulis) prove that chimpanzees are, to a certain extent, able to learn, comprehend, meaningfully apply, and transmit human language. Jensvold therefore pleads for continuity.

When primatologists, ethologists, or developmental psychologists observe, study, or theorize on the nature of communicative interactions in primates, including mother–infant relationships or non-verbal and verbal social communication, these researchers have to have a minimal working definition of "communication," "social interaction," and "information." What are these working definitions; how do they influence methodology; and how do they relate to a larger theoretical paradigm from wherein the results of observation and analysis are interpreted? These meta-theoretical questions are addressed by **Maria Botero** in her chapter on "How Primate Mothers and Infants Communicate: Characterizing Interaction in Mother–Infant Studies". Taking mother–infant interactions as exemplar, she historically and paradigmatically contextualizes how theoretical assumptions have shaped our understanding of social non-verbal communication. She distinguishes between two alternative methodological approaches to study primate mother–infant interactions: the ecological-linear model as it was introduced by Jeanne Altmann to study social mother–infant interactions in non-human primates; and the interactional model advanced by Jerome Bruner to study mother–infant interactions in humans. Because it was introduced for investigating communicative interactions in non-human primates, the ecological-linear model avoids assumptions on the presence or absence of belief states such as intentionality or theory of mind, but focuses on the immediate and observable outputs of communicative behavior (action–reaction schemes or does behavior x in the mother chimpanzee trigger a behavior y in the infant). Information, Botero demonstrates, is therefore understood from within Shannon's classic information theory as a message that is transmitted between a sender and a receiver. From within the school of ethology, the message (a vocal call or behavioral pattern) is often understood as "instinctive" or fixed, and thus informed neither by environmental context or social situation, nor learning. This in turn reinforces a methodological emphasis on the transmission mode (the onset, duration, and ending) of communicative messages, rather than an emphasis on the context and the meaning of the message. This epistemic stance, Botero points out, limits research on semantics of the message and the sociocultural situatedness wherein communication takes place. In contrast, the interactional model, designed to investigate human mother-infant communication, works from within the explicit assumption that there exists a functional sociocultural and communicative context. Such communication is understood as interactional, because it is assumed to be the outcome of ritualized sociocultural practice at both the level of the mother–child dyad, and at higher levels such as the group. The mother–infant dyad is therefore evaluated as embedded in a set of other communicative group interactions that affect and result in social, cognitive, and cultural learning. Such a stance, Botero demonstrates, also does not require assumptions on intentionality of the communicative process. Rather, the messages are considered semantically meaningful when one can observe that they are communicatively functional. Such functional assessments can only be made when one contextualizes the communicative behaviors as a sociocultural dyad. Botero demonstrates how this interactional approach to communication can be implemented into the study of chimpanzee mother–infant interactions.

Facial expressions are the most outstanding examples of non-verbal communication. How facial expressions are means of communication and how they in particular enable the articulation of emotions in both humans and other primates has long been the subject of social communication studies. Charles Darwin (1872) pioneered the study in his seminal work *On the Expression of the Emotions in Man and Animals*. He first linked specific expressions to certain emotions, and secondly, he contended that such expressions are instinctive and innate within a variety of species. Expressions of emotions are exemplar of adaptive behavior that evolved by means of natural selection to enable recognizable social communication among members of the same and different species. This work was revived and expanded by Paul Ekman, a clinical psychologist who introduced the facial affects theory. In humans, he differentiated between 7 "archetypical" and universal expressions for emotions of joy, fear, anger, sadness, surprise, disgust, and contempt. Ekman also devised several methodologies and testing devices with the goal to standardize comparative cross-cultural research on both expressions and emotions within the human lineage. In the chapter “On Prototypical Facial Expressions versus Variation in Facial Behavior: What Have We Learned on the “Visibility” of Emotions from Measuring Facial Actions in Humans and Apes”, **Augusta Gaspar, Francisco Esteves, and Patrícia Arriaga** give a historical review on how the facial affect theory became the standard paradigm within comparative psychology. The authors critically assess the basic tenets of the theory. First, they demonstrate that although humans have a distinguishable set of emotions and expressions and associated action recognitions, scholars active in ethnological/anthropological and cross-cultural psychological research discovered quite some cross-cultural variability in how certain emotions link to specific expressions. Secondly, their own comparative work on the development of expressions and emotions in human children and adults shows that both expressions and emotions are much more variable and behaviorally flexible during the course of development than assumed by adherents of the facial affect theory (in a state of anger, for example, people sometimes smile). Gaspar and co-authors identify many more non-verbal clues that accompany the behaviors, and both the expressions and their linkage to certain emotions are the outcome of considerable learning of the behaviors in a sociocultural context. This also has consequences for evolutionary comparative research that often straightforwardly assumes the universality of emotions and expressions. Gaspar and co-authors demonstrate significant difference between which emotions are linked to specific expressions in humans, and how such linkage is expressed in our phylogenetic cousins. The authors emphasize the continued need for observational descriptive, comparative ethological research to further flesh out how enculturation, learning, and the general sociocultural environment help shape both expressions and emotions and the linkage between the two.

Any type of social communication, be it linguistic, emotional, expressive, or gestural, requires a capacity for the communicative partners to share attention toward the communicative signal. **Timothy Racine, Tyler Wereha, Olga Vasileva, Donna Tafreshi, and Joseph Thompson** contribute a chapter on “The Evolution of Joint Attention: A Review and Critique”. Reviewing the literature, they list the various cognitive, behavioral, and social capacities that have been proposed either to enable or to associate with joint attention. These include intersubjectivity, attentional focus through gaze following or pointing, exchange of emotions, shared beliefs, and mind reading (also discussed in **Nagataki**, this volume). They show that most scholars therefore agree that joint attention is a form of social cognition that ontogenetically precedes the development of either social communication or language. This gives credibility to the idea that joint attention also evolutionary precedes social communication of any kind. Considerable debate arises, however, on whether non-human primates and human infants are able to share attention intentionally. In the second part of their chapter, Racine and co-authors demonstrate that the various theoretical and empirical problems associated with comparative developmental research on joint attention also hinder evolutionary research on its origin. The authors critically assess several theories on the evolutionary origin of social behavior and how they can or cannot be applied to examine the evolutionary origin of joint attention. These theories include the secondary intersubjectivity theory that was developed by Trevarthen, evolutionary psychology approaches as they were introduced by scholars such as Cosmides and Tooby, Pinker and Buss, and Tomasello's (2000) shared intentionality hypothesis. Trevarthen assumes intersubjective behavior to be innate and thus unlearned, but Racine et al. demonstrate that such a claim adds little explanatory value. Evolutionary psychologists are notorious for countering classic social science models that emphasize learning and instruction to the neglect of phylogenetic research. But the phylogenetic research such scholars deem necessary involves assuming that social behavior evolved by means of natural selection, as an adaptation to a Pleistocene environment. They also content that the social cognition required for joint attention is domain-specific and modular. Racine and co-workers point out that such a stance requires evolutionary psychologists to assume that "core knowledge" exists and that it is readily available during ontogeny, which again eliminates learning as a potential mechanism underlying the behavior. Racine and co-workers therefore evaluate these theories to contribute little explanatory insight into the onset of joint attention during ontogeny. Tomasello's theory on joint attention does integrate and rely on comparative ontogenetic research of joint attention in chimpanzees and human infants; and he concludes from these experimental observations that chimpanzees do not have the same number of mental states as humans and do not understand them in the same way as humans do. But here too, the difference is argued to be shared intentionality which is contended to be an adaptation, which again contributes little to overall theory formation or experimental research. Racine and co-workers criticize these adaptationist accounts on several grounds. For one, the latter is associated with the modern synthesis, which in biology is more and more discredited in favor of an extended synthesis. Secondly, the scholars involved do not take ontogeny into account. Evo-devo and epigenetics are evolutionary theories that developed within the extended synthesis and that do take both development as well as evolution into account when advancing evolutionary theories (see also **Boeckx**, **Tattersall**, **and Tamariz**, this volume). The authors therefore suggest that psychological research on joint attention is conducted from within evo-devo in order to further flesh out the origin of joint attention, during both ontogeny and phylogeny.

What is Theory of Mind? How does it unfold in humans? And how can scholars study it? In his "Describing Mental States: From Brain Science to a Science of Mind Reading", **Shoji Nagataki** guides us through distinct psychological theories and associated methodologies that enable researchers to investigate mind sets, especially emotional and mental states, in oneself and other humans. Historically, investigations into theory of mind have been conducted by simply asking the research subjects to describe their mental and emotional states (the method of introspection), by observing the outcomes of cognized behaviors such as language in experimental settings (the behavioristic approach), and by mapping the types of brain activity that are associated with certain states of mind (neuroimaging). Nagataki demonstrates how all these approaches fall short. Introspection is based upon subjective interpretations of the research subject and assumes an immediate reciprocal and intuitive understanding of the belief states by the investigator. Behaviorism merely focusses on the outputs of behavior and does not investigate the underlying mechanisms that enable cognizing. And neuroimaging techniques are troubled by the explanatory gap for how do the brain activation patterns relate to the mental states. Nagataki further demonstrates that all of these methods are conducted within a social experimental context where it is assumed that both the researcher and the subjects under investigation already understand one another's mental states. Whether it involves asking for personal descriptions, interpreting results from observation, or asking a patient to respond to, or to imagine mental states during PET or fMRI scans, such experiments are done within a social context where "mind reading" already and necessarily occurs in order to be successful. Where does this "common ground" stem from? Answering this question involves an inquiry into folk psychology. Folk psychology takes this "common ground" as a given, and there exist three theories that investigate its origin: rationality theory, simulation theory, and theory–theory. Rationality theory goes back to Kant and assumes that there is a psychic unity, all humans share a universal reason as well as universal mental categories that enable one to understand others as intentional agents. Simulation theory is more empiricist-informed and assumes that all humans share a common sense apparatus, which enables one to "simulate" or take on the perspective of others, via empathy, imitation, or imagination, and as such, to infer understanding. Questions for both theories then become whether the assumed common sense apparatus or universal reason is innate or acquired, and how either evolved. Theory–theory assumes a less universal and fixed state, and contends that humans develop theories on the mental states of others through learning and enculturation, which enables perspective taking. Nagataki demonstrates that all these theories assume that mind reading, whether it is through reason or through investigating bodily expressions, assume either inferential or intuitive learning, and the question then becomes how we are to interpret the latter. The author pleads for a "detranscendentalized" form of intuition and points to verbal communication and especially language as the locus of this folk psychology. (Linguistic) Communication itself is an empirically acquired behavior that infers meanings from bodily actions such as sound production, and this learning occurs within and across generations. According to Nagataki, languages therefore provide a "third-person" perspective of folk psychology, and folk psychology itself is a means to communicate socially. Once acquired, language is often a means to understand and interpret others' mental states. Language itself is a socially evolved means that enables intuitive understanding. He ends his chapter with an application of his approach in an experimental setup, where therapists were evaluated to read anger states in normal individuals.

3.3 Evolutionary Transitions from Social Communication Systems to Language

How can primate social communication be understood as an evolutionary precursor to human language, and how did the transition occur? For years, scholars have been debating what the anatomical–physiological modalities are wherefrom language evolved. Traditionally, two schools can be distinguished: scholars either defend that speech evolved first and thus that the evolution of language proceeded from vocal–auditory adaptations (for an overview, see Fitch 2000); or scholars defend that vocal language evolved from gestural (proto-)language, in which case the evolution of human language required both visual–gestural and later vocal–auditory modifications (Armstrong et al. 1995; Corballis 2002; MacNeilage 1998). Today, synergies of both ideas can be found, and several scholars suggest that the postulated protolanguage might have been prosodic, where rhythmic sounds and gestures that holistically conveyed semantically complete messages transitioned into fractioned, compositional language (for a discussion, see Arbib and Bickerton 2010; Mithen 2006; Tallerman 2007). In this part, and from within primatology, psychology, and linguistics, scholars investigate how gesture-first theories can be expanded to include other types of bodily communication, how non-human primates have more voluntary control over their vocalizations than originally presumed, and how both gestural and vocal communication might have transitioned into human language.

Given that there exist so many means by which primates can communicate non-linguistically, and accepting these communicative modes as evolutionary precursors to human language, how then, did human language evolve from these non-verbal modalities? This is the question raised by **Jordan Zlatev**, in his chapter on "Bodily Mimesis and the Transition to Speech". Zlatev expands on the work of Donald (1991), a neuropsychologist who in his *Origins of the Modern Mind*, defends that symbolic thought, language, and our human-specific consciousness primarily evolved from sociocultural stimuli. Rather than evolve from modular brain adaptations, the capacity for complex culture evolved from changes in how different brain regions functionally connected and associated with the expanding neocortex. He distinguishes between three consecutive stages of cognitive-cultural evolution in our species: the mimetic, mythic, and technology-supported stage. These stages are differentiated based upon the predominant means in which culture is individually and collectively expressed and memorized: first in bodily expressions, then in language, and eventually cultural inventions such as art and written language allow for external information storage of symbolic thought. Donald's theory has been adapted and expanded by many scholars interested in the evolutionary origin of language. Especially, adherents of a gesture-first origin of language investigate how a mimetic, primarily gestural expressive communication systems, could have originated from the communication systems present in other primates. Zlatev synthesizes gesture-first theories with Donald's concept of mimesis developing the bodily mimesis hypothesis. The bodily mimesis hypothesis assumes that our ancestors acquired more volitional control over bodily actions such as expressions and gestures enabling them to consciously initiate and rehearse such behavior, as well as imitate or re-enact bodily actions. Such rehearsals and re-enactments enable a more developed form of empathy and intersubjectivity and thus increase group cohesion and sociocultural learning where pantomime and ritualized behavior become a means to intentionally communicate non-linguistically. Zlatev's chapter first reviews the evidence in favor of the bodily mimesis hypothesis. For Zlatev, the first hominins who possessed a more advanced control over bodily actions were *Homo ergaster* and *Homo erectus*. These species invented symmetric Acheulean artifacts and fire, and the production of both require fine motoric skills if not craftsmanship (indicating rehearsal and re-enactment). In comparison, the action recognition that occurs in monkeys during the activation of mirror neuron systems might enable basic empathy, but not the development of systems of shared symbolic meaning (also see **Nolan**, this volume). In order to make such a transition, Zlatev contends that actions need not merely become recognized, but volitionally and intentionally imitated. Comparative psychology demonstrates that chimpanzees and other great apes already possess voluntary control over manual gestures. Nonetheless, Zlatev suggests that chimpanzee's cognitive abilities for hand–eye coordination, causal reasoning, executive control, social learning, teaching, social intelligence, and functional representation are different from our own. Evidence for a gestural, mimetic origin of language can also be found in the fact that in humans, vocal language learning is often preceded by mimetic types of communication and also adult vocal language use is often accompanied by gestures. From this, Zlatev concludes that a mimetic form of communication evolved first and later co-evolved with vocal language. The transition from mimetic to vocal language is clarified by referring to Cognitive Semiotics, a field that investigates the symbolism of vocal, gestural, or artistic signs in relation to the cognitive capacities that enable them. Vocal language is less iconic and more arbitrary than manual communication systems, which enables more conventionalization and standardization of the message across larger groups. It is is also less costly to learn vocal arbitrary sounds than to repeat whole sets of ritualized behavior to convey a message, which is why, according to Zlatev, speech at the expense of gesture became selectively favored, resulting in the multimodal communication system of modern humans.

David Leavens, Jared Taglialatela, and William Hopkins combine experimental and observational data on voluntarily produced vocal–auditory communicative signals in primates, with manual and gestural origin theories of social communication, into a multi-modal theory on the origin of human language. In their "From Grasping to Grooming to Gossip: Innovative Use of Chimpanzee Signals in Novel Environments Supports both Vocal and Gestural Theories of Language Origins", the authors take an ecological and epigenetic approach to both communication and language. This implies adhering to the following two tenets: first, the many means there exist for non-human primates to communicate, both vocally and gesturally, are deemed sufficient for communication, so neither modality is maladaptive; and secondly, although human language requires enhancements in the physiological capacity to communicate both manually or vocally, language did not solely evolve from these features. Rather, in line with Dunbar (1996), they assert that human language evolved particularly to enhance social communication in larger groups, and they present empirical evidence that supports this hypothesis. The structure of non-human primates' supralaryngeal vocal tract and breathing apparatus limits the types of vocal calls they can produce, and it is the main reason why non-human primates cannot speak. One of the major contributions of Leavens et al.'s work is that they demonstrate that non-human primates have more voluntary control over their supralaryngeal vocal tract than traditionally conceived. Their spectrum of vocal calls is also much larger. Sounds produced in the front of the mouth, such as lip smacks, sputters, kisses, and teeth chomps and also pants and grunts that are produced lower in the tract, can be understood not merely as behavioral actions, but as vocal-auditory communicative signals. The authors review both their own experimental research as well as reports on observations in the wild, and they demonstrate that primates possess voluntary control over such vocal signals; that such signaling behaviors are often co-opted innovatively and voluntarily to communicate (to acquire attention, for example); and that there is significant cultural variation in how and which vocal signals are used to communicate, which in turn implicates social learning. Leavens, Taglialatela, and Hopkins' work furthermore provides functional and neurobiological evidence that proves that it is primarily the left hemisphere that is active during such lip-produced sounds and that the activated brain regions show considerable homologous overlap with the regions active during speech in humans. They conclude that the evidence necessitates us to accept that our hominin ancestors, just as non-human primates, already possessed significant voluntary control over the production of vocal sounds and that such sounds were readily available to be co-opted to create novel intentional communicative signals (see also **Tattersall** and **Tamariz'** contributions). The consequence of these findings is that manual and vocal origin of language theories can be combined: language evolved simultaneously from both vocal as well as gestural communication systems. The reason vocal language is the dominant mode for human language today is explained by referring to Dunbar's theory: Evidence supports that vocal communication increases with group size, in humans as well as other primates, for it enables more rapid social bonding.

Chimpanzee's ability to produce voluntary and intentional vocal signals in socially communicative settings is also the theme of **Adam See's** chapter on "Reevaluating Chimpanzee Vocal Signals from the Ground Up". Problems he touches upon include the following: When do vocal sounds become communicative signals? Do communicative signals require learning? Does that learning need to be social and if so, when does learning become social? And when does communication become intentional? See reviews the recent scientific data obtained on chimpanzees' ability to voluntarily produce vocal sounds (including the work of Leavens et al. (2005), and this volume), and compares it with the criteria Tomasello introduced to understand chimpanzee's manual gestures as both communicative and intentional. Tomasello (2000) has asserted that, unlike in their vocalizations, non-human primates demonstrate significant evolutionary innovation in their voluntary control over manual gestures. They are often intentionally produced as communicative signals during dyadic reciprocal relationships, and many of these gestures can be learned. From a careful reading of Tomasello's work, See deduces that his main criteria for understanding several of these manual gestures as communicative signals are threefold: they are socially learned, voluntarily produced and used flexibly depending on the context, and they invoke attention to the attention of others. By using these criteria, he analyzes the literature on chimpanzee vocalizations and provides evidence for all three criteria being equally applicable to certain vocalizations. A specific category of chimpanzee vocalizations, namely those that are produced to acquire attention, differ from other vocal displays and vocal calls (such as uncontrollable alarm calls or food grunts), because they are less associated with emotional or environmental stimuli and more the result of socially learning to direct the behavior of others. For Tomasello, there must be intentional usage of gestures before they can become understood as signals: they must be directed toward others in a social context in order to trigger behavioral responses or mental states (such as acquiring attention). Tomasello therefore relates research on gestural signals to debates on the presence or absence of theory of mind in chimpanzees. See asserts that although there is reason to debate the issue whether primates' manual attention-getting gestures demonstrate second-order intentionality, there is no more reason to doubt its presence in the production of vocal signals than there is in regard to manual gestures. See thus concludes that the criteria Tomasello uses to understand manual gestures as communicative and intentional signals are met in regard to the voluntarily uttered attention-getting vocal sounds. See consequently agrees with scholars such as Leavens, Taglialatela and Hopkins that human language evolved from both vocal as well as manual signaling.

3.4 Evolutionary Origins of Human Language

How does human language differ from social communication as it unfolds in non-human primates? When did human language originate in time? Can the archeological and hominin fossil record provide insight into the origin of language? And by which evolutionary, biological, and cultural mechanisms did human language evolve? In the final part of this volume, authors analyze these questions from within anthropological, archeological, evolutionary, and linguistic sciences.

In the chapter on "Communication and Human Uniqueness", **Ian Tattersall** explains how, or better how little inferences we can make on the origin and evolution of language by studying either hominin fossil remains or archeological finds. To examine the onset of vocal linguistic sounds, anthropologists draw inferences on the length of the supralaryngeal vocal tract (the sound box) from the position of the cranial vault (the skull base) and how flexible the basicranium is. Inferences on the larynx are drawn from the hyoid bone (a floating bone supported by muscles located above the thyroid in the neck, in turn positioned above the larynx). Both basicranial flexion and hyoid bones enable scholars to calculate the position of the tongue and the air space available to produce the spectrum of sounds. Middle ear bone remains allow inferences on hearing capacities. But, Tattersall demonstrates, comparison of these anatomical features across later-evolving *Homo* species (especially Neanderthals and humans) displays both significant variation within species and similarity between species, thereby providing inconclusive evidence for or against these species' capacity to produce human(-like) vocal language. Recent insights coming from the newly emerging field of paleogenetics have also proven that Neanderthals share the specific mutations of the *FOXP2* gene. Fixed in our species, it is associated with articulate speech. But even if one postulates from these findings that Neanderthals possessed the capacity for articulate speech, the archeological record shows no compelling evidence they did actually speak a symbolic and syntactic language. The archeological record demonstrates a transition between four types of tool-technologies: the Oldowan (mode 1), Acheulean (mode 2), African Middle Stone Age and European Middle Paleolithic (mode 3), and African Later Stone Age and European Upper Paleolithic tools (mode 4). These archeological remnants provide behavioral proxies, i.e., scholars can speculate upon the cognitive and behavioral capacities that are required to produce these artifacts. Tattersall endorses the likelihood that from the onset, the hominins who produced these artifacts had an emotional, manual, and vocal communicative system, but there is little evidence that they possessed a linguistic mind. Only during the life span of *Homo heidelbergensis*, who is presumably the direct predecessor of our kind, archeologists find more complex and composite tools associated with the Middle Stone Age technological complex. Even Archaic *Homo sapiens*, first found in Ethiopia and 200,000 years old, produced artifacts not much different or more elaborate than the tools associated with older species. It is only with the appearance of anatomically modern humans, which happened around 100,000 years ago in Africa, that the archeological record undeniably demonstrates the presence of symbolic artifacts and ornamentation that include the 70,000-year-old geometric artwork and personal ornaments found in the South African Blombos Cave. The newly evolved skeletal features of anatomically modern humans demonstrate a major developmental reorganization of several bodily structures. According to Tattersall, this plausibly extended to the synaptic wiring of the neocortex, providing our species with new cognitive capacities for symbolization and abstract thought. These capacities remained dormant until they were triggered by cultural stimuli such as the described symbolic artifacts, which enabled a rapid evolution of semantic and syntactic language. For vocal language to emerge, several anatomical structures, including the vocal apparatus, became exapted and language evolved rapidly and exclusively in our species. Tattersall emphasizes that the vocal apparatus was already in place and of functional use in existing communicative systems, but that cultural stimuli triggered such features to become exapted for new use: language, which was vocal–auditory in kind. For Tattersall, the emergence of Nicaraguan Sign Language demonstrates another contingent evolutionary scenario where this time, due to other sociocultural circumstances, existing anatomical and cognitive structures became exapted for gestural language.

In the chapter on "How did Humans become Behaviorally Modern? Revisiting the "Art First" Hypothesis", **Rita Nolan** battles the standard philosophical model of human cognition that understand symbolization as the result of logic and computation, and language as that what uniquely features semantic–syntactic structures, recursion, unbounded productivity, and displacement. Nolan provides an alternative to this view, which goes back to philosophers such as Cassirer and Langer and she makes a case for art, or deliberate symbolic artifacts of material culture as she calls it, sharing these features. This has consequences for the origin of modern behavior in humans. Traditionally, language, art, and abstract thinking are assumed to distinguish us from other animals, and in our evolution toward modern behavior, language has been theorized to have evolved first, thereby enabling art, abstract thought, and other behaviorally modern traits such as navigation over water. Based upon the recent Blombos archeological findings—which include deliberately engraved ocher that is presumed to be of a symbolic nature as well as shell beads that served as personal ornaments and presumably symbolized status, gender, or other sociocultural features of the person who wore it—Nolan says that the capacity to produce deliberate symbolic material artifacts evolved first. Art facilitated the evolution of displacement, abstract thought or symbolization, recursion, and unbounded productivity, and eventually language. In making her case, Nolan provides a rich philosophical contextualization and critical assessment of many of the theories that have developed in regard to human language, human uniqueness, and modern behavior. Although agreeing with Chomsky's poverty of stimulus argument that in turn functioned as a critique against Skinner's verbal imitation and operant learning theory, Nolan criticizes Chomsky by making reference to Tomasello's theory that, following Grice, emphasizes that language requires and enables shared intentionality. But Tomasello's theory in turn is balanced against Grice's second requirement for language, which is shared semantics: all members of a language community more often than not attribute the same meaning to the same linguistic constructions. Gestural origin theories of language, which find proof of the evolutionary origins of shared intentionality and imitation in the discovery of mirror neuron systems (MNS), cannot by themselves explain the origin of abstract symbolization or features such as displacement, because actions and perceptions triggered and activated by the MNS require the presence of the object, the observer, and the performer of the behavioral scheme. If what makes language unique is its features such as semantic–syntactic structures, recursion, unbounded productivity, and symbolic displacement, then Nolan emphasizes that these features first and foremost have no immediate perceptual aspect. Following Harnad, language origin theories need to answer the symbol grounding problem: i.e., symbolization occurs without immediate empirical grounding or associating of the symbolized objects or behaviors to the physical or sociocultural world. Such decoupling or movement away from present and visible objects and behaviors allows for displacement, a term defined by Sterelny as the ability to talk about the past, present, or future. Combining these ideas and critiques, Nolan demonstrates how the late Middle Stone Age Art of Blombos already possesses these features traditionally attributed exclusively to language, and how the deliberate manufacture of material symbolic artifacts triggered and facilitated human language which is consequently assumed to have evolved later in time. She ends her contribution with guidelines on how to empirically test the theory.

How much of language evolution can be explained by referring to cultural rather than biological evolutionary processes? How does one define cultural evolution? And how does cultural evolution underlie the evolution of language? These are the questions raised by **Mónica Tamariz** in the chapter titled "Experiments and Simulations Can Inform Evolutionary Theories of the Cultural Evolution of Language". Tamariz elucidates how a combination of new experimental and computational techniques developed in the fields of evolutionary psychology, anthropology, and linguistics provide research tools that can complement cross-species comparative psychological research on cultural transmission (e.g., Gaspar et al., Botero, Jensvold, Leavens et al., this volume). Tamariz illustrates several of the iterated learning experiments and mathematical and computational agent-based simulations as they were developed by, among others, the James Hurford/Simon Kirby and the Luc Steels/Bart de Boer groups, which demonstrate that key linguistic features can evolve by means of cultural evolutionary mechanisms. More specifically, the creation of communicative symbols and conventions among individuals, the spread and stabilization (faithful sharing) of linguistic conventions within and across populations over time, the emergence of linguistics systems such as vowel systems, and compositionality of linguistic structure can be experimentally modeled and simulated by solely invoking cultural evolutionary dynamics. In the second part of the paper, she links these experiments and simulations to existing theories on babbling, social and imitation theories, and musical and gestural protolanguages, and she puts forward a theoretical framework for the cultural evolution of language. In this framework, she breaks with some of the long-standing ideas on how we are to conceptualize cultural evolution (how similar or different is it to biological evolution), how we can define cultural and linguistic evolutionary units (replicators), and how they are inherited and transmitted over generations. Tamariz advances that languages are complex adaptive systems that exist not so much because of biological, neuro-cognitive, or genetic adaptations for recursion or abstract thought, rather they emerge from contingent cultural evolutionary processes operating at both local and population levels. Following Maynard Smith and Szathmáry's (1995) work on the major evolutionary transitions, she reasons that the cultural evolution of language involved two key transitions: a selection for replicability and a selection for function. Babbling phases or presumed protolinguistic phases in language development or language evolution occur because there exists selection of replicators: Sounds or gestures are faithfully replicated through imitative social learning. Eventually, these replicating cultural units become co-opted or exapted (see **Tattersall's** contribution) for function: They are attributed communicative value between individuals at a local level. Innovative to Tamariz' approach is that she suggests that individual concepts or perceptions (such as memes or linguemes) are not constant entities, because they result from "statistical" social learning: they can change during the course of an individual's life time, and for this reason, they cannot be replicating units. What does replicate are social constructs that emerge as a consensus at the population level. At the population level, social interaction patterns that influence imitative learning and teaching, as well as the easiness with which certain symbols are (re-)produced, provide selection pressures on the type of linguistic features that evolve. Cumulative learning and transmission to new generations is necessary for linguistic systems to emerge. And such systems evolve more by random drift than natural selection and eventually demonstrate remarkable stability and complex features such as compositionality and arbitrariness of signs, due to constraints imposed by how we learn socially. In short, language is not so much a biological capacity that evolved in our species, but an emerging outcome of individual and group behavior that includes social learning and cultural transmission. By demonstrating how cultural transmission systems itself are both evolving and evolutionary systems, she concludes that we need to prioritize cultural evolutionary explanations of language over biological ones.

What is it that makes human language specific and unique, and how exactly does it differ from animal and primate communication? That is the main question tackled by **Antonio Benítez-Burraco, Ana Mineiro, and Alexandre Castro-Caldas** who, in their chapter "The Emergence of Modern Communication in Primates: a Computational Approach", present a summary and review of the Biolinguistic tradition. The authors recognize the diverse means there are in the animal and primate taxa to communicate. Nonetheless, in line with Chomsky, they contend that such communicative systems cannot be understood as a semantic or symbolic coding system that is so typical of human language. They assert that human language neither evolved from animal cognition nor animal communication systems. Rather, human language evolved from our species' unique cognitive and neurologically underlain abilities to "compute" complex structures. Language is typified by symbols that function as displaced conceptual representations and that are syntactically stringed together into semantic compositional structures. Such computational binding is recursive and enables humans to make an infinite series of meaningful linguistic structures that enables us to demonstrate abstract knowledge processes such as creative thought. Again in line with the Chomskyan tradition, they emphasize that social communication is but one function of human language. Other functions of human language, in particular these cognitive–neurological computations that underlie both the competence and performance of language, are what characterizes and defines its uniqueness. After having outlined the theoretical framework they work from, Benítez-Burraco and co-workers review how recursion, binding, compositionality, and displacement are evidenced in current human, vocal, and gestural languages. Afterward, they present archeological and paleontological fossil evidence for the evolutionary emergence of these types of computational behavior. Problems investigated include how the tying of knots and the production of complex composite tools can allow scholars to draw inferences on the presence of recursive cognitive thought; and how endocasts hint at the origin of brain lateralization, brain size, and increased neural connectivity required for language. They end with inferences on the evolution of language from recent genetic studies on the *FOXP2* gene (in humans correlated to the rise of articulate speech) and the *MICROCEPHALIN* gene (associated with brain size).

Cedric Boeckx, in his "What Can an Extended Synthesis do for Biolinguistics: On the Need and Benefits of the Eco-Evo-Devo Program", zooms in on the distinction Chomsky made, together with Marc Hauser and Tecumseh Fitch, between the faculty of language in the broad sense (FLB) and the faculty of language in the narrow sense (FLN) (Hauser et al. 2002). The author demonstrates that the reception of this article has mostly resulted in a focus on FLN and how it contributes to human uniqueness, while Boeckx' thinks through the consequences of recognizing that language is decomposable into subunits and accepting that many of these subunits (FLB) are shared with other primates. His main argument is that recognizing FLB enables, if not altogether necessitates the introduction of a subbranch of biolinguistics, namely "comparative" biolinguistics. Originally, the field of biolinguistics focused on the human neurocognitive and ontogenetic capacity to develop I-language or language competence. By acknowledging, first of all, that there is no discrete and homogenous "language organ" in the brain; and secondly, by recognizing that multiple components of the FLB are shared with other primates and thus have an evolutionary history and continuity, biolinguistics now has to move beyond ontogeny and investigate how the subcomponents of the FLB evolved phylogenetically. From this, it also follows that one needs to investigate how the various elements of the FLB combined into our current language capacity, as well as how the unique properties of FLN (especially recursion and displacement), evolved. Boeckx argues that the Eco-Evo-Devo program holds potential to address these questions (see Racine et al., this volume). The Evo-Devo field originated exactly with the goal to reconcile phylogenetic evolutionary research with development. This field also takes on a more systems-theoretical approach, and studies on self-organization, exaptation, and emergence are key concepts in theory formation. Such concepts, Boeckx contends, enable a synthesis between synchronic, diachronic, and evolutionary linguistics. E-language is an emergent property of language users and self-organizes outside human individuals in language communities (see **Tamariz** contribution), while I-language appears to be an emergent property that results from the co-optation of various brain structures and cognitive capacities (see **Tattersall**, this volume). In pleading for the adoption of an Evo-Devo perspective, and following pioneering scholars such as Gould, Boeckx emphasizes that a functional-adaptationist approach, so characteristic of mainstream evolutionary linguistics and evolutionary psychology, is insufficient to explain the origin of anatomical form or the neurocognitive brain structures that underlie I-language. Without making the claim explicit, he argues that not only the ultimate (functional-adaptationist) causes of E-language, but also the proximate causes of I-language, or the physiological neurocognitive brain structures that underlie our human capacity for language, need to be the subject of scientific investigation. Here too, Boeckx hopes that taking on an Evo-Devo perspective will provide the answers to the questions long asked. Boeckx' paper is a programmatic one, wherein he most of all pleads for a moderation of several of the traditional, synchronic, and biolinguistic claims, in favor of the incorporation and integration of evolutionary and comparative studies in order to examine both the ontogenetic and phylogenetic emergence of I-and E-language. The incorporation of the Evo-Devo perspective will, according to the author, enable evolutionary pluralistic accounts that can complement the functionalist-adaptationist, modular and gene-centered theories of our time.

4 Current Trends, Future Questions, and Concluding Remarks

Every school of thought comes with scientific leaders, and all scientific leaders are visionaries that oppose the standard thoughts of their time and propose new ideas as well as methodologies to test them. These methodologies then serve as a canon for subsequent generations of researchers who, when following outlined ideas and methodologies, either confirm the postulated hypotheses and contribute to theory formation; or they find anomalies, contradictions, or new methodologies that lead to criticism and new ideas, which in turn enables theory formation by more clearly delineating the research problem (Kuhn 1962).

Behaviorists such as Skinner and Watson battled the phenomenological and nativist philosophers of their time by focusing on visible behavior and learning. With the goal to extend evolutionary theory to behavior and taking an instructionist approach in their behavioral studies, primatologists and experimental psychologists sought the observable continuation and boundaries of animal and human communication. Chomsky and Piaget battled Skinner and Watson by emphasizing the importance of cognition, and evolutionary psychologists and evolutionary linguists opposed both instructionist and cognitive thought by evolutionizing the existing theories. None of these advances could have been made if not for the previous work that had been done in the fields they eventually opposed, for theories do not come into existence de novo. Rather, scientific theories themselves are a shared learning experience of our species.

From within a variety of disciplines, scholars are now able to more clearly demarcate the means by which primates communicate gesturally, vocally, and linguistically. They have been able to point out many of the cognitive and anatomical requirements that enable communication and language. And they have been able to more scientifically ground that much of non-linguistic and linguistic communication is socioculturally embedded. The means used to communicate not merely depend upon anatomical structures and individual cognitive learning abilities, they also depend upon group behavior and sociocultural modes of information storage and teaching thereof. What anthropologists used to call the "super-organic structure" has equally become evolutionized, and today, the cultural dominion is either theorized to co-evolve with the biological realm or to form a complex emerging (adaptive) system (Steels 2002).

In sum, the various new epistemic frameworks that have been proposed to investigate the origin of social communication demonstrate that research is more successful when, rather than provide old answers, it creates new ways in which to formulate the questions and re-evaluate the obtained data. Having read the various chapters that make up this volume, we are happy to distinguish plenty new research avenues. How can we characterize the current trends in research on communication and language? And which directions should future scholars take?

What characterizes current research on emotions, expressions, vocal calls, or mental states is that the researchers involved first and foremost situate the sociocultural context in which these behaviors are displayed. And it is becoming more and more accepted that it is the sociocultural context that to a large extent facilitates if not initiates the rise of these communicative behaviors. The study of social communication therefore falsifies many of the original philosophical assumptions on the nature of perception, cognizing, and mental states. Expressions, perceptions, symbols, grammatical rules, and mental states become understood less as prototypical cognition-based entities, or static biologically or genetically determined instincts, and more as malleable and learnable sociocultural constructs. Consequently, what evolved on a biological level are not the various types of communicative behavior or specific cognitive traits, but the anatomical and cognitive means there are to initiate behavior that can become interpreted as communicative. As the various observations and experiments demonstrate, all primates, to a lesser or larger extent, are able to transform random behavior into functional and communicative behavior; to attribute meaning to meaningless gestures, expressions, sounds, or eye gazes; and to reach some level of mutual understanding. But questions that remain pertinent are whether, and to what extent, the biological evolution of anatomical modalities and neurocognitive features that facilitate communication are sufficient to explain the origin of social communicative behavior, and how sociocultural (evolutionary) mechanisms provide additional explanations. Where do we put epistemic importance? What is causally relevant? Anatomical structure, behavior, cognition, or culture?

Another pertinent question that remains is how human language differs from social communication. Does sociocultural communication suffice to define human language and explain its evolutionary origin? Volitional control, levels of arbitrariness, types of symbolization, intentionality and awareness of course vary in gradation between human and other primates, and our species indeed by far surpasses the capacities of other primates in how we attribute meaning and communicate on the past, present, and future of events. When does primate communication end, and when does language begin, or is there no difference? Stated otherwise, what is non-linguistic about primate communication? And what is non-communicative about language? If there is no difference, then distinguishing between primate communication and human language is unnecessary from an epistemic point of view. If there is a difference, one can ask whether that is sufficiently characterized by referring to recursivity, compositionality, and displaced symbolism; or by referring to the larger outreach human language appears to have.

A distinguishable trend is that while linguistic, cognitive, and behavioral studies used to focus on ontogenetic research, and even oppose evolutionary research, it is safe to say that today, all these fields have taken on an evolutionary perspective to examine the sources and the means by which communicative behavior evolved. For many years, evolutionary scholars merely had the modern synthesis framework at their disposal, and so they examined how social communication could have evolved by means of natural selection. Today, with the recognition that evolution can proceed by a myriad of evolutionary mechanisms, scholars are just beginning to investigate how epigenetics and evo-devo frameworks, drift theory, exaptation and niche construction theory, and cultural evolutionary processes can be put to use to make sense of the evolutionary origin of social communication.

Future research on social communication will be greatly advanced by examining how the myriad of evolutionary mechanisms, both biological and cultural, that are currently identified within an extended synthesis, can identify the proximate and ultimate causes of social communication in primates.

Finally, it is good to keep in mind that shifting epistemic outlooks and applying new methodologies might call for a reinterpretation of the data and the theory that explains it, but it does not falsify the data obtained. None of the theories that have developed through the years demonstrate less or more scientific rigor. A behavioral experiment that quantifies empirically measurable actions is not methodologically more or less impaired than a neurocognitive experiment that examines the presence or absence of voluntary control over these actions. Scientists expand on the different means there are to theoretically conceptualize, as well as methodologically and experimentally delineate research problems, and it is through this expansion that science progresses. It enables comparison of conceptual frameworks, methodologies, and experimental setups, and most of all, it enables combining the results of the latter. Both comparison and combination can only lead to a richer demarcation as well as understanding of the problem. Both are highly necessary, for neither of the theories on their own are able to answer the question how social communication or language evolved in primates. What adherents of these various frameworks have done, every single one of them, is provide perspectives on how the problem can be defined, examined, and experimentalized. The result in the end is that more data, theories, and methodologies are available to the researcher, which enables epistemic plurality. A future question therefore becomes how the various data and theories can be combined into fuller and more encompassing theories that explain the biological, developmental, cognitive, neurological, and cultural evolution of social communication in primates.

Acknowledgements We thank the Portuguese Fund for Science and Technology (SFRH/BPD/89195/2012) as well as the American John Templeton Foundation (Grant ID 36288) for their support. We are grateful to Slawomir Wacewicz for proofreading the text for English orthography.

References

Arbib M, Bickerton D (eds) (2010) The emergence of protolanguage: holophrasis vs compositionality. John Benjamins, Amsterdam

Armstrong D, Stokoe W, Wilcox S (1995) Gesture and the nature of language. Cambridge University Press, Cambridge

Axelrod R (1981) The evolution of cooperation. Science 211:1390–1396

Baars BJ (1986) The cognitive revolution in psychology. Guilford Press, New York

Barrett L, Dunbar RI, Lycett J (2002) Human evolutionary psychology. Palgrave, Hampshire

Bickerton D (1984) The language bioprogram hypothesis. Behav Brain Sci 7(2):173–188

Boyle KV, Bar-Yosef O, Stringer C, Mellars P (2007) Rethinking the Human Revolution. McDonald Institute for Archaeological Research: Mcdonald Institute Monographs, Cambridge

Byrne R, Whiten A (eds) (1988) Machiavellian inheritance: social expertise and the evolution of intellect in monkeys, apes, and humans. Oxford University Press, Oxford

Chomsky N (1959) A review of BF Skinner's verbal behavior. Language 35(1):26–58
Chomsky N (1965) Aspects of the theory of syntax. The MIT Press, Cambridge
Chomsky N (1972) Language and mind. Harcourt, Brace, Jovanovich, New York
Corballis MC (2002) From hand to mouth: the origins of language. Princeton University Press, Princeton
Cosmides L, Tooby J (1994) Beyond intuition and instinct blindness: toward an evolutionary rigorous cognitive science. Cognition 50:41–77
Darwin C (1872) The expression of the emotions in man and animals. John Murray, London
Donald M (1991) Origins of the modern mind. Harvard University Press, Cambridge
Dunbar R (1996) Grooming, gossip and the evolution of language. Harvard University Press, Cambridge
Fitch WT (2000) The evolution of speech: a comparative approach. Trends Cogn Sci 4(7):258–267
Fossey D (1983) Gorillas in the mist. Houghton and Mifflin Company, Boston
Fouts R, Mills S (1997) Next of kin, what chimpanzees have taught me about who we are. Perennial Currents, HarperCollins Paperbacks, New York
Gardner RA, Gardner BT (1969) Teaching sign language to a chimpanzee. Science 165:664–672
Gardner RA, Gardner BT, Van Cantfort TE (eds) (1989) Teaching sign language to chimpanzees. State University of New York Press, New York
Goodall J (1986) The chimpanzees of Gombe: patterns of behaviour. Harvard University Press, Cambridge
Gontier N (2006) Pathologies and the origin of language: an epistemological reflection. Cogn Syst 7(1):35–62
Gontier N (2009) The origin of the social approach in language and cognitive research exemplified by studies into the origin of language. In: Pishwa H (ed) Language and social cognition: expressions of the social mind. Mouton de Gruyter, Berlin, pp 25–46
Gontier N (2010) How to identify the units, levels and mechanisms of language evolution. In: Smith ADM, Schouwstra M, de Boer B, Smith K (eds) The evolution of language: proceedings of the 8th international conference (EVOLANG 8). World Scientific, London, pp 176–183
Gontier N (2012a) Applied evolutionary epistemology: a new methodology to enhance interdisciplinary research between the human and natural sciences. Kairos J Philos Sci 4:7–49
Gontier N (2012b) Selectionist approaches in evolutionary linguistics: an epistemological analysis. International Studies in thePhilosophy of Science 26:67–95
Haldane JBS, Huxley J (1927) Animal biology. Clarendon, Oxford
Hamilton WD (1964) The genetic evolution of social behavior, I and II. J Theor Biol 7:1–52
Hauser MD, Chomsky N, Fitch WT (2002) The faculty of language: what is it, who has it, and how did it evolve? Science 298:1569–1580
Heinroth H (1910) Beiträge zur Biologie, namentlich Ethologie und Psychologie der Anatiden. Berichte des V. Int. Ornithologen Kongresses, Berlin: S. 559ff
Hurford J, Studdert-Kennedy M, Knight C (eds) (1998) Approaches to the evolution of language. Cambridge University Press, Cambridge
Jenkins L (2000) Biolinguistics: exploring the biology of language. Cambridge University Press, Cambridge
Kellog WN, Kellog LA (1933) The ape and the child. Hafner Publishing Company, New York
Knight C, Studdert-Kennedy M, Hurford J (eds) (2000) The evolutionary emergence of language. Cambridge University Press, Cambridge
Kuhn TS (1962) The structure of scientific revolutions. University of Chicago Press, Chicago
Leavens DA, Hopkins WD, Bard KA (2005) Understanding the point of chimpanzee pointing: Epigenesis and ecological validity. Curr Dir Psychol Sci 14(4):185–189
Lorenz K (1937) Biologische Fragestellungen in der Tierpsychologie. Zeitschrift für Tierpsychologie 1:24–32
Lorenz K (1941) Kant's Lehre vom Apriorischen im Lichte gegenwärtiger Biologie. Blätter für Deutsche Philosophie 15:94–125

Lorenz K (1958) The evolution of behaviour. Sci Am 199(6):67–78
MacNeilage P (1998) Evolution of the mechanism of language output: comparative neurobiology of vocal and manual communication. In: Hurford J, Studdert-Kennedy M, Knight C (eds) Approaches to the evolution of language. Cambridge University Press, Cambridge, pp 222–241
McBrearty S, Brooks A (2000) The revolution that wasn't: a new interpretation of the origin of modern human behaviour. J Hum Evol 39:453–563
Mellar PA, Stringer C (1989) The human revolution: behavioral and biological perspectives on the origins of modern humans. Princeton University Press, Princeton
Mithen S (2006) The singing Neanderthals: the origin of music, language, mind and body. George Weidenfeld & Nicholson, London
Morris D, Collett P, Marsh P, O' Shaghnessy M (1979) Gestures: their origin and distribution. Stein and Day, Minnesota
Patterson FG (1978) The gestures of a gorilla: language acquisition in another pongid. Brain Lang 5(1):72–97
Piaget J (1972) The principles of genetic epistemology. Basic Books, New York
Piattelli-Palmarini M (ed) (1980) Language and learning: the debate between Jean Piaget and Noam Chomsky. Harvard University Press, Cambridge
Pinker S, Bloom P (1990) Natural language and natural selection. Behav Bran Sci 13(4):707–784
Puppel S (ed) (1995) The biology of language. John Benjamins Publishing Company, Amsterdam
Rumbaugh DM (ed) (1977) Language learning by a chimpanzee: the Lana project. Academic Press, New York
Savage-Rumbaugh ES (1986) Ape language: from conditioned response to symbol. Columbia University Press, New York
Skinner BF (1957) Verbal behaviour. Appleton-Century Crofts Inc, New York
Skinner BF (1986) The evolution of verbal behaviour. J Exp Anal Behav 45:115–122
Steels L (2002) Language as a complex adaptive system. In: Brisard F, Mortelmans T (eds) Language and evolution. Universiteit Antwerpen: Departement Germaanse Talen, Afdeling Linguïstiek, pp 79–88. [Antwerp Papers in Linguistics, 101]
Tallerman M (2007) Did our ancestors speak a holistic protolanguage? Lingua 117:579–604
Thorndike E (1911) Animal intelligence: experimental studies. The Macmillan Company, New York
Terrace HS (1979) Nim. Knopf, New York
Tinbergen N (1963) On aims and methods of ethology. Zeitschrift für Tierpsychologie 20(4):410–433
Tomasello M (2000) The cultural origins of human cognition. Harvard University Press, Cambridge
Tomasello M, Call J (1997) Primate cognition. Oxford University Press, New York
Whiten A, Byrne R (eds) (1997) Machiavellian intelligence II: extensions and evaluations. Cambridge University Press, Cambridge
Tyson E (1699) Orang-Outang, sive *Homo sylvestris*: or the anatomy of a Pygmie compared with that of a monkey, an ape and a man. To which is added, a philological essay concerning the pygmies, the cynocephali, the satyrs, and sphinges of the Ancients. Wherein it will appear that they are all either apes or monkeys, and not men, as formerly pretended. London
von Uexküll JJ (1909) Umwelt und Innenwelt der Tiere. Springer, Berlin
Watson JB (1913) Psychology as the behaviorist views it. Psychol Rev 20:158–177
Wilson EO (1975) Sociobiology: the new synthesis. Harvard University Press, Cambridge
Winsor M (1991) Reading the shape of nature, comparative zoology at the Agassiz museum. The University of Chicago press,Chicago

Part I
Philosophical and Historical Roots of Social Communication Studies

Lord Monboddo's *Ourang-Outang* and the Origin and Progress of Language

Stefaan Blancke

Abstract During the Enlightenment, the great apes from Africa and Southeast Asia sparked an intense debate about whether these animals should be considered human or not. Language played an important part in these discussions. Not only did the protagonists (anatomists, taxonomists, and philosophers) differ in their opinion over whether language should be regarded an essential part of human nature, but they also thought differently about the linguistic competence of the great apes. After briefly sketching this debate, I will focus on one eccentric voice, Lord Monboddo. This Scottish judge claimed that the *Ourang-Outang* were humans living in a primitive state and that the study of these creatures could tell us many things about the nature of man, his origins, and the progress of language. Monboddo was convinced that the *Ourang-Outang* had both the physical and mental capacities to acquire language and at one point even suggested an experiment in which a young ape would be taught to speak. Monboddo's worldview was built upon ancient Greek philosophy and the Great Chain of Being. Nevertheless, his ideas about the great apes still sound familiar to modern ears.

Keywords Great apes • Enlightenment • Nature of man • Language

1 Introduction

By the turn of the eighteenth century, the great nations of Europe were exploring large parts of the globe, thereby discovering an unanticipated richness in fauna and flora and revealing ever more exotic creatures. New specimens were constantly transferred to the homeland to tickle and satisfy the curiosity of both the intellectual

S. Blancke (✉)
Department of Philosophy and Moral Sciences, Ghent University, Ghent, Belgium
e-mail: st.blancke@gmail.com

M. Pina and N. Gontier (eds.), *The Evolution of Social Communication in Primates*, Interdisciplinary Evolution Research 1, DOI: 10.1007/978-3-319-02669-5_2,

elite and the vulgar and evoke awe for the wonderful world God had created. One type of animal in particular attracted attention. Adventures and tradesman brought back stories of human-like, hairy creatures that lived in the woods of Africa and on the islands of Java and Sumatra where they were called "men of the woods." Such reports also often claimed that when captured at a young age, these peculiar creatures could be taught to behave somewhat like humans. In 1698, the first young chimp arrived in England. It was consequently put on display in a freak show, but died soon after because of an infection. Its body was dissected by England's foremost anatomist at the time, Edward Tyson, a famous doctor and member of the Royal Society. His anatomical findings and the philosophical considerations he drew from them marked the beginning of a debate that would engage several important taxonomists, physicians, and philosophers in the Enlightenment focusing on two questions: (1) Are these creatures human and (2) why (not)? Much of this debate revolved around language, whether the great apes had it and whether it was a necessary property of a human being (Wokler 1995).

In this chapter, I will first sketch the great ape debate of the Enlightenment, thereby briefly highlighting the main actors and arguments. Next, I will zoom in on one particular and somewhat eccentric voice within that debate, the Scottish judge Lord Monboddo, and focus on his ideas about the origin and development of language and the role the great apes play therein. I conclude by comparing Monboddo's views with the modern study of communication of nonhuman primates and its importance for our understanding of the evolution of human communication.

2 *Ourang-Outang* in the Enlightenment

Before we proceed to the Enlightenment discussion about the human status of the great apes, it is important to realize that eighteenth century Europeans could not rely on the scientific information of these animals that we have today. Most of our modern knowledge about nonhuman primates has only been gathered in the last 50 years anyway from longitudinal field studies, repeatable lab experiments and close observations in zoos. Instead, they largely depended on reports from travelers of divers plumage who in turn had most of their information from locals and on some rare occasions from personal encounters with young animals. Unsurprisingly, these stories comprised a mash of real information, distortions, and exaggerations. The great apes were regarded as exotic instances of the *Homo sylvestris*, who in medieval mythology were the men and women of the woods who indulged in sexual excesses. They were believed to abduct boys and girls from the villages who they kept for both business and pleasure. Moreover, the word "orangutan" is Indonesian for "man of the woods," allegedly the product of a lustful woman and an ape, which strengthened the connection between reality and fiction. In addition, they were associated with Plinian human races such as the cynocephali, who were humans with a dog's head, or the sciapods who only had one foot that they used as an umbrella (Corbey 2005; Nash 1995).

Both the inaccurate and fanciful information and the mythologies in which the stories about the great apes were framed explain why European intellectuals struggled desperately to get their facts straight. They tried to distinguish one species from the other, label, and categorize them. These attempts to get solid grip on their subject were further hindered by the fact that the locals often ascribed very different names to these animals ranging from Jockos over Itsena to Barris. At some point, Europeans were able to discern smaller from bigger species (probably the chimp and the gorilla), but they did not make the modern distinction between chimpanzee, bonobo, gorilla, and orangutan. Instead, they used the term *"Ourang-Outang"* in a somewhat generic way, loosely referring to all great apes. To capture the uncertainty and mythological aspects of the Enlightenment debate about non-human primates and avoid modern connotations, I will employ the term *"Ourang-Outang"* in the rest of the chapter.

3 *Ourang-Outang*: Animal or Man?

Unsurprisingly, the most significant feature of the *Ourang-Outang* was its human-like appearance. From the very first observations, Europeans could not resist making morphological and behavioral comparisons between the creature and themselves. In 1641, Nicolaas Tulp (1593–1674), the Dutch anatomist who takes central stage in Rembrandt's famous painting *The Anatomy Lesson*, described a young chimp or bonobo from Angola, which he had seen alive in the menagerie of the Prince of Orange. "Orang-outang: sive homo sylvestris" he named the animal, and, believing that it had provided the source of inspiration for the ancient stories about satyrs, he decided to give his particular specimen the name of *Satyrus indicus*. Tulp noticed that this creature, like an animal, walked on all fours (although, he admits, it often walked upright). The face, however, resembled that of a human, with a flat and crooked nose of an old woman. Furthermore, the ears and chest were of the human form, and the shape and physiology of its limbs were identical to those of a human. Moreover, wiping its mouth after drinking from a can, and sleeping with its head on a pillow and modestly covered under a sheet, it also behaved remarkably human (Tulp 1641). Tulp's fellow countryman, Jacobus Bontius (1592–1631), who worked as a physician in Java, even grotesquely exaggerated the resemblance with humans by depicting his *Ourang-Outang* as a furry version of a human female. He also mentioned that "the Javanese claimed that the *Ourang-Outangs* could talk, but that they did not want to because they did not want be forced to work" (Bontius 1658, my translation).

Although these accounts clearly hinted at the close relationship between humans and the *Ourang-Outangs*, the authors did not make any attempts to establish the particular nature of that relationship. This changed 50 years later. In 1698, a young chimp was brought to England and was consequently put upon display in a freak show in London attracting the attention of both the common people and the intellectual elite, including Edward Tyson (1650–1708), at that time England's foremost anatomist (for a detailed biography, see Montagu 1943). The animal, however, suffered from a serious infection on a wound it had incurred during the long journey from Angola and died three months after its arrival. Tyson seizes the opportunity to dissect the body of this peculiar creature with both hands. One year later, he publishes his results of his anatomical research in a volume with the title *Orang-Outang, sive Homo sylvestris: or the anatomy of a Pygmie compared with that of a monkey, an ape, and a man. To which is added, a philological essay concerning the pygmies, the cynocephali, the satyrs, and sphinges of the Ancients. Wherein it will appear that they are all either apes or monkeys, and not men, as formerly pretended* (Tyson 1699).

Acknowledging the similarities between his subject and Tulp's satyr, Tyson decides to adopt the name *Ourang-Outang*. However, uncertain whether both specimens actually belong to the same species—he slightly distrusts the Dutchman's account—he distinguishes both creatures by calling his ape a *Pygmie*. Like Tulp before him, Tyson believes that the ancient stories of strange human races including satyrs and pygmies are founded upon inaccurate and fanciful accounts of real, human-like creatures such as his *Ourang-Outang*, an idea that he elaborates in the philological essay attached to the part containing his anatomical findings.

In his anatomy, Tyson does not only relate the results of the dissection on his *Pygmie*, but he also contrasts them with the available data on the anatomy of a man and a monkey. In conclusion of this comparison, Tyson constructs two lists: in the first, he enumerates the traits in which the *Pygmie* is more similar to a monkey than a man. In the second, he tallies the properties in which the creature resembles a human more than a monkey. The reason why he introduces this methodological novelty—Tyson is regarded as the father of comparative anatomy—lies in his ardent belief in the Great Chain of Being. This cosmological view, which had its origins in Greek philosophy, had become most popular in Britain in Tyson's days. It depicted the universe as a static whole created by God in which each species was hierarchically and gradually ordered in a large chain from the simplest of minerals to the most perfect creature, God himself. One of the main implications of the Great Chain of Being was that no real gaps could exist between the different realms in nature (Lovejoy 1936). With his *Pygmie*, Tyson believed he had bridged the gap between humans and the rest of earthly life. The lists proved this very point:

> T is a true remark, which we cannot make without admiration; that from minerals to plants; from plants to animals; and from animals to men; the transition is so gradual, that there appears a very great similitude, as well between the meanest plant, and some minerals; as between the lowest rank of men, and the highest kind of animals. The animal of which I have given the anatomy, coming nearest to mankind; seems the nexus of the animal and the rational (Tyson 1699).

Although in the chain the animal takes its place between animals and humans, it is no more than an animal. Tyson does not arrive at this conclusion after meticulously evaluating the two lists he composed: in the end, the animal had more features in common with humans than with monkeys. For instance, he assumes that the animal walked upright and had it represented as such supported by a stick, in the wonderful drawings of his colleague William Cowper. Somewhat to his own surprise, Tyson also found that the *Pygmie*'s brain looked quite similar to a human brain and that the animal was endowed with speech organs. However, Tyson nor anyone else had heard the animal speak when alive which leads him straight to the conclusion that the *Pygmie* lacked the metaphysical principle that enables humans to reason and communicate their thoughts by language. As a true missing link, the *Pygmie* possessed the necessary bodily requirements to talk, but missed the spiritual qualities that would allow the animal to actually use them. Based on this Cartesian, dualistic criterion, in which language, or rather the absence of it, plays central stage, Tyson concludes that the *Pygmie* is not a man.

More than half a century later, the highly respected and internationally renowned French naturalist, Georges-Louis Leclerc de Buffon (1707–1788) endorsed Tyson's conclusion. Being the head of the royal gardens, Buffon had set himself the task of assembling, clarifying, and categorizing the then available knowledge about the natural history of life on earth. The result of this immense enterprise was his *Histoire Naturelle*, of which the first volume was published in 1749. Forty-three more volumes would follow, some of which were published posthumously. In the fourteenth volume, published in 1766, Buffon attempts to get some grasp on the ever growing but increasingly confusing amount of data concerning the *Ourang-Outang*. He is also familiar with Tyson's work of which he provides an extensive summary and he agrees with the English anatomist that the animal comes the closest to man in bodily form and structure, both externally and internally. However, resemblance is not everything, because these properties do not resort the same effect. Buffon explains:

> For example, the tongue and all the speech organs are the same as in a man, and yet the *Ourang Outang* does not speak; the brain is absolutely of the same form and of the same proportion, but yet he doesn't think; can it be more evident that matter only, despite its perfect structure, cannot produce thought or language, unless it is animated by a superior principle?

From this, one can draw only one conclusion: The *Ourang-Outang*'s extraordinary human-like properties "do not bring it any closer to the nature of man, nor do they elevate it above the nature of animals."

As such, Buffon disagrees with the classification of Carolus Linneaus (1707–1778), the Swedish taxonomist who is famous for introducing the modern scientific method of labeling species with two Latin names. In 1758, Linneaus caused quite a shock by categorizing humans within the realm of animals, instead of above it. More particularly, he put the genus *Homo* in the order of Primates and distinguished two human species, *Homo sapiens* (*H. diurnus*) and *Homo troglodytes* (*H. nocturnus*) with the latter including Bontius' *Homo sylvestris Orang-Outang* (Chazan 1995; Linneaus 1758). His contemporaries reacted appalled as they considered his classification an attack on human dignity (Corbey 2005). Linneaus believed that the *Ourang-Outang* dwells in caves during the day and forages for food at night and has, like cats, a membrana nictitans, a third eyelid.

He also mentioned that the creature can speak, albeit in a hissing tone (Linneaus 1758). However, Buffon very much doubts Linnaeus' account as he had personally seen a small *Ourang-Outang* (which he called *Jocko* to distinguish it from the larger type called *Pongo*) alive on several occasions, and never heard it speak or hiss. In fact, the animal behaved no better than a well-trained dog. Moreover, Buffon suspected that Linnaeus' animal did not exist and that it was probably based on a poor and distorted description of a "white negro" (Buffon 1766). As such, Buffon again widened the gap between man and animal, a gap that Linnaeus had dared to close (Corbey 2005).

4 Lord Monboddo's *Ourang-Outang*

Because of his authority, Buffon seemed to have settled the matter definitely in following Tyson's suggestion that the *Ourang-Outang*, although it had been endowed with the appropriate anatomy, could not speak and therefore lacked a metaphysical principle that belonged exclusively to the human species. In Scotland, however, an important jurist, James Burnett, better known as Lord Monboddo (1714–1799) (for a detailed biography, see Cloyd 1972), disagreed wholeheartedly with this conclusion and opposed it vehemently and repeatedly in the course of his two great works, *Of the origin and progress of language* (OPL, 1773–1792) and *Antient Metaphysics* (AM, 1779–1799), six volumes each. Most relevant for a study of his thoughts on the *Ourang-Outang* are the first volume of OPL, in particular the second edition which has an entirely new chapter on the *Ourang-Outang*, and volumes three, four, and five of AM. The first edition of the first volume of OPL, however, contains the clearest and most concise statement of his ideas on the history of man and the origin of language.[1]

4.1 Why Language is Not Essential to Man

Monboddo finds it startling that Tyson, but particularly Buffon, deny the *Ourang-Outang* the human status because they have never heard the animal speak. By that criterion, he reasons, we can no longer consider young children and dumb people to be human as well. They too have the necessary physical properties, yet they do not talk. Would Buffon and Tyson really be willing to draw such harsh conclusions? Obviously, there is something wrong with their ideas about human nature. The main problem, Monboddo argues, is that they form their ideas about man only on the basis of their knowledge about modern man. To get at human nature, however, one needs to look at man in his natural state, uncorrupted by civilization. Only then can a man of science deduce the properties essential to man. Monboddo

[1] I here follow Barnard (1995).

indeed believes that man has not always existed in his current state. Man in his natural state was a hairy animal, crawling on all fours and solitary. However, he disagreed with Hobbes that this natural state could be described as a continuous battle of one man against the other. Rather, Monboddo pictured natural man to be a peaceful animal, enjoying a warm and hospitable climate and living of the fruits of the earth, probably somewhere in Asia. As the animal lived the largest part of its live alone, it had no need for language and therefore could not speak. In fact, it could not do much anything.

Another reason why Tyson and Buffon ended up on the wrong track, Monboddo claims, is that they neglected much of the philosophy of the ancient Greeks. A great admirer of Greek civilization, language, and thought, Monboddo thinks very little of the novel ideas proposed by modern thinkers such as John Locke, George Berkeley, and David Hume. He also opposes Newtonian physics in defense of the Aristotelian alternative it had replaced. The same goes for his anthropological views. According to Monboddo, Aristotle provided us with the most beautiful and accurate description of the essence of man. The teacher of Alexander the Great had defined man as "a rational animal, mortal, capable of intellect and science." (Monboddo 1774) As such Aristotle gets at the core of what it means to be human, but the definition also situates us in the larger scheme of things. Like Tyson, Monboddo endorses the cosmological view of the great chain of being, stating for instance that "[t]his is the scale of being, rising by proper gradations, from mere matter and sense to intellect." (Monboddo 1773) Aristotle's definition perfectly determines man's place within that chain. The first part, "rational animal" situates humans among the higher animals, including horses, elephants and dogs, which master at least some basic reasoning processes. The second part, the fact that humans are mortal, distinguishes the human species from the purely spiritual beings and the last part "capable of intellect and science" separates us from the rest of the material world. What is important about the definition is that it does not determine the essence of man in terms of actual properties, but in terms of capacities. As such, the possession of language by itself does not constitute a necessary condition to be called human, one only needs to have the potential of acquiring that particular property. Moreover, language, albeit a wonderful human faculty, is but one of many signs of the intellectual capacities of man, next to using sticks for weapons, building huts, making fire, and many other cultural habits. With Monboddo, in a sense, language looses the significance it had for Tyson and Buffon in determining the nature of man.

4.2 Language and the History of Man

Nonetheless, language still plays an important part in Monboddo's anthropological views as it allows us to track the intellectual development of the human species. Following ancient Greek philosophy, Monboddo argues that language consists of a material part, articulation or speech, and a formal part, the concepts expressed by language. As humanity progresses from its primitive animalistic toward the intellectual state of modern Europeans, language follows suite and becomes more and more complex and abstract. As long as man lived in his natural state, alone or simply herding together, he had no need for language. Only when circumstances or the "necessities of human life" (Monboddo 1773), which were mainly the need for sustenance and defense against other conspecifics, induced humans to start living in societies and to collaborate, the need for communicative tools became pressing and language started to develop. At first, when interactions were only quite basic, inarticulate cries and signs would suffice. However, as their wants increased, and they needed to communicate increasingly complex information, articulation became necessary, until finally a proper language would emerge.

> [T]he first sounds articulated were the natural cries of men, by which they signified their wants and desires to one another, such as calling one another for certain purposes, and other such things as were most necessary for carrying on any joint power. Then in process of time other cries would be articulated, to signify that such and such actions had been performed, or were performing, or that such and such events had happened relative to common business. Then names would be invented of such objects as they were conversant with. This increase of words would make more articulation necessary. And thus the language would grow by degrees; and as it grew, it would be more and more broken by consonants; but still the words would retain a great deal of their original nature of cries: and thus things would go on, words still multiplying, till at last the language became to cumbersome for use; and then art was obliged to interpose, and form a language according to rule and method. (Monboddo 1773)

This process in which both intellect and language ascend on the mental stairs toward ever more abstractive thinking accumulates in the ancient Greek culture and language, which are the most perfect culture and language humanity has ever produced. Since that glorious period, however, things have been going gradually and steadily downward. The development of our intellectual potential also leads to luxury and unnatural habits and diets that go to the detriment of our animal nature and make us weaker and weaker. Languages suffer likewise and deteriorate. To a certain extent, man is able to counter the degradation of his body and mind by exercise and extra care—Monboddo was renowned for the naked sunbaths he took on his estate. Nonetheless, in the end, our intellectual part would inevitably break free from our bodies and in a grand cosmological drama, the human race as a whole would occupy a higher rank in the great chain of being. As such, the history of man mimics the life of an individual:

> [I]f it be true [...] that this scene of man is to have an end, as well as the present system of nature, and that man is to appear again in some other form, as we are told the heavens and the earth will do, it is according to the order of nature that this change of this state should not happen at once, but should come by degrees, and, consequently, that the species should decline, degenerate, and become old, as we see the individual does, before its extinction.

Hence, the current, miserable state of modern man is but a necessary step in the progress from the mere animal to a higher spiritual being. As such, man has come a long way from being the natural creature as defined by Aristotle.

4.3 *Ourang-Outang* and the Natural State of Man

Monboddo realizes that in order to substantiate his views about the history of man he needs to demonstrate to his contemporaries that man indeed once lived in a natural state. To prove his point, he turns to the several alleged instances of individuals and people living in such a state. He is rather ambiguous as to whether the purely natural man still exists. Nonetheless, several proxies are available to him which are sufficiently convincing that man indeed once lived in a natural state and consequently progressed from that state in subsequent stages. Closest to the natural state come the feral or wild children—neglected or abandoned youngsters who were believed to live in the woods like animals—epitomized by Peter the Wild Boy. Peter was found in Germany and consequently transferred to the court of King George spending the rest of his life in England. He was found living solitary, crawling on all fours, and feeding upon grass and the moss of the tree. Monboddo is fascinated by this character. He visits him twice and features him repeatedly throughout his publications, including an account of one of his own visits (Monboddo 1784). Although Peter had lived almost like an animal and had only learned to speak a few words, no one would deny that he was human. As such, Peter's existence beautifully supported Monboddo's anthropological views:

> If these facts concerning Peter be true, and the inferences I have drawn from them just, such a living example of the state of nature will, I hope, satisfy even the men of experiment, who will believe nothing but what falls under the evidence of their senses. (Monboddo 1784)

The *Ourang-Outang* represents the next stage in which the natural men lived together, performed some joint actions but yet had no need for language. Monboddo saw a stuffed *Ourang-Outang* in the French king's cabinet of curiosities (Monboddo 1773) and encountered two specimens alive in London (Monboddo 1795), but most of the information he relies upon is secondhand. The Scottish judge proves to be quite gullible in response to the data of ancient historians and modern travelers on the *Ourang-Outang* and other strange types of man. Men with tails, Satyrs with feet of goats and with horns upon their heads, men with heads of a dog and men without heads, but with eyes in their breast, and mermaids, Monboddo accepts their existence without hesitation (Monboddo 1784, 1795). And why should he? According to Aristotle, "everything exists which is possible to exist," nor can Monboddo imagine

> that a benevolent and omnipotent Being, infinite in production as in everything else, should not have produced every sensitive being that is capable of pleasure, and can enjoy a happiness suitable to its nature, whose existence is possible, that is, implying no contradiction; for otherwise there would be something wanting in the System of Nature, which would not be perfect and complete, as, I think, of necessity it must be. (Monboddo 1784)

The same credulity Monboddo applies to the stories about the *Ourang-Outang*. These stories often exaggerate the human-like features of this creatures but Monboddo sees no reason to be skeptical. The humanity of the *Ourang-Outang* fits perfectly within his philosophy. Besides, he was not the first to believe that

Ourang-Outang was of the human species. Rousseau, one of the few modern thinkers who Monboddo admired, had suggested as much in the tenth footnote of his *Discourse on inequality*. The French philosopher, however, remained suspicious of the travelers' reports he relied upon (Rousseau 1984 [1755]). Monboddo on the other hand regarded them a trustworthy source of valuable information. And they clearly demonstrate that the *Ourang-Outang* is human. Remarkably, Monboddo also refers to the creature's emotions as evidence of its human status:

> He has the sense of what is descent and becoming, which is peculiar to man, and distinguishes him from the brute as much as anything else. And he has a sense of honour [...]; for he cannot bear to be exposed as a show, nor to be laughed at; and travelers mention examples of some of them having died of vexation, for being so treated. He has also the feeling of humanity in a strong degree; and a sense of justice [...]. Further, he has made some progress in the arts of life; for he builds huts, and he has got the use of a stick for attacking or defending [...]. He has also the use of fire [...]; and lastly, he buries his dead. (Monboddo 1795)

Although these men of the woods do not (yet) talk, they appear to have sufficient skills and display clear signs of their intellectual capabilities to include them, under Aristotle's definition, within the human species. To Monboddo, the humanity of the animal is evident. In the third volume of *Antient Metaphysics,* he asks rhetorically:

> If an animal, who walks upright, is of the human form, [...] who uses a weapon for defence and attack, - associates with his kind,—makes huts [...],—is tame and gentle, [...] [who] is capable of great attachments to particular persons, [...];—who has so much of the docility of man, that he learns, not only to do the common offices of a menial servant, [...] but also to play upon the flute; [...]—and lastly, [...] has the organs of pronunciation, and consequently, the capacity of speech, though not the actual use of it;—if, I say, such an animal is not a man, I should desire to know in what the essence of man consists. (Monboddo 1784)

Because of these characteristics and traits, the *Ourang-Outang* can be considered even more human than Peter the wild boy. But, maybe the skeptics need more evidence?

4.4 Ourang-Outang and Language

The fact that the *Ourang-Outang* shows clears signs of intelligence and has the organs of pronunciation, as Tyson had demonstrated, lead Monboddo to assume that if taught properly an Ourang-Outang could definitely learn how to speak. If that would happen, the skeptics would have to admit they were wrong. Being absolutely certain about the possibility, Monboddo proposes a bold experiment:

> It will naturally occur that this controversy might be easily decided, by trying the experiment upon an Ourang Outang, whether he could be taught to speak, and I should be very glad the experiment were tried and if the creature was young [...] I should think it was probable the experiment would succeed. (Monboddo n.d.)

However, even if the experiment would fail, Monboddo would not have been proven wrong. He explains:

> But if it should not I would not therefore give up my hypothesis. For in the first place [...] articulation is a business of very great difficulty, and it is well known that all savages are naturally very indolent, it's not improbable, that the Ourang Outang would not be at the trouble necessary to acquire language. (Monboddo n.d.)

The manuscript probably lay at the basis of the added chapter on the *Ourang-Outang* in the second edition of the first volume of *Of the origin and progress of language*, in which he reiterates the proposal (Monboddo 1774). Several years later, however, in 1779, the Dutch physician Peter Camper published a report in the form of letter to the president of the Royal Society, John Pringle, in which he sums up his findings on the anatomy of the ape and the *Ourang-Outang*. Unlike Tyson, Camper had not found any speech organs, which to him adequately explained why the monkey nor the *Ourang-Outang* could speak. They simply did not have the appropriate equipment to do so (Camper 1779). Monboddo, however, is not impressed. Tyson's *Ourang-Outang* came from Angola, whereas Camper's was brought from Borneo. Clearly, these were two entirely different creatures (Monboddo 1784). As such, Monboddo could simply discard Camper's anatomical data and continue to maintain that the *Ourang-Outang* could be taught to speak in the third volume of *Antient Metaphysics* (Monboddo 1784). Understandably, Monboddo was not going to give up on the humanity of the *Ourang-Outang* that easily. The human status of the creature was simply too important to prove several essential points in his philosophy. The man of the woods demonstrated that once humans had lived in a natural state, like an animal, and that language was not, contrary to the opinion of many of his contemporaries, essential to man. As a result, Monboddo never ceased to refer to the *Ourang-Outang* in support of his views on the essence and the history of man and the proper role of language in both.

5 The Modern Monboddo

Despite his very traditional worldview and his ancient views on the essence of man, some of the ideas Monboddo proposes come across as quite contemporary. Particularly, the idea that a close connection pertains between humans and the great apes in their intellectual and emotional repertoire rings very familiar to a modern ear—although Monboddo often exaggerates the similarities. Of course, today we know that this connection obtains because of our shared ancestry, and in recognition of this evolutionary history, humans and the other great apes are categorized within the family of the *Hominidae*. Several researchers even argue that humans should be considered a third type of chimpanzee (Diamond 1993) or, on the basis of genetic similarities, that chimps should be included within the genus *Homo* (Wildman et al. 2003). Moreover, according primatologists and philosophers have suggested that the gap should not only be bridged taxonomically but also morally, meaning that the great apes, because they display such a richness in emotional and cognitive capacities, should be granted basic rights (Cavalieri and Singer 1993). Monboddo also closed the gap, albeit in a somewhat different way. Most importantly, he believed that the human species had a special ontological status in relation to all other animals. As such, by including the great apes within the human species, Monboddo did not remove the gap, but replaced it. Nonetheless, by emphasizing our animalistic origins and employing the *Ourang-Outang* as a proxy for that natural state—and thus explicitly associating humans with animals—, Monboddo certainly infringed on the dignity of his contemporaries. In 1802, Martinus Stuart, a Dutch protestant parson and publicist wrote: "Let Camper's memory be blessed whose comparative dissections have liberated you from the unbearable humiliation to which a Monboddo intended to bring you—that you should have to call the disgusting Orang-Outang your brother." (quoted in Corbey 2005) 50 years later, Darwinian theory would raise similar concerns that still play up today, particularly in religious circles.

Another important aspect in which Monboddo appears to be a forerunner, and perhaps more relevant to the theme of this volume, is his suggestion that we can use data on the social communication and linguistic capacities of the great apes to learn more about (the evolution and development of) human language. On the basis of the available data, he surmises that language started out as signals and inarticulate cries, as the *Ourang-Outang* produced, and that the *Ourang-Outang* has the required intellectual capacities to learn how to speak. Monboddo, however, had been misinformed by Tyson's anatomy about the presence of speech organs in the great apes and he restricted language to articulate sounds. Today, the concept of language has been broadened to include sign language as well, which has allowed researchers to study the mental capacities of primates by testing their ability to signal particular items and concepts. By trying to teach American Sign Language to some individual great apes, they have taken up Monboddo's challenge and show that these animals are indeed capable to master some basic language. The results, however, remain quite controversial. Nevertheless, even if those experiments would have failed completely, as the skeptics believe, the natural communication of primates can still tell us a great deal about the evolution of language. For instance, Pollick and de Waal (2007) claim that "the natural communication of apes may hold clues about language origins, especially because apes frequently gesture with limbs and hands, a mode of communication thought to have been the starting point of human language evolution." This sounds like a true modern Monboddo.

6 Concluding Remarks

The Enlightenment debate about the *Ourang-Outang* show that almost from the very first encounters, these remarkable creatures have raised fundamental questions about our humanity. Moreover, the debate also reveals that language has always constituted an important ingredient within the discussions concerning the relationship between the great apes and ourselves. In that regard alone, the Enlightenment debate still resonates within the modern discussions and is therefore certainly worthy of our attention. However, what is perhaps even more surprising and fascinating is that the answers that some of the protagonists provided align beautifully with modern approaches to the study of the great apes. Monboddo in particular seem to have been right on the spot on a number of occasions which is quite ironic in light of his contempt for the modernist ideas of his contemporaries. In addition, Monboddo is no evolutionist, let alone a scientist in the modern meaning of the term. An armchair philosopher, he founds his ideas mainly upon secondhand information, which allows him to mold the *Ourang-Outang* into a creature that beautifully serves his purposes. Nevertheless, even with these constraints in place he is able to pick out essential criteria for establishing the close connection that obtains between ourselves and the great apes and to make astute observations about the intellectual capacities of the latter, some of which were later confirmed by professional and evolutionarily informed scientists. Finally, he is one of the first people to appreciate the importance of studying the great apes in order to improve our understanding of who we are.

Acknowledgements The research for this chapter was funded by Ghent University (BOF13/24 J/089).

References

Barnard A (1995) Monboddo's Ourang Outang and the definition of Man. In: Corbey R, Theunissen B (eds) Ape; man, apeman: changing views since 1600. Department of History of Leiden University, Leiden

Bontius J (1658) Historiae naturalis et medicae Indiae orientalis libri sex in Gulielmus Piso, De Indiae utriusque re naturali et medica, libri quatuordecim. Elsivirium, Amsterdam

Buffon GL (1766) Histoire Naturelle, vol. XIV. Paris

Camper P (1779) Account of the organs of speech of the Orang Outang in a letter to Sir John Pringle. Phil Trans R Soc LXIX:139–59

Cavalieri P, Singer P (eds) (1993) The great ape project. Equality beyond humanity. St. Martin's Griffin, New York

Chazan M (1995) The meaning of Homo sapiens. In: Corbey R, Theunissen B (eds) Ape; man, apeman: changing views since 1600. Department of History of Leiden University, Leiden

Cloyd EL (1972) James burnett, Lord monboddo. Clarendon Press, Oxford

Corbey R (2005) The metaphysics of apes. Negotiating the animal-human boundary. Cambridge University Press, Cambridge

Diamond J (1993) The third chimpanzee. In: Cavalieri P, Singer P (eds) The great ape project. Equality beyond humanity. St. Martin's Griffin, New York

Linneaus C (1758) Systema Naturae, 10th edn. Laurentius Salvius, Stockholm

Lovejoy AO (1936) The great chain of being. The history of an idea. Harvard University Press, Cambridge

Monboddo L (n.d.) Of the Ourang Outang, and whether he be of the human species. Monboddo papers, National Library of Scotland

Monboddo L (1773) Of the origin and progress of language, vol 1. Balfour and Cadell in the Strand, Edinburgh and London

Monboddo L (1774) Of the origin and progress of language, vol 1, 2nd edn. Balfour and Cadell in the Strand, Edinburgh and London

Monboddo L (1784) Antient metaphysics, vol 3. Cadell in the Strand, London

Monboddo L (1795) Antient metaphysics, vol 4. Bell and Bradfute, Edinburgh

Montagu A (1943) Edward Tyson, M.D., F.R.S 1650–1708. American Philosophical Society, Philadelphia

Nash R (1995) Tyson's Pygmie: the Ourang-outang and Augustan 'Satyr'. In: Corbey R, Theunissen B (eds) Ape; man, apeman: changing views since 1600. Department of History of Leiden University, Leiden

Pollick AS, de Waal FBM (2007) Ape gestures and language evolution. Proc Natl Acad Sci 104:8184–8189

Rousseau J (1984 [1755]) A discourse on inequality. Penguin Group, London

Tulp N (1641) Observationes medicae. Elsivirium, Amsterdam

Tyson E (1699) Orang-Outang, sive Homo sylvestris: or the anatomy of a Pygmie compared with that of a monkey, an ape and a man. To which is added, a philological essay concerning the pygmies, the cynocephali, the satyrs, and sphinges of the Ancients. Wherein it will appear that they are all either apes or monkeys, and not men, as formerly pretended, London

Wildman DE, Uddin M, Liu G, Grossman LI, Goodman M (2003) Implications of natural selection in shaping 99.4 % nonsynonymous DNA identity between humans and chimpanzees: enlarging genus Homo. Proc Natl Acad Sci 100:7181–7188

Wokler R (1995) Enlightening apes: eighteenth-century speculation and current experiments on linguistic competence. In: Corbey R, Theunissen B (eds) Ape; man, apeman: changing views since 1600. Department of History of Leiden University, Leiden

Ferality and Morality: The Politics of the "Forbidden Experiment" in the Twentieth Century

Sandra Swart

Abstract In the first decades of the twentieth century, several experiments were conducted to compare the development of human and non-human animals. These investigations were premised on "cross-fostering"—raising animal and human babies together in human families. Such experiments were themselves loaded with difficult ethical questions, grappled with in the fledgling fields of primatology, ethology and, particularly, psychology. Yet the antithetical case study, where a human baby was raised by apes—the so-called Forbidden Experiment—violated too many social norms and ethical principles to be conducted. Thus, academics embraced the possibility of "natural" case studies. This essay tracks the various cases of "real Tarzans" that materialised, with special focus on the "Baboon Boy" of South Africa. Researchers used the feral children to bolster their own arguments about nature and nurture, particularly with reference to early child development. The essay delineates the contesting contentions which erupted in international academic circles in the 1930s and 1940s, which created the intellectual context for ensuing cases of feral children. Thus, the essay traces change over time in the shifting understanding of the "Forbidden Experiment".

Keywords Feral child • Primatology • Baboon • Forbidden experiment

1 Introduction

Two babies were raised by a young couple anxious not to miss a moment. From seven and a half months and from 10 months, respectively, the little girl and the little boy had every nuance of their development recorded. The baby boy

S. Swart (✉)
History Department, Stellenbosch University, Stellenbosch, Republic of South Africa
e-mail: sss@sun.ac.za

M. Pina and N. Gontier (eds.), *The Evolution of Social Communication in Primates*, Interdisciplinary Evolution Research 1, DOI: 10.1007/978-3-319-02669-5_3,

was the couple's biological child, and the baby girl was fostered. From 26 June 1931, over a 9-month period, the mother and father watched closely as their infants' emotions, physical abilities and intellectual acumen developed. They filmed the siblings obsessively, taking reel after reel of 16 mm, black-and-white film. Like all parents, they could not help comparing their progeny. Unlike most parents, however, they published their results.

The other difference was that only one of their babies was human. The baby girl was a chimpanzee named Gua, temporarily adopted from Yale's Anthropoid Experiment Station, in order to be compared with her human foster-brother, Donald, by parent-scientists Winthrop and Luella Kellogg.[1] Kellogg devised the experiment in 1927, while still a graduate student, amidst the publicity surrounding the Indian "wolf children", debated in the *American Journal of Psychology* [Squires (1927), for this hypothesis, see Benjamin (1982)]. Like the classic wolf-suckled children from Herodotus's twins to Rudyard Kipling's Mowgli, "Amala" and "Kamala" were purportedly nursed by a loyal she-wolf, in Midnapore (or Midnapur) in Bengal (for discussion, see Benzaquén 2001) who had to be shot in order to separate her human cubs from her. The wolf-girls were captured or rescued (depending on a lupine or human viewpoint) in 1920 by a Reverend J.A.L. Singh, a Bengali-born, pro-Empire missionary, who cared for them at his orphanage, until Amala's and Kamala's early deaths in 1921 and 1929, respectively. The series of articles they inspired recorded that the girls retained wolf-ways, crawling on all fours, "pouncing on and devouring" small mammals and birds, howling into the night, rather startling the other orphans (Kellogg 1931b). While Squires and some others argued that the girls could not learn to be human owing to their being born with subnormal intelligence, Kellogg, by then Associate Professor of Psychology at Indiana University, argued that they were born normal but *learned* to be wolves because that is what their environment demanded of them (Kellogg 1931a, b). He followed the behaviourist line in arguing for the enduring impact of early experience, which rendered the feral child impossible to draw back into human society. This followed the robust "objective" behaviourism of John B. Watson, epitomised in his 1925 *Behaviorism*, which argued that environment so impacts on heredity that *any* child can be shaped into *any* adult. Kellogg then drew the Watsonian links to the feral:

> Instead of supposing that the 'wild' children were inherently feeble-minded, as is usually done, the proponent of the environmental doctrine would hold that originally such children were probably normal... Those placed with animals may actually have learned... to be wild themselves, in the same way that a Caucasian child reared among the Chinese grows into the Chinese customs and language, or a baby that has been kidnapped by Gypsies knows in later years only the Gypsy manner of living. (Kellogg and Kellogg 1933: 5–7).

---

[1] Gua was born in Cuba on 15 November 1930, delivered to the Kelloggs in June 1931, while Donald was born 31 August 1930. There is nothing unusual in a scientist's observing his or her own children—both Darwin and William Prayer watched their offspring with researchers' eyes.

2 Deprived or Depraved?

Kellogg himself toyed with the idea that a definitive way to test his theory of the power of early environmental conditioning would be to place a normal infant in a deprived environment. This was a reference to a long-held idea, dubbed the "Forbidden Experiment" by Roger Shattuck, that children could be raised in highly unusual circumstances to test certain theories (Shattuck 1980). It has a long lineage, dating to seventh century BC when, Herodotus insists, an Egyptian pharaoh, endeavouring to determine the first language of humanity, raised an infant in isolation. He discovered that the original language was Phrygian—as the child's first word sounded like *bekos* (Phrygian for "bread").[2] The Enlightenment philosopher Montesquieu, calling it a "beautiful experiment", suggested raising "three or four children like animals, with goats or with deaf-mute nurses. They would make a language for themselves. Examine this language. See nature in itself, and freed from the prejudices of education; learn from them, after they are instructed, what they had thought; exercise their mind by giving them all the things necessary to invent; finally, write the history of the experiment" (Montesquieu quoted in Benzaquén 2006: 113). In the early twentieth century, the clandestine attraction of such an experiment remained. Kellogg, however, conceded that his deprivation experiment would be both "illegal" and "immoral", so he conjured up another way—both moral and legal—to test his premise. He would perform the mirror image of that experiment—he would take an animal into his home and raise it as a human (Kellogg and Kellogg 1933).

Of course, others had tried. On some level, the Victorian era's insistent and sentimental anthropomorphising by the middle classes of their pets was efforts to create furry little humans. Animals were increasingly portrayed and understood as beloved family members (Kean 1998: 13). Lavish Victorian funerals for pets were not uncommon. In other anthropomorphic incidences, some South American indigenous women have traditionally breastfed baby monkeys, although the cultural significance of this is debated (Morris and Morris 1966: 84). Moreover, others raised apes as faux-children with no thought of science, merely for fashion, amusement or simple curiosity. An A list of celebrities, historical and contemporary, including Catherine de Medici, Frida Kahlo, Jiang Qing (Madame Mao), Michael Jackson and Justin Bieber, have all kept primate pets.

However, for Kellogg, the point was to raise Gua as a child—not a laboratory animal or, in some ways worse, a pet. He introduced a further fresh variable by raising the animal alongside his own human child, as a control subject, over the objections of his wife. Although popular reaction was excited, collegial responses were tepid.[3] Academics objected that it was inhumane towards the

---

[2] Herodotus (2001).

[3] *Time* magazine popularised the experiment in June 1933.

human baby. Very few (but some) objected to Gua's separation from her mother and fellow chimpanzees (Benjamin and Bruce 1982: 468, and for a more pragmatically critical review, see Krout 1934). Some Christian fundamentalists protested, feeling that Kellogg was inverting the "natural order". The experiment survived, although the Kellogg marriage almost did not (Kellogg and Kellogg 1933: ix).

On a fundamental level, the experiment was intended to discover how "human" an animal could be if raised as one—but, perhaps more importantly, it was supposed to wage a definitive battle in the war between nurture and nature. Its general believed environmentalism would emerge triumphant. While the experiment certainly did reveal the enhanced effect of an augmented environment on development, it also exposed the curb nature (heredity) places on an organism despite its environment. So, instead of ending the war, it showed how powerful both armies remain.

There was little reason given for why the experiment ended abruptly—Gua was simply returned to the primate colony and underwent "rehabilitation" to be a non-human again. Perhaps the Kelloggs were worried that sibling rivalry would become dangerous for Donald as Gua matured physically. There is some evidence to suggest they were concerned that Donald was becoming slightly retarded by not seeing human children and he was acquiring some distinctly chimpanzee traits, like barking with excitement at the sight of food (Kellogg and Kellogg 1931: 281). While the Kelloggs consciously humanised Gua, she was effectively subversively simianising Donald (for discussion, see Desmond 1980: 81–82). So, on some level, the Kelloggs feared their control subject—their own son—was becoming a feral child.

3 The Animal Human and the Human Animal

As this essay will show, several scientific comparisons were made between the infant development of human and non-human animals (overwhelmingly primates) in the early decades of the twentieth century. These experiments in "cross-fostering" meant raising animal and human babies alongside each other in human homes, as with Gua and Donald. Such experiments were themselves fraught with difficult moral questions, evinced in the young disciplines of primatology, ethology and, particularly, psychology. Yet the inverse case study, the so-called Forbidden Experiment—where a human baby was raised by apes—could not be conducted, as discussed. It breached too many ethical principles.

Thus, scientists fell with feverish excitement upon the possibility of such "natural" case studies. Zingg (1940: 514) declared that these feral humans were frequently authentic and had "the greatest significance to students of human psychology". They seemed to provide the empirical data no ethical university research could generate. This essay tracks the various cases of "real Tarzans" that emerged sporadically through the twentieth century, with particular emphasis on the "Baboon Boy" in South Africa validated by Raymond Dart from the University of the Witwatersrand, as the first "authenticated case of a human child adopted and raised by infrahuman primates". Over the next decade, academics in psychology and primatology used the cases to buttress their own developing arguments about the gamut between nature and nurture, with reference to early child development. This chapter discusses the suite of arguments unleashed in international academic circles in the 1930s and 1940s, by locating it within the disciplinary context of the consciously cross-fostered chimpanzees. These debates are then contrasted with subsequent cases of feral children, in order to track change over time in the shifting academic understanding of the "Forbidden Experiment".

4 Fostered by the Wild

The notion not only of feral people but of their affinity with animals has a long pedigree. Frequently, the feral children in these tales have been associated with primates: John, the so-called monkey boy, allegedly discovered in Burundi in the early 1970s and investigated by Harlan Lane, a pupil of behaviourist B.F. Skinner; Saturday Mthiyane (or Mifune), discovered in 1987 who purportedly spent a year with vervet monkeys in KwaZulu-Natal in South Africa and the case of John Sesebunya, ostensibly discovered with a troop of monkeys in Uganda in 1991.[4] But other animals have acted—or been perceived to act—as adoptive family. In 2004, South African health authorities stumbled onto a shocking case of a family who were locally understood as feral. The father was a farm labourer, the mother was mildly retarded, and their four children lived with livestock as their only companions. Further afield, Mirny in the Ukraine, which before the collapse of the Soviet Union had been a thriving naval town, became a grey place of empty blocks of flats, bleak prospects—low employment and high crime. Wild dogs—once, pets in better times—roam the streets in packs. In 2001, a four-year-old boy was found with such a pack, the pack having adopted him as a useful intermediary between them and human society. In return for the food the little boy begged from passers-by and shared with his pack, they protected him and warmed him in the long nights. Recently, a newborn human baby was abandoned in a wooded area in Nairobi and rescued by a stray dog, who—witnesses claimed—dragged the baby across a dangerous road, under a barbed-wire fence, and then placed her gently in with her own litter. Authorities took the infant to Kenyatta National Hospital. The dog was baptised *Mkombozi*—meaning "saviour". The list of human waifs fostered by animals goes on: a little boy cared for by a dog he understood to be his mother in Machakos in the 1980s, a little girl found with

[4] http://spiritrambler8.blogspot.com/2003/12/homeferal-children-found-in-1991-john.html, Accessed 1 May 2013.

primates in the swamps of southern Sumatra in 1983, and "Ivan" adopted by feral dogs in Moscow in the 1990s. Most fostering appears to be by primates or by domestic animals: both understandable, as the primates offer the virtue of approximating human morphology and "body language", while domestic animals (particularly dogs) offer proximity and several thousand years of breeding to see humans as allies. Whether or not these animals are capable of fostering a human child, these reasons made them *appear* as possibly viable parents to human observers.

Cases of "fostering" by non-human animals appear sporadically since Herodotus recorded the legend of Romulus and Remus, founders of Rome, suckled by a wolf. Linnæus himself took feral humans seriously, classifying "*Loco ferus*" under "*Homo*", describing them as *tetrapus*, *mutus* and *hirsutus*. Linnæus provided case histories of a Lithuanian bear-boy, the Hesse wolf-boy, the Irish sheep-boy and the Bamberg calf-boy. Stories abound of human babies adopted by wolves, bears or apes, and reared to supra- or, sometimes, super-humanity. Among the palpable legends, there are, however, several quasi-authenticated cases of human children reared under extremely atypical circumstances. The first famous feral children were "Peter the wild boy", "a naked, brownish, black-haired creature", captured in Hanover in 1724; the wild boy of Aveyron, found in a French forest in 1797, survived by appropriating animal lifeways and Kaspar Hauser, who arrived at the gates of Nuremberg castle in 1828, had been reared in isolation from human society (Candland 1993). In mid-nineteenth century, several cases erupted in newly colonised corners—in 1852, six cases were reported from India alone (Sleeman 1848). This precipitated a new "scientific" acceptance of the validity of animal-fostered wildlings by the English anthropologist Tylor (1863). The weight of this worthy's support for the likelihood of "beast-children", as a subspecies of the feral, prompted the concomitant endorsement by Darwin's cousin, Francis Galton, a polymath anthropologist and eugenicist, and James Frazer, one of the founders of modern anthropology (Galton 1865; Frazer 1929). Some of these feral cases were obvious hoaxes, and others quasi-legitimated—all became phenomena that obsessed intellectuals and the public.[5]

5 Primacy of Primates?

There has existed a tradition of the "epistemological child study" in philosophy. Perhaps epitomised by John Locke's writing, it has examined the "nature of the human" and the ontology of mind through the lens of early childhood, which

[5] For an obvious fake, see the story of Misha, a Jewish orphan, who purportedly drifted through occupied Europe during World War II, cared for by wolves (Defonseca 1997). For discussion of both the frauds and the more authentic cases, see Douthwaite (2002), Newton (2002), Shattuck (1980).

sometimes overlapped with more pedagogical studies (Kessen 1965: 112). Both were swamped by the 1920s and 1930s by "scientific child studies" (Prochner and Doyon 1997: 103). Right up until the present, publications urge the significance of researching non-human primates to understand human child development (Weiss and Santos 2006). But it is not a new idea that the other primates can help understand what might be loosely termed "the human condition" (see Tomasello and Call 1997). Darwin himself mooted the idea. Yerkes (1943, 1971) in the USA and Köhler (1925) in Germany used primate studies as transparent windows into the human psyche, unobscured by the curtains of culture.

Attempts to teach animals to speak date back into antiquity—by the Enlightenment, philosophers made frequent allusion to the prospect of communicating apes (Bishop 2010: 350). In the eighteenth and early nineteenth centuries, thinkers turned their attention to the "nature of the human". Intellectuals such as Hobbes, Rousseau and Locke contemplated the "natural" condition of "man", and its relation to the earliest forms of human society. The "almost human" (feral children, monsters and the other primates) had already wrenched at the seventeenth-century understanding of the self, intimating Hobbesian fears about human identity—the beast beneath the fiction of the civilised self. Eighteenth-century thinkers sustained and transformed this tradition. Some thought feral children the unfortunate hybrid of human–animal breeding, perhaps between women and apes (Douthwaite 2002: 19). Most, however, slotted them into the crude, often unconscious, hierarchy of existing beings. The sense that the mental abilities of the world's creatures could be ranked is not new. A fluid but powerful idea of a "Great Chain of Being", articulated by Aristotle and Plato, has guided Western thought for over two thousand years. The shifting understanding of the permutations of this Great Chain of Being, coupled to newly discovered primates like "Orang Outangs",[6] opened a window into the question of the *humanity* of other primates. Lord Monboddo in fact suggested in the late eighteenth century that orangutans were a breed of wild men who had not (yet) learned to speak (Monboddo 1773, 1970, and see Blancke, this volume).

Twentieth-century attempts to rectify this reticence included those of Lightner Witmer, expert in the psychology of special education at the University of Pennsylvania, who tried (and failed) to teach an ex-performing chimpanzee to speak in 1909. Witmer ventured that such efforts would be more fruitful if the animal were raised like a human child: "within a few years chimpanzees will be taken early in life and subjected for purposes of scientific investigation to a course or procedure more closely resembling that which is accorded the human child" (Witmer 1909: 205). Many studies continued to focus on language acquisition by primates. Before the Kelloggs (but unknown to each other), a study was conducted in Russia by Ladygina-Kohts (1935), an isolated psychological pioneer in Stalinist Moscow. Her study was of particular interest to historians, as it was unaffected by American behaviourism, yet drew similar conclusions. Ladygina-Kohts raised the

6 In the eighteenth century, "Orang Outang" meant great apes generally chimpanzees and orangutans (gorillas were then unknown in Europe), while "ape" usually meant the baboon.

chimpanzee, Joni, for two and a half years until he died, aged four, from respiratory failure in 1916. Nine years later, their son Roody was born, and she essayed a comparative study, prefiguring that of the Kellogg's, by comparing her human son with her memories of Joni. The study was published in 1935 by the Museum Darwinianum in Moscow, which was run by her husband.

Following the Kellogg's, Catherine and Keith Hayes raised Viki, as a chimpanzee "daughter" for almost seven years, attempting to teach her words, of which she purportedly added four to her repertoire: "mama", "papa", "cup" and "up" (shaped initially by lip manipulation) (Hayes and Hayes 1952).[7] In the 1960s, psychologists Allen and Beatrice Gardner taught American Sign Language to Washoe, a chimpanzee whom they had raised in their own home, in the tradition of the Kelloggs (Gardner and Gardner 1969). Others studies on communicating apes followed on the chimpanzees Sarah, Lana and Nim Chimpsky,[8] the gorilla Koko, the orangutan Chantek and the bonobos Kanzi and Panbanisha (Terrance 1979; Savage-Rumbaugh et al. 1998). While these were not studies in the Kellogg mould and were not attempting to humanise their subjects, most of these primates were taught the behaviour deemed apposite to a human child (wearing nappies and using crockery). There followed a robust body of critique of this kind of humanisation of the animal, including the accusation of sensationalism and the performance of an "academic Barnum and Bailey act" (Desmond 1980: 59).

Apes have been understood as ur-human, creatures capable of becoming—through a civilising process—human. Bishop makes an intriguing argument for similarities between contemporary ape language projects and historical discourses of human childhood, which link children to a state of animality (Bishop 2010: 350).[9] Both women and children have been (at various times and places) located as closer to the animals on the man–beast continuum. Like the great apes themselves, women and children have been seen as "almost human", existing in the gamut between the rational and irrational, mind and body, and articulate and voiceless. Bishop points out that those liminal figures at the "edge of humanness" have been drawn into a pedagogical ideology that claims to train its subject by disciplining the body and the mind. Thus, the humanised ape of Kellogg's imagining produces a creature Goodall (1990: 10), once likened to "a changeling", wistfully classifying it "some other kind of being".

The assumption that these creatures of cultural limbo could learn to be a bit more human has a history. It was critical to the eighteenth (and some nineteenth)-century notions of socialising feral children. Jean-Jacques Rousseau essayed an attempt to cast feral man in a romantic light in his disquisition on "natural man". Post-Revolutionary France's Victor de l' Aveyron (discovered 1797), brought out of the woods and taken to Paris in 1799, was placed in the care of Jean Itard. Itard described Victor as "purely animal", teaching him to control his bodily habits and the use of a few distinguishable words (Candland 1993). This was part of the

[7] For popular reading of raising the "child ape", see Hayes (1951).

[8] A pun on the name Noam Chomsky.

[9] For pioneering research in the gendered history of primate studies, see Haraway (1991).

understanding of wild children (along with apes) as creatures that could overcome their own "animality", and through undergoing a humanising process (especially through learning to speak) could progress towards a "state of civilised sentience" (Bishop 2010: 353). Gender played a particular role, as illustrated in early eighteenth-century France visualising the taming of the human shrew Marie-Angélique le Blanc (discovered 1731), who had to be taught not only to be human but how to be female. Le Blanc was restricted to a convent and taught the feminine skills of "silence, immobility and physical constraint" (Douthwait 2002: 31).

This analysis may be extended further: certain human groups have been more closely linked (by other groups—especially by those in a position of socio-economic or political power above them) to a "state of animality". For example, after the Anglo-Norman invasion of Ireland, the barbarous "wild Irish" label stuck. This designation shifted in the nineteenth century to a racialised biological category—the apelike Irishman. Following the famines of the 1840s, there was a mass migration to England, which was followed by increasing resistance to English dominion in Ireland through sabotage. It became socially and politically necessary to demean the enemy. Previously, the Irish had long been described as "child-like", feckless and immature (De Nie 2004). They had also been described as feminine, sensuous and quick to love (Renan 1896). Later in the century, once the idea that humans had "descended from apes" (rather than Darwin's more subtle articulation) gained traction, public imagination became stimulated by evolutionary notions filtering down and the display of the first gorilla in the London Zoo (from 1860). It has been argued that the Victorians displaced their own anxiety about their newly discovered descent by claiming close kinship between apes and "primitive peoples", dissociating "civilized human beings" of the near relatedness (of either) [see Haller (1971, 1995)]. The Irish were dubbed the "Missing Link" by Punch in 1862, described as a creature between "the gorilla and the negro", that usually confined itself to its own colony but sometimes "sallied forth in states of excitement, attacking civilized human beings" (Pieterse 2003: 112; Valente 2011: 16). Caricatured in parodies, working-class Irish (especially those subversive of English authority) were routinely depicted as monkey-men by the 1860s and 1970s.[10] So the fenian became depicted as simian.

Not only gender, class and ethnicity, but also the understanding of race in the discourse of the "feral child" was affected by this imprecise but widespread "Great Chain of Being", coupled to crude Social Darwinism. Racial thinking shifted from the ill-defined but functional universalism of the late eighteenth century to a cultural racism, predicated on stereotyping the "native" (or, as we have seen women, other ethnicities or the "lower" classes like the Irish immigrants) as brutish, closer to nature and to the other primates (for good discussions, see Stepan 1982 and Dubow 1995). Popular taxonomies of race were conveyed in the literature—especially the adventure fiction of Empire—which reached a mass audience and

[10] Particularly interesting for our argument is that the Irish were depicted either as simian or as feminine, which further reinforces the intellectual connection we trace between the two almost-human creatures, woman and ape, in historical discourse (Valente 2011: 16).

was probably much more influential than the abstruse writings of academics. Fictionalised feral children raised by apes became briefly fashionable. Bryden in 1907[11] and H. Rider Haggard in 1913[12]—and Edgar Rice Burrough's *Tarzan of the Apes* (1912)—elevated the trope to international populist fame. Unlike the more vulnerable, damaged "real" examples, Mowgli and especially Tarzan were not subhuman, but super-human. In fact, the latter demonstrated that being an aristocrat—indeed, an Englishman—was so genetically ingrained that even being raised by apes could not dent it—Lord Greystroke remained quite literally a "Noble savage". Tarzan's story provided readers with the reassurance that despite the sunset of Empire, an English gentleman remained an English gentleman. Moreover, imperial soldier-traveller Ivor Thord-Gray (Thord Ivar Hallstrom 1878–1964), who served in South Africa, averred to having been interviewed by Edgar Rice Burroughs. He narrated that he had once blundered across a troop of baboons, one of whom died in the mêlée. The victim transpired to be a little human boy.

This story was echoed eerily by that of "Lucas the Baboon Boy", c.1890 to 1948 (from Swart 1914). Purportedly, thirty years before, two members of the Cape Police were riding through a remote region of the south-eastern Cape. On spying a troop of baboons, they fired a few shots into the group for fun. One of the troop was wounded, which—to the policemen's amazement—turned out to be "a native boy between 12 and 14 years of age", "unable to speak any language but mimicked like a baboon". They made inquiries at surrounding African homesteads, tracking down a rumour that a baboon had stolen a local baby. Eventually, they took him to Grahamstown Mental Institution.

A local man, George Henry Smith, employed him as a farm labourer.[13] Smith claimed that he believed Lucas had indeed been raised by baboons because of his "profile, mannerisms, long arms, constant jerking and nodding of the head, the scratching of parts of his body, and his peculiar and frightened-looking grin…" Smith maintained that Lucas had confirmed that baboons had kidnapped him. Smith enjoyed concluding his narration by explaining that he himself had taught Lucas not to be "wild" anymore, by "thrashing him", especially "as regards his dirty animal habits".[14]

International newspapers—the *New York Times* and *Morning Post*—and the South African national press—*The Star* and *The Outspan*—carried the reports,

---

[11] Bryden (1907) in *The Gold Kloof* wrote of white heroes stumbling on an African boy living feral with a troop of baboons. For discussion, see Jenkins (2003).

[12] In Haggard's *Allan's Wife*, Hendrika, a white woman, is brought up by baboons in the mountains. Again buttressing this essay's argument, women are closer to being beasts in Haggard's view: "for if women are more jealous than men, small-hearted men are more jealous than those of larger mind and wider sympathy; and animals are the most jealous of all" Haggard (1915, 2007: 78).

[13] *Grocott's Daily Mail*: 14 November 1938.

[14] See Albany History Museum (1927–1938), SMD 194, GH Smith statement, 16 November 1927.

Fig. 1 A posed "Tarzanlike" depiction of Lucas, captioned "Baboon Boy", by the press [Albany History Museum (1927–1938), *Farmers' Weekly*, 19 May 1948]

illustrated with obviously posed photographs that tried to emphasise Lucas's supposed ferality—consciously trying to portray him as Tarzan (Lewis 1930; Smith 1931) (Fig. 1).[15]

There were intermittent attempts to authenticate Smith's story.[16] Then, the middlebrow travel writer Lawrence Green (1900–1972) adopted the tale in 1937 (Green 1937: 61–67). Writing strictly for a white South African *bourgeois* audience, Green popularised a nebulous but nevertheless palpable twentieth-century version of the Great Chain of Being, including ostensibly "true tales" of human children raised by baboons. "The chimpanzee is the most intelligent animal in the world.... [H]e flourishes in captivity and adopts human ways with delight. ... It is often said that a chimpanzee, educated side by side with a black child, would make faster progress". "I am convinced that rare cases of the adoption of children by animals do occur, and that the widespread belief in 'baboon boys' in South Africa is justified". His ersatz research both mirrored and, with its best-seller status, actually contributed to the intellectual atmosphere that rendered the racialised taxonomy acceptable. The white reading public to a degree believed Lucas's story because they believed Africans more closely related to the other primates.

Academia interest was aroused—Professor Gesell at Yale, Dr Foley at George Washington, Dr Zingg at Denver and Dr Reginald Ruggles Gates at Bristol University had visited South Africa in 1929, photographing African peoples (Foley 1940).

[15] See also "Believe Baboons brought up Native South African Boy", *The Gazette Montreal*, 23 August 1927. Albany History Museum (1927–1938), SMD 194, Prof E Mountain, "Nurtured by Baboons...", *Morning Post*, 9 August 1927 copy by E Mountain 17-7-1948 and *New York Times* 21 August 1927.

[16] Albany History Museum (1927–1938), SMD 194, District Commandant, letter to Fort England re: "Native Lucas" 21 November 1927; Physician Superintendent, letter to District Commandant, SAP, 25 November 1927; LB Stent, Dept of Interior, letter to Dr E Moon, Fort England Hospital, 14 October 1930; Physician Superintendent, letter to LB Stent, Dept of Interior 24 October 1930; Dr George E Stratton, letter to the Superintendent, Fort England Hospital from University of California, 16 December 1935; Physician Superintendent, letter to Dr JC Twomey, Port Alfred Mental Hospital, 3 August 1938.

Raymond Dart, South Africa's famous palaeoanthropologist, called upon to deliver a verdict, endorsed Lucas as the world's first modern authenticated case of a real feral child, raised by "infrahuman primates".[17]

But as the real details of the case were revealed, Dart (1940: 293–295), Zingg, Foley and Gesell et al. slowly dropped the case.[18] It transpired that there was no official police report and witnesses from the Grahamstown Mental Hospital could remember no "baboon boy". However, records confirm that Lucas was admitted in 1904, after probably working as a herd boy, and then (perhaps after his injury), he was incarcerated as an "indoor pauper". An emaciated, bewildered boy who spoke only English was on record, with a severe scalp wound, which affected his speech and his perception of himself. He was certified as he claimed (while interned) to be a sheep and, at other times, a goat—never a baboon, which was an identity imposed by Smith, it turned out, who entertained the idea of exhibiting Lucas on the stage. The identity that stuck was that of "baboon boy" because of the zeitgeist that encouraged a toxic taxonomy of race.

6 Feral Fictions

Thus, in some ways, animal–children narratives did not offer their scientist-storytellers new knowledge, simply a chance to validate the theories they already claimed as truth. Benzaquén (2001: 61) dismisses this process of ostensible scientific discovery, as rather a form of intellectual colonisation. By mid-twentieth century, Zingg and others condemned strong environmental determinism: "They overlook that mentality is a bioneurological mechanism, and mind is the environmentally conditioned content organized by that mechanism. …we have seen well-attested cases of human beings reduced to wolf-conditioning, the radical thesis still needs a case of a wolf raised to human behaviour and language". Zingg (1940: 515) concluded that "Deprived too long of human association, or animal-conditioned too strongly, the sensitive potentialities of human development are permanently inhibited and the traces of animal conditioning are never completely lost".

There was an inner circle of scholars of the animal-fostered child: Foley (1940), a scholar at George Washington University, was known for his work on conditioned responses in primates and abnormal behaviour of humans, and Zingg (1940), scholar at the University of Denver, was familiar with the writings of Squires and Kellogg (1938) on the wolf children and was in contact with Gesell of Yale

[17] Albany History Museum (1927–1938), SMD 194, Dart to Dr Gillis, 20 July 1939; SMD 194, "scientists study the Baboon Boy…Professor Dart convinced…", *Daily Dispatch* (Article, typed copy) 5 July 1939; "Science: Baboon Boy", *Time Magazine*, 1 April, 1940.

[18] "Story of 'Baboon Boy' Now Thrown in Doubt" in *Science Newsletter* Society for Science and the Public. 1940, 350–351; Tryers, Rev (SJ) refutes criticism of Prof Dart (unidentified incomplete article, dated London 8 July 1939, entitled "King George's 'Tarzan'"—copy-pasted into Zingg's article.

and R.R. Gates of the University of Bristol, although Zingg (1940) called him Gasell. Davis, a scholar at Pennsylvania State College, in studying isolated children drew on Squires' and Kellogg's work and disagreed with the long-held theory of such subjects' innate feeble-mindedness, concurring with Kellogg that they possessed normal intelligence which permitted their adaptation to the singular environments (Davis 1940).

Zingg was entrusted with Singh's manuscript by the late 1930s, for Zingg intended to use the diary to raise the subject of feral man to legitimate scientific status. His *Wolf-Children and Feral Man*, published in 1942, is a defensive ***cri de coeur***, with four forewords by Gates, Gesell, Francis Maxfield and Kingsley Davis (Zingg 1942). Gesell was one of the first Western scientists to contact Singh. He published his psychobiography of Kamala and Amala, *Wolf Child and Human Child* in order, as he explained to Zingg to help overcome the "profound, unfathomed resistance to the acceptance of Feral Man" (Gesell to Zingg, 30 September 1940, quoted in Benzaquén 2001: 67). Gesell strongly countered environmentalism's claims on beast-children, by defending the idea that Kamala was born normal and stayed normal, following key developmental stages. Gesell (1941: xii) filled in the obvious lacunae with his insistence of knowledge of "the child" in general: he traces the story of a "normal infant" in a narrative that simply showed the "reactions of normal human potentialities under extremely abnormal stress".[19] Gesell (1941: 21) concluded that although she acquired "wolf-ways", "by no stretch of the imagination can we say that she became a wolf creature". He took this as a Panglossian lesson about the faith one could still have in human nature at a time when the Second World War was making wolves of men.[20]

7 Experimental Ethics?

Gua, Joni and Nim all died young.[21] What the Kelloggs and others did not know—in fact, really only came to light from Jane Goodall's fieldwork in the 1960s—is that Chimpanzees are weaned at 4 or 5 and until then live closely by their mother's side. They tend to die if their mothers die before they are 3, despite attempts by other family members to care for them. They frequently live with their mothers until 7 or even 10 years old for females.

Modern experiments on primates frown on the "humanisation" so blithely accepted as ethically acceptable by Kellogg, Ladygina-Kohts, Hayes and others. They are required to receive correct nutrition (to circumvent the feeding of solely human food which, for example, resulted in Joni's early death from malnutrition). They are required to receive veterinary care (that might have prevented such premature deaths).

---

[19] This work is a narrative interpretation of the life history of Kamala, the wolf girl.

[20] For the phrase "wolves of people" see Benzaquén (2001: 70).

[21] Nim was the oldest survivor—he was 27 when he died, chimpanzees usually die around 40.

Most significant change, for this essay, is that they are now required to have the companionship of the same species. Cross-fostering was beset with other ethical problems: Ladygina-Kohts referred to her chimpanzee as "our little prisoner", and Joni was whipped as punishment.

But one must be careful to eschew the teleological lens that condemns these scientists from modern-day perspectives. Few experiments at the time considered issues of paramountcy today—privacy and informed consent. In fact, if contextualised with contemporary experiments on children, Gua and the others did not suffer as much as they could have. Several key researchers seem, from our perspective, to flirt with what can only be dubbed scientific child abuse, as in Watson's 1920 study in terrifying of the 11-month "Little Albert" to see whether he could be conditioned into fearing a harmless object. These repeated efforts reduced him from a cheerful little baby to a whimpering wreck (Blum 2002: 72–73). Equally, as the true feral child was too rare, researchers looked in orphanages and pauper's hospitals for quasi-feral characteristics, which produced some key studies in the 1930s (for example, see Skeels 1936: 1–5). But this should be seen in the light of little respect accorded society's underclasses, in a period when eugenics was still deemed a method of social reform (see Dubow 1995: 166–196). Aside from moral dilemmas, there was a practical problem: what could be usefully learned from simply a few individuals—could one extrapolate from one idiosyncratic creature and make generalisations about chimpanzees or humans as a *species*? Certainly, many, like Gesell (1934: 10), attempted to prove that one could draw valuable insights from the intense study of a single individual.

8 Conclusion

Animal sensitive historians have urged researchers to add the analytical category of "species" to the holy trinity of the social sciences—race, class and gender (Swart 2010). Yet, ironically, historians of the feral and the cross-fostered have focused on species to the exclusion of other categories. Race, class and gender are critical in understanding the construction of the humanised animal and the animalised human. Thus, this essay has shown the importance of understanding not only the taxonomies of species, but also the taxonomies of power.

These difficult debates tell us about the porous membrane between the human and the animal, particularly our closest animal relatives, and the shifting ideas about what it means to be human. The beast-children roamed the scientific imagination just as they had prowled forests and veld. The hope of forbidden knowledge continued to haunt. Harlan Lane, biographer of Victor de l' Aveyron, and Richard Pillard searched for the truth in the narrative of supposedly feral "John" in Burundi. Lane asked "How much more could we discover about what it means to grow up in society from this terrible experiment of nature, which chance had designed and which science could exploit?" (Lane and Pillard 1978: 5)

Historically, however, the Forbidden Experiment and its inverse have largely failed. First the philosophers and then the scientists have left records of frustration. The grand questions about human nature have been mainly unanswered and remain unanswerable by such experiments. The experimental subjects (both human and simian) have often died young, or been discarded, and others have been revealed as innocent (or witting) frauds, frequently exploited due to their own incapacity. Mostly, they have acted as blank screens upon which people project their issues—as breathing inkblots in humanity's Rorschach test.

References

Albany History Museum. South Africa (1927–1938) Clippings, documents, letters, statements

Benjamin L, Bruce D (1982) From Bottle-fed chimp to bottlenose dolphin: a contemporary appraisal of Winthrop Kellogg. Psychol Record 32:461–482

Benzaquén AS (2001) Kamala of Midnapore and Arnold Gesell's Wolf Child and Human Child: Reconciling the extraordinary and the normal. Hist Psychol 4(1):59–78

Benzaquén AS (2006) Encounters with Wild Children: temptation and disappointment in the study of human nature. McGill-Queens University Press, Montreal

Bishop R (2010) Some other kind of being: human nature and animal subjects in ape language research. Feminism Psychol 20(3):350–364

Blum D (2002) Love at goon park—Harry Harlow and the science of affection. Basic Books, New York

Bryden HA (1907) The Gold Kloof. Thomas Nelson, London

Candland D (1993) Feral children and clever animals—reflections on human nature. Oxford University Press, Oxford

Dart R (1940) The legend of Lucas 'the Baboon Boy'. Sci Afr J 37:293–295

Davis K (1940) Extreme social isolation of a child. Am J Sociol 45(4):554–565

De Nie M (2004) The eternal paddy: Irish identity and the British press, 1798–1882. University of Wisconsin Press, Madison

Defonseca M (1997) Misha: a Mémoire of the Holocaust Years. Mount Ivy Press, Boston

Desmond A (1980) The Ape's reflexion. Quartet, London

Douthwaite J (2002) The wild girl, natural man and the monster. University of Chicago Press, Chicago

Dubow S (1995) Scientific Racism in modern South Africa. Cambridge University Press, Cambridge

Foley JP (1940) The Baboon Boy of South Africa. Am J Psychol 53(1):128–133

Frazer J (1929) Fasti of Ovid, II, Translation with a commentary by Frazer. Macmillan, London

Galton F (1865) The domestication of animals. Trans Ethnological Soc London 3:136

Gardner RA, Gardner BT (1969) Teaching sign language to a chimpanzee. Science 165:664–672

Gesell A (1934) Infant behaviour: its genesis and growth. McGraw-Hill, New York

Gesell A (1941) Wolf Child and Human Child, being a narrative interpretation of the life history of Kamala, the wolf girl. Harper, London and New York

Goodall J (1990) Through a window—My thirty years with the Chimpanzees of Gombe. Houghton Mifflin, Wilmington

Green L (1937) Great African Mysteries. Stanley Paul, London

Haggard HR (1915, 2007) Allan Quatermain #3: Allan's Wife. A & D Publishing, Radford, Virginia

Haller JS (1971, 1995) Outcasts from evolution: scientific attitudes of racial inferiority, 1859–1900. University of Illinois, Urbana

Haraway D (1991) Simians, Cyborgs and Women—The Reinvention of nature. Free Association Books, London

Hayes C (1951) The ape in our house. Harper and Row, New York
Hayes KJ, Hayes C (1952) Imitation in a home-raised Chimpanzee. J Comp Physiol Psychol 45:450–459
Herodotus (2001) An account of Egypt: being the second book of his histories called Euterpe. In Voyages and travels: ancient and modern, vol XXXIII. P.F. Collier & Son, New York, 1909–1914; Bartleby.com, 2001. Available at www.bartleby.com/33/ (Accessed 1 May 2013)
Jenkins E (2003) Nudity, clothing and cultural identity in some South African Children's books. Engl Afr 30(1):87–101
Kean H (1998) Animal rights—political and social change in Britain since 1800. Reaktion Books, London
Kellogg WN (1931a) More about the wolf children of India. Am J Psychol 43:508–509
Kellogg WN (1931b) Humanizing the ape. Psychol Rev 38:160–176
Kellogg WN, Kellogg LA (1933) The Ape and the child. McGraw-Hill, New York
Kellogg WN (1938) A further note on the wolf children of India. Am J Psychol 46:149–150
Kessen W (1965) The Child. Wiley, New York
Köhler W (1925) The mentality of apes. Kegan, Trench, London
Krout MH (1934) Review of the Ape and the Child. J Abnorm Soc Psychol 29(2):247–249
Ladygina-Kohts NN (1935) Infant Chimpanzee and Human Child—A classic. In: de Waal F (ed) 2002 Comparative study of Ape emotions and intelligence. Oxford University Press, Oxford
Lane H, Pillard R (1978) The Wild Boy of Burundi: a study of an outcast child. Random House, New York
Lewis E (1930) A native Romulus of modern times. The Star, Johannesburg
Monboddo JB (1773, 1970) Of the origin and progress of language. Garland, New York
Morris R, Morris D (1966) Men and Apes. Hutchinson, London
Newton M (2002) Savage girls and wild boys: a history of feral children. Faber and Faber, London
Pieterse JN (2003) White Negroes. In: Gail D, Humez JM (eds) Gender, race, and class in media: a text-reader. Sage, Thousand Oaks
Prochner L, Doyon P (1997) Researchers and their subjects in the history of child study: William Blatz and the Dionne quintuplets. Can Psychol 38(2):103–110
Renan E (1896) The poetry of the celtic races, and other essays. Walter Scott Publishing, London
Savage-Rumbaugh S, Shanker S, Taylor TJ (1998) Apes, language and the human mind. Oxford University Press, Oxford and New York
Shattuck R (1980) The forbidden experiment. Quartet, London
Skeels HM (1936) The relationship of foster home environment to the mental development of children placed in infancy. Child Dev 7(1):1–5
Sleeman W (1848) A journey through the Kingdom of Oude. London, Richard Bentley. Available at http://www.gutenberg.org/files/16997/16997-h/16997-h.htm (Accessed 1 May 2013)
Smith GH (1931) The truth about the Baboon Boy. Outspan, South Africa
Squires PC (1927) Wolf children' of India. Am J Psychol 38:313–315
Stepan N (1982) The idea of race in science: Great Britain 1800–1960. Macmillan, London
Swart S (2010) Riding high—horses, humans and history in South Africa. Witwatersrand University Press, Johannesburg
Swart S (1914) Lucas's story. The invention of the South African Baboon Boy. In Speitkamp W, Zehnle S (eds) Afrikanische Tierräume: Historische Verortunge, Rüdiger Köppe Verlag
Terrance H (1979) Nim: a chimpanzee who learned sign language. Knopf, New York
Tomasello M, Call J (1997) Primate cognition. Oxford University Press, Oxford
Tylor EB (1863) Wild men and beast-children. Anthropol Rev 1:21–32
Valente J (2011) The myth of manliness in Irish national culture, 1880–1922. University of Illinois, Urbana
Weiss DJ, Santos LR (2006) Why primates? The importance of nonhuman primates for understanding human infancy. Infancy 9(2):127–140
Witmer L (1909) A monkey with a mind. Psychol Clin 3(7):205
Yerkes RM (1943, 1971) Chimpanzee: a laboratory colony. Johnson Reprint Corporation, New York
Zingg R (1940) Feral man and extreme cases of isolation. Am J Psychol 53(4):487–517
Zingg R (1942) Wolf-children and feral man. Harper, New York

Part II
The Elements of Social Communication in Primates and Humans

Experimental Conversations: Sign Language Studies with Chimpanzees

Mary Lee Jensvold

Abstract Sign language studies of chimpanzees are a tool for studying the continuity between human behavior and behavior of other animals and between verbal behavior and other intelligent behavior. Cross-fostered chimpanzees paralleled children in their acquisition and use of signs and phrases. These procedures occurred under rigorous and systematic record keeping and experimental paradigms. The study of Wh-questions and the use of remote videotaping (RVT) are examples of experimental procedures. These revealed chimpanzee-to-chimpanzee signing and private signing. Face-to-face interactions between the chimpanzees and an interlocutor who presented various systematic probes is another experimental procedure. The chimpanzees adjusted to the interlocutor with revisions, new signs, or no response when appropriate. The hallmark of the sign language studies is that caregivers treated the chimpanzees as conversational partners socially motivated rather than experimental subjects bribed or forced into participation. These findings confirm continuity with differences of degree among species.

Keywords Sign language experiments • Project Washoe • Chimpanzees • Immersion learning

M. L. Jensvold (✉)
Department of Anthropology and Museum Studies, Central Washington University,
Ellensburg, WA 98926-7544, USA
e-mail: jensvold@cwu.edu

M. Pina and N. Gontier (eds.), *The Evolution of Social Communication in Primates*, Interdisciplinary Evolution Research 1, DOI: 10.1007/978-3-319-02669-5_4,

1 Cross-Fostering

Nearly 40 years of research using various methodologies in a group of chimpanzees reveals discoveries about the remarkable capacities of this species. In these studies, creating comparable conditions between chimpanzees and humans has allowed valid comparisons between the two species.

Sign language studies of chimpanzees are a tool for studying the continuity between human behavior and behavior of other animals and between verbal behavior and other intelligent behavior. Gardner and Gardner used cross-fostering to study infant chimpanzees' acquisition of signs (Gardner & Gardner 1969). Cross-fostering is a procedure to study the interaction between environmental and genetic factors by having parents of one genetic stock rear the young of a different genetic stock. It seems as if no form of behavior is so fundamental or so distinctively species-specific that it is not deeply sensitive to the effects of early experience (Stamps 2003). Ducklings, goslings, lambs, and many other young animals learn to follow the first moving object that they see, whether it is their own mother, a female of another species, or a shoebox. The mating calls of many birds are so species-specific that an ornithologist can identify them by their calls alone without seeing a single feather. Distinctive and species-specific as these calls may be, they, too, depend upon early experience (Slater and Williams 1994; West et al. 1997). The development of the individual is dependent upon the interaction between both genes and environment.

How about our own species? How much does our common humanity depend on our common human genetic heritage and how much on the equally species-specific character of a human childhood? The question is as traditional as the story of Romulus and Remus and so tantalizing that even alleged but unverified cases of human cross-fostering, such as the wolf children of India (Singh and Zingg 1942) and the monkey boy of Burundi (Lane and Pillard 1978) attract serious scholarly attention. An experimental case of a human infant cross-fostered by nonhuman parents would require an unlikely level of cooperation from both sets of parents. In a few cases, however, chimpanzees have been cross-fostered by human parents (Kellogg and Kellogg 1933; Hayes and Hayes 1951).

2 A Conversational Laboratory

2.1 Chimpanzees as Cross-Fosterlings

In making discoveries about human behaviors, chimpanzees are an obvious first choice for cross-fostering, as they look and act remarkably like human beings and recent research reveals close and deep biological similarities of all kinds (Goodall 1986). In blood chemistry, for example, chimpanzees are not only the closest species to humans, but chimpanzees are closer to humans than chimpanzees are to gorillas or to orangutans (Stanyon et al. 1986; Ruvolo 1994) and 98 % of human and chimpanzee DNA shares the same structure (Sibley and Ahlquist 1984; The Chimpanzee Sequencing and Analysis Consortium 2005). Cross-fostering a chimpanzee is very different from keeping one in a home as a pet. Many people keep pets in their homes. They may treat their pets very well, and they may love them dearly, but they do not treat them like children. True cross-fostering—treating the chimpanzee infant like a human child in all respects, in all living arrangements, 24 h a day every day of the year—requires a rigorous experimental regime that has rarely been attempted.

2.2 Sign Language Immersion

Allen and Beatrix Gardner reared the infant chimpanzees, Washoe, Moja, Tatu, Dar, and Pili, in a cross-fostering laboratory at the University of Nevada Reno. With the similarities between chimpanzees and humans, the Gardners hypothesized that the chimpanzees would acquire aspects of human culture if immersed in those aspects.

> Washoe at about 9-months, followed by Moja, Pili, Tatu, and Dar within days of birth, entered a laboratory of rigorous cross- fostering. They had all the usual toys, games, and picture books that human infants get. They lived in quarters in which humans had lived, ate human food at tables from dishes with forks and spoons, and drank from cups. They also helped set and clear their tables, clean their quarters, and put away their playthings. They dressed and undressed themselves and used human toilets, wiping themselves, flushing, even asking to go potty during lessons and naptimes. Most important, at least one human member of their foster families stayed in close attendance throughout their waking hours. Never caged, they were as free as human infants to move about the world with supervision. When they slept, human family members listened on intercoms to comfort troubled infants during the night. (Gardner 2002: 624)

Caregivers integrated American Sign Language (ASL) into the procedure so that the chimpanzees were immersed in a sign language environment much like a human child is immersed in a speech environment. In teaching sign language to Washoe, Moja, Pili, Tatu, and Dar, caregivers imitated human parents teaching young children in a human home. They called attention to everyday events and objects that might interest the young chimpanzees, e.g., THAT CHAIR, SEE PRETTY BIRD, and MY HAT. Caregivers often molded the chimpanzees' hands in the shape of new signs as deaf parents do (Maestas and Moores 1980). The cross-fosterlings learned many signs by watching adults sign about interesting objects and activities (Gardner and Gardner 1989: 17–19). Caregivers asked probing questions to check on communication, and they always tried to answer questions and to comply with requests. They expanded on fragmentary utterances using the fragments to teach and to probe. They also followed the parents of deaf children by using an especially simple and repetitious register of ASL and by making signs on the youngsters' bodies to capture their attention (Maestas and Moores 1980; Marschark 1993; Schlesinger and Meadow 1972; Gardner et al. 1989a).

2.3 Deleterious Effect of Operant Techniques

While many experimental psychologists were putting rats and rewards in boxes to study learning, Gardner and Gardner discovered that operant techniques had detrimental effects on the desired behavior and often interfered with the task at hand (Gardner and Gardner 1989, p. 20).

> ...Washoe, Moja, Pili, Tatu, and Dar both learned and used the signs of ASL in an environment modelled after the living and learning conditions of a human household. We did not have to tempt them with treats or ply them with questions to get them to sign to us. They initiated conversations on their own, and they commonly named objects and pictures of objects in situations in which we were unlikely to reward them. (Gardner et al. 1989b: 23)

Rewards have a negative effect on children's behavior as well. For example, rewards suppress drawing in a classroom setting (Lepper et al. 1973).

2.4 Acquisition of Signs

Size of vocabulary, appropriate use of sentence constituents, number of utterances, proportion of phrases, and inflection all grew robustly throughout five years of cross-fostering. The growth was patterned growth and the patterns were consistent across chimpanzees. Wherever there are comparable measurements, the patterns of growth for cross-fostered chimpanzees paralleled the characteristic patterns reported for human infants (Gardner and Gardner 1994, 1998). Both chimpanzees and children used immature forms of the signs, generalized the early meaning of the signs, used negation (Gardner et al. 1989a), and inflected signs in questions and expression of person, place, and instrument (Chalcraft and Gardner 2005; Gardner and Gardner 1974, 1978; Rimpau et al. 1989).

3 Systematic Records

These procedures occurred under rigorous and systematic record keeping and experimental paradigms. The caregivers doubled as human observers and made systematic records of the chimpanzees' development in the form of diary records, inventories of phrases, and samples of utterances. There was rigor in these records; for example, signs were recorded with descriptions of their shape and use, each sign in the chimpanzees' vocabulary had to meet a criteria to be included, and periodically old signs were reassessed to make sure they were still part of the vocabulary (see Gardner et al. 1989a for review). Under meticulous conditions, Gardner and Gardner video recorded conversations between the chimpanzees and caregivers for later analysis (Gardner and Gardner 1973; Chalcraft and Gardner 2005).

Besides naturalistic observation, Gardner and Gardner also used systematic experiments (Gardner and Gardner 1984; Gardner et al. 1989b, 1992). These procedures including systematic controlled manipulation of independent variables were embedded in the daily routine making the chimpanzees' participation in the experiment no more forced than American children saying "trick-or-treat" on Halloween.

3.1 *Wh-Questions*

The study of Wh-questions is an example of an experiment in the cross-fostering laboratory. In Wh-questions, the critical word begins with wh such as whose, where, who, what, and why. "…in English, questions of this type correspond with individual declarative sentences in which there is one unknown, or *x* element, corresponding to some major grammatical constituent" (Brown 1973: 14). Wh-questions are used to study grammatical structure in the early linguistic development of children. Periodically throughout the cross-fostering project, systematic samples of replies to Wh-questions were recorded.

> We embedded the sample of questions and replies into the normal stream of everyday conversation…. We avoided concentrated series of questions, as in a battery of tests. Sometimes, of course, a series of questions of different types fitted unobtrusively into the normal conversational context. (Gardner et al. 1992: 33)

Later analysis of these questions and responses revealed that the Washoe, Moja, Tatu, Dar, and Pili replied to Wh-questions with appropriate sentence constituents (Gardner and Gardner 1975; Gardner et al. 1992). Longitudinally, Moja, Tatu, and Dar followed the same developmental sequence in their responses to Wh-questions as human children (Van Cantfort et al. 1989). The Gardners made no sacrifice in rigor while testing grammatical skill in the chimpanzees.

3.2 *Vocabulary Test of Communication*

Since the chimpanzees frequently named objects, in another experiment Gardner and Gardner (1984) used naming to test whether the chimpanzees could communicate information otherwise unknown to a human observer. A hidden experimenter projected slides onto a screen that only the chimpanzee could see. There were two human observers: one with the chimpanzee and another hidden in a room with a view of the chimpanzee. The human observers could only see the chimpanzee signing and could not see the screen. The chimpanzee began a trial by sliding open a door or pressing a button, which made the slide appear. If the chimpanzee failed to sign, then the observer asked him or her to sign again. Gardner and Gardner (1984) described how the cross-fosterlings participated in the test.

> Washoe, Moja, Tatu, and Dar were awake and accompanied by a human member of their foster family approximately 70 hr a week. During those 70 hr, the exposure to objects and the ASL names for objects was very great compared with the brief periods spent in vocabulary tests. Moreover, these tests were as different from the routines of the rest of their daily lives as similar testing would be for young children. For caged subjects, a session of testing is probably the most interesting thing that happens in the course of a laboratory day. For Washoe, Moja, Tatu, and Dar, most of the activities of daily life were more attractive than their formal tests. Nor could we starve them like rats or pigeons and make them earn their daily rations by taking test. Getting free-living, cross-fostered chimpanzees to do their best under the stringent conditions of these test required a great deal of ingenuity and patience. The basic strategy was to establish the testing routine, by a regular program of pretests that were kept short, usually less than 30 min, and infrequent, rarely more than two sessions per week. (Gardner and Gardner 1984: 385)

The agreements between the two observers and between the signs reported by the two observers and the correct name of the slide were high ranging from 70 to 95 %. Using a procedure that required voluntary chimpanzee participation, Gardner and Gardner showed that the chimpanzees could communicate novel information and that their signs were distinct and intelligible. Additionally, this procedure provided a control for cuing. Interlocutors can unwittingly lead subjects to correct or incorrect responses as the horse Clever Hans famously demonstrated (Gardner et al. 2011 for review). Controls for cueing are essential in tests of language and intelligence, and this study provided such a control.

4 Signing Among Chimpanzees

4.1 Leaving Reno

In 1970, Washoe left Reno with Roger and Deborah Fouts for the Institute of Primate Studies (IPS) at the University of Oklahoma. While Washoe was wild-caught by the US Air Force and arrived in the cross-fostering laboratory at about 9 months of age, Moja, Pili, Tatu, and Dar were born in American laboratories and each arrived in Reno within a few days of birth. Moja arrived in November 1972 and cross-fostering continued for her until winter 1979 when she left for IPS. In 1980, Washoe and Moja moved with the Fouts to the Chimpanzee and Human Communication Institute (CHCI) on the campus of Central Washington University in Ellensburg, WA. Tatu arrived in Reno in January 1976 and Dar in August 1976. Cross-fostering continued for Tatu and Dar until May 1981 when they left to join Washoe and Moja in Ellensburg. Pili arrived in Reno in November 1973, and he died of leukemia in October 1975.

4.2 Project Loulis: A Case of Transmission

The Gardners' studies utilized naturally occurring behaviors, which eliminated any need to coerce the chimpanzees into participation. These principles and procedures continued after the chimpanzees left Reno. The first of these studies began in 1979 at IPS in Oklahoma when Washoe adopted a 10-month-old son, Loulis. To show that Loulis would learn signs from Washoe and other signing chimpanzees without human intervention, experimenters restricted human signing when Loulis was present except for seven specific signs, WHO, WHAT, WHERE, WHICH, WANT, SIGN, and NAME. Humans instead used vocal English to communicate in his presence. Loulis began to sign in 7 days and at 73 months of age his vocabulary consisted of 51 signs (Fouts 1994b; Fouts et al. 1982, 1989b). At 15 months of age, he combined signs and the development of his phrases paralleled the cross-fostered chimpanzees and children (Fouts et al. 2002). In June 1984, the signing restriction around Loulis ended. The control showed ASL was easily acquired from other chimpanzees.

5 Remote Videotaping: A Method to Record

RVT was an observational technique used to record the behaviors of the chimpanzees with no humans present at CHCI. Initially, three cameras were mounted in a chimpanzee enclosure and focused on the chimpanzees' enclosure. Later, a fourth camera was added. The cameras were attached to television monitors and a VCR in another room away from the chimpanzees. Only one camera recorded at a time and the VCR operator could control which camera recorded. During recording sessions, no one entered the chimpanzee quarters, surrounding hallways, and adjacent rooms. All humans stayed out of view and kept silent, and all activities in the laboratory that might be a distraction ceased.

5.1 Chimpanzee-to-Chimpanzee Conversations

Data collection with RVT initially occurred during a 15-day period at the end of July and the beginning of August 1983 in a study that focused on Loulis' use of signs with Washoe, Moja, Tatu, and Dar. Every day during the 15-day period, the video cameras were turned on for two 20-min recording periods between the hours of 9:00 a.m. and 5:00 p.m. There were 45 recording periods so that each hour of the day was sampled randomly without replacement either five or six times. On the videotapes, there were 189 chimpanzee-to-chimpanzee signed interactions. Dar and Loulis were the most frequent dyad with 167 signed interactions, and Loulis and Tatu were the second most frequent dyad with 76 signed interactions. Most of the interactions occurred in the Affinitive Social (33 %) and Play (38 %) contexts (Jensvold and Gardner 2000; Jensvold et al. 2014).

5.2 Private Signing

In the 45 h of videotape, Fouts (1994a) reported 115 private signs that Loulis made when his face and body were not oriented toward another chimpanzee. In a second analysis of the videotapes, Bodamer (1987) found 90 instances of private signing by the other chimpanzees. These were signs made in the absence of interactive behaviors such as looking toward another individual. Bodamer classified these into categories of private speech that humans use (Furrow 1984). A later study recorded 56 more hours of RVT and found 368 instances of private signing (Bodamer et al. 1994). In both samples, one of the most common categories of signing was Referential (59 % in the 56 h sample). In this category, the chimpanzee signed about something present in the room for example naming the pictures in a magazine. The informative category, an utterance that refers to an object or event that is not present, accounted for 12 % in the 56 h sample and 14 % in the 45 h sample. An example of this category was when Washoe signed DEBBI to herself when Debbi was not present.

One category of private signing was imaginative (Furrow 1984) and accounted for 17 instances in the 56 h of RVT. A later study recorded 15 h of RVT while the chimpanzees' enclosure was filled with toys. There were six instances of imaginary play classified into categories of imaginary play that human children use (Matthews 1977). There were four instances of Animation in which the chimpanzee treated an object as if it was alive. For example Dar signed PEEKABOO to a stuffed bear. There were four instances of substitution in which the chimpanzee treated one object as if it were another. For example, Moja wore a shoe and signed SHOE. She then removed the shoe, put a purse on her foot, and zipped it up (Jensvold and Fouts 1993).

RVT was a systematic data collection technique that controlled for cuing and provided a way to discover what the chimpanzees did without any human influence. It revealed various functional uses of signs, socially between chimpanzees, privately, and creatively. Other research manipulated independent variables within the context of the chimpanzees' typical daily signed interactions with their human caregivers.

6 Systematic Interactions with Human Interlocutors

6.1 A Bottom-up Approach to Linguistics

The tradition in theoretical linguistics is to examine syntax and semantics using a top down approach. Yet successful face-to-face interactions involve the orchestration of pragmatics and context as well as syntax and semantics evoking a bottom-up approach to questions in linguistics. More recent research in human adults and children systematically explores pragmatic and contextual appropriateness during the stream of conversation in a variety of ways including initiation of conversation, topic introduction and maintenance, turn taking, responses to questions, conversational repair and changes in conversational register (Abbeduto and Hesketh 1997; Ferguson 1998; Galski et al. 1998; Ripich et al. 2000; Duncan 2000; Pickering and Garrod 2004; Mol et al. 2012; Benus et al.

2012). Interlocutors systematically vary input in conversations to examine conversational skill in children.

6.2 Experimental Conversational Breakdowns: Human Studies

Communication depends on the interaction between two speakers or signers. In the give-and-take of conversation, conversational partners must respond appropriately to the communicative actions of each other. However, communication breakdown between partners is not uncommon and partners must make contingent adjustments in their responses for the conversation to continue. Conversational contingency is evident in behaviors of very young human children and develops gradually and systematically on into adulthood (Wilcox and Webster 1980; Golinkoff 1986; Wootton 1994; King and Gallegos-Santillan 1999; Most 2003). Systematic studies of children's early responses to conversational breakdown show they initially repeat their original utterance (Gallagher 1977). Later, they begin to add more information by revising the original utterance and adding new words (Brinton et al. 1986a). Researchers have examined this during natural conversations (Garvey 1977; Golinkoff 1993; King and Gallegos-Santillan 1999) and in paradigms where the interlocutor presents systematic probes in response to a child's utterance (Brinton et al. 1986a, b; Ciocci and Baran 1998; Most 2003; Wilcox and Webster 1980; Wooton 1994) or request (Marcos 1991; Marcos and Knrnhaber-le Chanu 1992; Ryckebusch and Marcos 2004).

Anselmi et al. (1986) examined responses of children to general questions such as "What?" versus specific questions such as "What banana?" Wilcox and Webster (1980) presented questions versus statements to children. They found that children were more responsive to questions than statements. Brinton et al. (1986a) presented a succession of general questions to children 2–7 years old. In the succession, the interlocutor followed the child's response to the first general question with a second general question for a total of three questions. For example,

Child: Gimme ball.
Probe 1: Huh?
Child: Gimme ball.
Probe 2: What?
Child: Gimme that ball.
Probe 3: I don't understand.
Child: That ball there, gimme it.

Younger children had more difficulty responding to questions occurring later in the series than older children. Older children provided more information to the interlocutor than the younger children. General question series shows differences between normal versus language impaired children (Brinton et al. 1986b) and among developmentally disabled adults (Brinton and Fujiki 1991). By systematically varying the response of the interlocutor, research showed that responses of children were contingent on the questions of the interlocutor.

6.3 Systematic Conversational Experiments with Chimpanzees

In a study with the chimpanzees, the interlocutor systematically varied her input during casual interactions with the chimpanzees. A single interlocutor presented one of four types of probes: general requests for more information, on-topic questions, off-topic questions, or negative statements. At the time of data collection, she had 8 years of experience caring for and interacting with this group of chimpanzees and 10 years of experience communicating in ASL.

When she arrived at the interaction area MLJ either approached a chimpanzee or waited for a chimpanzee to approach her as she normally did in the course of a day. The interlocutor then attempted to engage the chimpanzee in a typical conversational interaction on a subject such as looking at a book, eating a meal, playing a game, or some other common activity (Fouts et al. 1989a). When ready, the camera operator signaled that the camera was ready and prompted the interlocutor by indicating which condition to present on that trial. The next time that the chimpanzee signed to the interlocutor, she replied with the first probe in the series specified by the condition for that trial. When the chimpanzee signed in response to the first probe, the interlocutor probed again, and so on until the interlocutor completed the series of three planned probes specified for that trial. The chimpanzees were free to interact with the interlocutor or to ignore her. After each probe, the chimpanzee was free to answer with any signs or phrases in his or her vocabulary, to continue to face the interlocutor, to look away, or to leave the scene entirely.

6.4 Conditions

There were four conditions of probes. The general conditions were questions such as HUH? And WHAT? indicating a general misunderstanding. The on-topic condition was Wh-questions that were related to the chimpanzee's previous utterance. For example when Washoe signed THAT the interlocutor asked WHO WANT THAT? The off-topic condition was Wh-questions that were unrelated to the chimpanzee's previous utterance. For example, when Washoe signed RED THERE, the interlocutor asked WHO FUNNY? The negative condition was three statements indicating the interlocutor could not comply with probes such as CAN'T, SORRY CAN'T and I MUST GO.

An experimenter transcribed the signs on the videotape. She then categorized replies in relationship to the chimpanzees' previous utterance and again in relationship to the interlocutor's previous utterance. Reliability scores ranged from 87 to 93 %.

6.5 Results

Like older children (Brinton et al. 1986a), the chimpanzees were as responsive to the later probes as they were to the initial probes with in a trial. This was the result in all four conditions.

Human children react to general probes by expanding across turns as in Brinton et al. (1986a: 77).

Child: A girl's playing on the swing
Adult: Huh?
Child: A girl and boy are playing on the swing.

Expanding across turns maintains the topic (Garvey 1977; Wilcox and Webster 1980) and also adds information. As human children develop, they are more likely to expand across turns (Brinton et al. 1986a; Anselmi et al. 1986). The cross-fostered chimpanzees often reacted to probes by expanding across turns. This maintained their original topic while adding more information.

Trial # 3/1:04:40
1:04:51 Washoe: ME GIMME (toward edible plants beyond her reach)/
1:04:54 Probe: NOT UNDERSTAND/
1:04:56 Washoe: FOOD GIMME/

It was also appropriate to react to a general probe by repeating the signs in the chimpanzee's previous utterance, thus emphasizing or clarifying something that the interlocutor may have missed.

Trial # 4/0:22:27
0:22:28 Dar: FLOWER THERE/
0:22:29 Probe: WHAT?/
0:22:30 Dar: FLOWER THERE/

Yet Washoe, Moja, Tatu, and Dar were more likely to expand across turns to general probes.

Incorporation was a category of response in which the chimpanzee used some of the signs in the interlocutor's previous utterance. Expansion was a category in which the chimpanzee used some signs from the interlocutor's previous utterance and added new signs. When the interlocutor asked relevant on-topic questions, the chimpanzees responded with many incorporations and expansions. These were absent in the general condition. In conversation, both human adults and human children often incorporate all or part of the utterance of a partner into their own next rejoinder. Keenan (1977), Keenan and Klein (1975), and Wilbur and Petitto (1981) suggest that incorporation indicates assent in the utterances of children. They interpreted examples of incorporation as a pragmatic device indicating positive response as in Keenan (1977: 130):

Adult: And we're going to have hot dogs.
Child: Hot dogs! (excitedly)

In Reno, Tatu and Dar incorporated signs from the interlocutor's previous utterance and incorporations were more likely to occur in response to announcements of positive events than to neutral or negative announcements. For example, "in response to the statement TIME ICECREAM NOW, Tatu signed ICECREAM ICECREAM ICECREAM ICECREAM ICECREAM ICECREAM" (Gardner et al. 1989b: 47).

Adults and older children interact in conversation by expanding as well as incorporating while younger children tend to incorporate without expanding (Bloom et al. 1976; Keenan 1977), as in Bloom et al. (1976: 528).

Adult: Take your shirt off.
Child: Shirt off

In this experiment, Washoe, Moja, Tatu, and Dar often expanded on the interlocutor's utterance when they responded to probes adding information to the interlocutor's previous utterance as in the following example:

Trial # 2/0:19:23.
0:19:24 Tatu: SMELL/
0:19:26 Probe: WHO SMELL?/
0:19:27 Tatu: TATU SMELL YOU/

As cross-fosterlings in Reno, Washoe, Moja, Tatu, and Dar replied to Wh-questions with expansions (Van Cantfort et al. 1989: 210). In Reno as in Ellensburg, the conversation of these chimpanzees resembled older children rather than younger children.

When the interlocutor asked off-topic questions, the chimpanzees often failed to respond and when they did respond they used few incorporations and expansions. On-topic and off-topic probes had the same Wh-signs combined with signs that were either contingent on or irrelevant to the signs in the previous utterance of the chimpanzee.

On-topic Trial # 2/1:39:59
1:39:57 Moja: EAT/
1:39:59 Probe: WHO EAT?/
Off-topic Trial # 3/0:30:15
0:30:12 Moja: FLOWER/
0:30:15 Probe: WHO IN THERE?/

With on-topic probes, the interlocutor followed the conversational lead of the chimpanzee. Off-topic probes created a situation like a dual monologue; the interlocutor's responses ignored the conversational utterances of the chimpanzee.

The cannot condition evoked the least response of all the conditions of this experiment. Moja and Tatu responded slightly more often than Washoe and Dar showing more persistence since they also expanded on their previous utterances, thus staying with their original topic. Marcos and Bernicot (1994) examined reactions of 18- to 30-month-old human children to an interlocutor who refused to cooperate with requests for objects. Like the chimpanzees in this experiment, the children sometimes persisted in their original request; sometimes, they switched to a different topic, but more often, they failed to respond.

6.6 Systematic Responses to Chimpanzee Requests

Interlocutor nonverbal behavioral responses to chimpanzees' utterances are another type of independent variable. The chimpanzees depended on their caregivers to fulfill many of their needs, and often used signs to request objects and activities of humans. Typical interactions between caregivers and the chimpanzees included games, such as chase and peek-a-boo; activities, such as coloring and looking at books; chores, such as cleaning; and meals (Fouts et al. 1989a). The objective of Leitten et al. (2012) was to experimentally manipulate caregiver responses to the chimpanzees' requests and determine whether changes in the chimpanzees' signing were contingent upon this interlocutor input. Following the chimpanzee's request, a human interlocutor responded systematically.

The daily activity of the chimpanzees contained a wide spectrum of events. A master list of potential requests was created before data collection began. Three of the chimpanzees' longtime caregivers compiled a list of 18 object signs or action signs that the chimpanzees often requested. Then, nine other longtime caregivers ranked each chimpanzee's preference for each object or action (after Gardner et al. 1989b). The five objects or actions that caregivers ranked as the least preferred for each chimpanzee were those offered during the misunderstand condition, described below.

6.7 Trials

On experimental days, the interlocutor entered the chimpanzee–human interaction area with a camera operator and waited for a chimpanzee to initiate a conversation. A trial began, when the chimpanzee signed for an object on the list. GIMME and THERE could also begin a trial if the referent of the sign was visible and was included on the list. The interlocutor then responded to the chimpanzee's request with a probe from one of four conditions, described below. Trials concluded either after the chimpanzee signed in response or after 30 s.

6.8 Conditions

There were four conditions of interlocutor probes in this study: comply, misunderstand, refuse, and unresponsive. The probe in the comply condition was when the interlocutor offered the chimpanzee the requested object or performed the requested action. Requested objects were readily available in a nearby area, but often not visible to the chimpanzees. For example, if Tatu requested MASK, the interlocutor would enter the adjacent enrichment room and return to Tatu with a mask. As per typical daily protocol, if the chimpanzee requested gum, toothbrushes, snacks, or other items, the interlocutor offered the item to all of the chimpanzees regardless of whether they had participated in a trial.

The probe in the misunderstand condition was when the interlocutor offered an object or activity that was not part of the initial request. For example, if Tatu requested MASK, the interlocutor would enter the adjacent enrichment room and return to Tatu with a brush. The interlocutor used a list of objects to determine which requests were candidates for misunderstand trials and what objects to offer in the misunderstand condition. The interlocutor only presented a probe to utterances that contained a sign for an object or action that was on the list for that chimpanzee.

The probe in the refuse condition was when the interlocutor refused to comply with the chimpanzee's request, with signs such as CAN'T. For example, if Tatu requested MASK, the interlocutor would respond by signing SORRY CAN'T. The refusals in this condition were like the refusals that typically occur in interactions between caregivers and their charges, be they children (Marcos and Bernicot 1994) or chimpanzees.

The probe in the unresponsive condition was when the interlocutor made no signed response to the chimpanzee's request, but continued to face the chimpanzee. For example, if Tatu requested MASK, the interlocutor refrained from responding and ignored the request.

6.9 Transcription

An experimenter transcribed the trials and then classified each of the chimpanzees' responses by comparing them to the chimpanzees' initial requests. Interobserver reliability ranged from 82 to 95 %. In the category repetition, the signs in the chimpanzee's response were the same as the signs in the chimpanzee's initial request. An example of a repetition is as follows:

Trial #1
0:31:17 Tatu: TOOTHBRUSHx/
0:31:38 LL: Offers Tatu a glove
0:32:07 Tatu: TOOTHBRUSHx/

In the category revision, the signs in the chimpanzee's response contained more or fewer signs than the chimpanzee's initial request, or the response contained completely different signs than the initial request. Some examples of revisions are as follows:

Trial #140
0:31:56 Moja: FOODx GUMx/
0:32:16 LL: Gives Moja a string
0:32:18 Moja: FOOD THERE GUM/
Trial #48
1:37:25 Washoe: GIMMEx TOOTHBRUSHx/
1:37:44 LL: Offers Washoe a hammer
1:37:58 Washoe: TOOTHBRUSHx HURRYx/
Trial #33
1:19:24 Tatu: TOOTHBRUSHx/
1:19:58 LL: Offers Tatu a ball
1:20:28 Tatu: MASKx/

In the category nonsign, the chimpanzee failed to sign within 30 s after the probe (Brinton et al. 1986a, p. 377).

6.10 Results

For each chimpanzee, there were 10 trials in each of the four conditions, yielding a total of 40 responses per chimpanzee. The 3 × 4 FET tests indicated that for four of the five chimpanzees the distribution of response types differed significantly across the four conditions, Washoe, $p < .001$, Tatu, $p < .001$, Dar, $p = .002$, Loulis, $p = .03$. There was no evidence that Moja's response types differed across the conditions, $p = .28$, FET.

Systematic differences in nonsigning contributed to the significance of the omnibus tests. Each chimpanzee ceased signing in the comply condition significantly more often than would be expected given the null hypothesis, Washoe, APR = 3.7, $p < .001$, Tatu, APR = 5.90, $p < .001$, Dar, APR = 3.70, $p < .001$, and Loulis, APR = 2.51, $p = .01$.

In the comply condition, Washoe, Tatu, and Dar used revisions significantly less often than expected, Washoe, APR = −2.01, $p = .04$, Tatu, APR = −2.53, $p = .01$, and Dar, APR = −2.25, $p = .02$. Also in the comply condition, Loulis exhibited less repetition than would be expected as indicated, APR = −2.68, $p = .007$. Finally, in the unresponsive condition, Washoe used significantly more repetitions than would be expected, APR = 3.65, $p < .001$.

Like human children (Brinton et al. 1986a, b; Wilcox and Webster 1980), the chimpanzees repeated their initial request more often in the refuse and unresponsive conditions than in either of the other conditions. For example,

Trial #182
0:34:05 Washoe: HUG x/
LL: No response
0:34:07 Washoe: HUG x/

Repetitions add no new information. In contrast the other category of response, revisions (also termed modifications) contain addition, deletion, or substitution of words or signs that change the original message (Halle et al. 2004). As described by Halle et al. (2004: 45):

> Whereas both repetitions and modifications can be effective from the child's point of view… modifications have been viewed as important indicators of children's development in perspective-taking skills needed for conversation. Modifications have obvious practical advantages as well. If a child's communication is not successful, then producing an alternative communication act that matches environmental conditions and partner behavior would seem to have adaptive value.

Revisions require persistence and elaboration which are indicators of intentionality (Bates et al. 1979; Golinkoff 1986, 1993). Golinkoff states that these behaviors in young children "indeed reveal an understanding … of the mind of another"

(p. 203). Adjusting responses to conversational partners is part of the shared negotiation and give-and-take of conversation. The chimpanzees in this study used revisions often when the interlocutor misunderstood the request, for example,

Trial #140
0:34:05 Moja: EATx GUMx/
LL: Gives Moja string
0:34:07 Moja: EAT THAT GUM/

If revisions are indicators of intentionality in human children, then the same can be said for chimpanzees.

6.11 Chimpanzees Initiate Conversations

In similar procedures, Bodamer and Gardner (2002) explored initiation of interactions. At the original CHCI facility, the chimpanzees had access to a suite of enclosures. One of the enclosures was across the hall from a human work room. When a caregiver was in the workroom, the chimpanzees often came to the nearby enclosure to request objects or activities. They often made noises if the human was not looking at them. Bodamer and Gardner systematically studied these initiations. The interlocutor sat in the work room with his back toward the chimpanzees' enclosure. When the chimpanzee made a noise, the interlocutor turned and faced the chimpanzee immediately or after a 30-s delay. When the interlocutor was not facing the chimpanzees, they made noises, such as Bronx cheers, and rarely signed. The few times the chimpanzees signed they used signs that made noise, such as DIRTY where the back of the hand hits the bottom of the jaw. With force this sign is noisy. In the delay condition, the noises became louder and faster. Once the interlocutor faced the chimpanzees, signing began and noises ended. Using a naturally occurring situation, this experiment showed the chimpanzees initiate interactions and sign spontaneously.

Experimental conversations with chimpanzees using systematic probes show their linguistic skill. The chimpanzees initiate and maintain conversations contingent on and appropriate to the interlocutor's input.

7 Conversations Versus Drill

The discoveries about chimpanzee signing occur with experimental trials presented in conversational contexts. The opposite of a conversational context is a drill context, which is characterized by requests to name objects and lacks in the natural give-and-take of conversations. Gardner and Gardner commented that drill tended to reduce the amount and quality of signing by the cross-fosterlings (Gardner and Gardner 1971: 140–141). This was later demonstrated experimentally when O'Sullivan and Yeager (1989) compared a conversational style of interaction to drill interactions with signing chimpanzee Nim (Terrace 1979). They found that in a conversational context, Nim interrupted less often and produced a high rate of spontaneous utterances. He behaved like a conversational partner when treated as one. Brown (1973: 178) noted children had deficient performance when mothers tried to elicit speech by asking their children to names things rather than interacting in a natural way.

The hallmark of the sign language studies is that caregivers treated the chimpanzees as conversational partners socially motivated rather than experimental subjects bribed or forced into participation. The chimpanzees were always free to leave the testing situation and free to respond to their world with their full repertoire of behaviors; these were often the dependent variables. Interlocutors nearly always doubled as chimpanzee caregivers. The tests were then embedded into the rich daily interactions that occurred between two friends. No rigor was lost, and an understanding of the remarkable similarities between human and chimpanzee behaviors is gained.

With comparable conditions, the research supports continuity with differences of degree among species as predicted by Darwin's theory of evolution.

References

Abbeduto L, Hesketh LJ (1997) Pragmatic development in individuals with mental retardation: learning to use language in social interactions. Ment Retard Dev Disabil Res Revs 3:323–333

Anselmi D, Tomasello M, Acunzo M (1986) Young children's responses to neutral and specific contingent queries. J Child Lang 13:135–144

Bates E, Benigni L, Bretherton I, Camaioni L, Volterra V (1979) The emergence of symbols. Academic Press, New York

Benus S, Gravano A, Hirschberg J (2012) Pragmatic aspects of temporal accommodation in turn-taking. Linguist Lang Speech 43:3001–3027

Bloom L, Rocissano L, Hood L (1976) Adult-child discourse: developmental interaction between information processing and linguistic knowledge. Cogn Psychol 8:521–552

Bodamer MD (1987) Chimpanzees signing to themselves. Thesis, Central Washington University

Bodamer MD, Gardner RA (2002) How cross-fostered chimpanzees (*Pan troglodytes*) initiate and maintain conversations. J Comp Psych 116:12–26

Bodamer MD, Fouts RS, Fouts DH, Jensvold MLA (1994) Private signing in chimpanzees. Hum Evol 9:281–296

Brinton B, Fujiki M (1991) Responses to requests for conversational repair by adults with mental retardation. J Speech Hear Res 34:1087–1095

Brinton B, Fujiki M, Loeb DF, Winkler E (1986a) Development of conversational repair strategies in response to requests for clarification. J Speech Hear Res 29:75–81

Brinton B, Fujiki M, Winkler E, Loeb D (1986b) Responses to requests for clarification in linguistically normal and language-impaired children. J Speech Hear Disord 51:370–378

Brown R (1973) Development of the first language in the human species. Am Psychol 28:97–106

Chalcraft VJ, Gardner RA (2005) Cross-fostered chimpanzees modulate signs of American Sign Language. Gesture 5:107–131

Ciocci SR, Baran JA (1998) The use of conversational repair strategies by children who are deaf. Am Ann Deaf 143:235–245

Duncan J (2000) Conversational skills of children with hearing loss and children with normal hearing in an integrated setting. Volta Rev 101:193–211

Ferguson A (1998) Conversational turn-taking and repair in fluent aphasia. Aphasiology 12:1007–1031

Fouts D (1994a) The use of remote video recordings to study the use of American Sign Language by chimpanzees when no humans are present. In: Gardner RA, Gardner BT, Chiarelli B, Plooij FX (eds) The ethological roots of culture. Kluwer, The Netherlands

Fouts R (1994b) Transmission of human gestural language in a chimpanzee mother-infant relationship. In: Gardner RA, Gardner BT, Chiarelli B, Plooij FX (eds) The ethological roots of culture. Kluwer, The Netherlands

Fouts RS, Hirsch AD, Fouts DH (1982) Cultural transmission of a human language in a chimpanzee mother-infant relationship. In: Fitzgerald HE, Mullins JA, Page P (eds) Psychobiological perspectives: child nurturance, vol 3. Plenum Press, New York

Fouts RS, Abshire ML, Bodamer M, Fouts DH (1989a) Signs of enrichment: toward the psychological well-being of chimpanzees. In: Segal EF (ed) Housing care and psychological wellbeing of captive and laboratory primates. Noyes, Park Ridge

Fouts RS, Fouts DH, Van Cantfort TE (1989b) The infant Loulis learns signs from cross-fostered chimpanzees. In: Gardner RA, Gardner BT, Van Cantfort T (eds) Teaching sign language to chimpanzees. SUNY Press, New York

Fouts RS, Jensvold MLA, Fouts DH (2002) Chimpanzee signing: Darwinian realities and Cartesian delusions. In: Bekoff M, Allen C, Burghardt G (eds) The cognitive animal: empirical and theoretical perspectives in animal cognition. MIT Press, Cambridge

Furrow D (1984) Social and private speech at two years. Child Dev 55:355–362

Gallagher TM (1977) Revision behaviors in the speech of normal children developing language. J Speech Hear Res 20:293–302

Galski T, Tompkins C, Johnston MV (1998) Competence in discourse as a measure of social integration and quality of life in persons with traumatic brain injury. Brain Inj 12:769–782

Gardner RA (2002) The proper study of chimpkind. Behav Brain Sci 25:624–625

Gardner RA, Gardner BT (1969) Teaching sign language to a chimpanzee. Science 165:664–672. doi:10.1126/science.165.3894.664

Gardner BT, Gardner RA (1971) Two-way communication with an infant chimpanzee. In: Schrier AM, Stollnitz HD (eds) Behavior of nonhuman primates, vol 4. Academic Press, New York

Gardner RA, Gardner, BT (1973) Teaching sign language to the chimpanzee, Washoe (Film). Available from Psychological Cinema Register, State College, PA

Gardner BT, Gardner RA (1974) Comparing the early utterances of child and chimpanzee. In: Pick A (ed) Minnesota symposium on child psychology, vol 8. University of Minnesota Press, Minneapolis

Gardner BT, Gardner RA (1975) Evidence for sentence constituents in the early utterances of child and chimpanzee. J Exp Psychol 104:244–267

Gardner RA, Gardner BT (1978) Comparative psychology and language acquisition. Ann NY Acad Sci 309:37–76

Gardner RA, Gardner BT (1984) A vocabulary test for chimpanzees. J Comp Psych 98:381–404

Gardner RA, Gardner BT (1989) A cross-fostering laboratory. In: Gardner RA, Gardner BT, Van Cantfort TE (eds) Teaching sign language to chimpanzees. SUNY Press, Albany

Gardner BT, Gardner RA (1994) Development of phrases in the utterances of children and cross-fostered chimpanzees. In: Gardner RA, Gardner BT, Chiarelli B, Plooij FX (eds) The ethological roots of culture. Kluwer, The Netherlands

Gardner RA, Gardner BT (1998) The structure of learning. Lawrence Erlbaum, Mahwah

Gardner BT, Gardner RA, Nichols SG (1989a) The shapes and uses of signs in a cross-fostering laboratory. In: Gardner RA, Gardner BT, Van Cantfort T (eds) Teaching sign language to chimpanzees. SUNY Press, New York

Gardner RA, Gardner BT, Drumm P (1989b) Voiced and signed responses in cross-fostered chimpanzees. In: Gardner RA, Gardner BT, Van Cantfort T (eds) Teaching sign language to chimpanzees. SUNY Press, New York

Gardner RA, Van Cantfort TE, Gardner BT (1992) Categorical replies to categorical questions by cross-fostered chimpanzees. Am J Psychol 105:27–57

Gardner RA, Scheel MH, Shaw HL (2011) Pygmalion in the laboratory. Am J Psychol 124:455–461

Garvey C (1977) Contingent queries and their relations in discourse. In: Ochs E, Schieffelin B (eds) Developmental pragmatics. Academic Press, New York

Golinkoff RM (1986) I beg your pardon? The preverbal negotiation of failed messages. J Child Lang 13:455–476

Golinkoff RM (1993) When is communication a meeting of minds? J Child Lang 20:199–207

Goodall J (1986) The chimpanzees of Gombe. Harvard University Press, Cambridge

Halle J, Brady N, Drasgow K (2004) Enhancing socially adaptive communicative repairs of beginning communicators with disabilities. Am J Speech-Lang Pathol 13:43–54. doi:10.1044/1058-0360(2004/006

Hayes KJ, Hayes C (1951) The intellectual development of a home-raised chimpanzee. Proc Am Philos Soc 95:105–109

Jensvold MLA, Fouts RS (1993) Imaginary play in chimpanzees (*Pan troglodytes*). Hum Evol 8:217–227

Jensvold MLA, Gardner RA (2000) Interactive use of sign language by cross-fostered chimpanzees. J Comp Psychol 114:335–346

Jensvold ML, Wilding L, Schulze SM (2014) Signs of communication in chimpanzees. In: Witzany G (ed) Biocommunication in animals. Springer, Dorderecht

Keenan EO (1977) Making it last: repetition in children's discourse. In: Ervin-Tripp S, Mitchell-Kernan C (eds) Child discourse. Academic Press, New York

Keenan E, Klein E (1975) Coherency in children's discourse. J Psycholinguist Res 4:365–380

Kellogg WN, Kellogg LA (1933) The ape and the child. Whittlesey House, Oxford

King JM, Gallegos-Santillan P (1999) Strategy use by speakers with dysarthria and both familiar and unfamiliar conversational partners. J Med Speech-Lang Pathol 7:13–116

Lane H, Pillard R (1978) The wild boy of Burundi. Random House, New York

Leitten L, Jensvold ML, Fouts R, Wallin J (2012) Contingency in requests of signing chimpanzees (*Pan troglodytes*). Interact Stud 13:147–164

Lepper MR, Greene D, Nisbett RE (1973) Undermining children's intrinsic interest with extrinsic reward: a test of the overjustification hypothesis. J Pers Soc Psychol 2:129–137

Maestas Y, Moores J (1980) Early linguistic environment: interactions of deaf parents with their infants. Sign Lang Stud 26:1–13

Marcos H (1991) Reformulating requests at 18 months: gestures, vocalizations and words. First Lang 11:361–375. doi:10.1177/014272379101103304

Marcos H, Bernicot J (1994) Addressee co-operation and request reformulation in young children. J Child Lang 21:677–692

Marcos H, Kornhaber-le Chanu M (1992) Learning how to insist and clarify in the second year: reformulation of requests in different contexts. Int J Behav Dev 15:359–376. doi:10.1177/016502549201500305

Marschark M (1993) Psychological development of deaf children. Oxford University Press, New York

Matthews WS (1977) Modes of transformation in the initiation of fantasy play. Dev Psychol 13:212–216

Mol L, Krahmer E, Maes A, Swerts M (2012) Adaptation in gesture: converging hands or converging minds? J Mem Lang 66:249–264

Most T (2003) The use of repair strategies by children with and without hearing impairment. Lang Speech Hear Serv Schools 33:112–123. doi:10.1044/0161-1461(2002/009

O'Sullivan CO, Yeager CP (1989) Communicative context and linguistic competence: the effects of social setting on a chimpanzee's conversational skill. In: Gardner RA, Gardner BT, Van Cantfort TE (eds) Teaching sign language to chimpanzees. SUNY Press, New York

Pickering MJ, Garrod S (2004) Toward a mechanistic psychology of dialogue. Behav Brain Sci 27:169–225

Rimpau JB, Gardner RA, Gardner BT (1989) Expression of person, place, and instrument in ASL utterances of children and chimpanzees. In: Gardner RA, Gardner BT, Van Cantfort TE (eds) Teaching sign language to chimpanzees. SUNY Press, Albany

Ripich DN, Carpenter BD, Ziol EW (2000) Conversational cohesion patterns in men and women with Alzheimer's disease: a longitudinal study. Intl J Lang Commun Dis 35:49–64

Ruvolo M (1994) Molecular evolutionary processes and conflicting gene trees: the hominoid case. Am J Phys Anthropol 94:89–113. doi:10.1002/ajpa.1330940108

Ryckebusch C, Marcos H (2004) Speech acts, social context and parent-toddler play between the ages of 1; 5 and 2; 3. J Pragmat 36:883–897

Schlesinger HS, Meadow KP (1972) Sound and sign: Childhood deafness and mental health. University of California Press, Berkeley

Sibley CG, Ahlquist TE (1984) The phylogeny of the hominoid primates indicated by DNA-DNA hybridization. J Mol Evol 20:2–15. doi:10.1007/BF02101980

Singh JAL, Zingg RM (1942) Wolf children and feral man. Shoe String Press, Hamden. Reprinted in 1966 by Harper & Row

Slater PJB, Williams JM (1994) Bird song learning: a model of cultural transmission? In: Gardner RA, Gardner BT, Chiarelli B, Plooij FX (eds) The ethological roots of culture. Kluwer, The Netherlands

Stamps J (2003) Behavioral processes affecting development: Tinbergen's fourth question comes of age. Anim Behav 66:1–13. doi:10.1006/anbe.2003.2180

Stanyon R, Chiarelli B, Gottlieb D, Patton WH (1986) The phylogenetic and taxonomic status of *Pan paniscus*: a chromosomal perspective. Am J Phys Anthropol 69:489–498

Terrace H (1979) Nim. Knopf, New York

The Chimpanzee Sequencing and Analysis Consortium (2005) Initial sequence of the chimpanzee genome and comparison with the human genome. Science 437:69–87. doi:10.1038/nature04072

Van Cantfort TE, Gardner BT, Gardner RA (1989) Developmental trends in replies to Wh-questions by children and chimpanzees. In: Gardner RA, Gardner BT, Van Cantfort T (eds) Teaching sign language to chimpanzees. SUNY Press, New York

West MJ, King AP, Freeberg TM (1997) Building a social agenda for the study of bird song. In: Snowdon CT, Hausberger M (eds) Social influences on vocal development. Cambridge University Press, Cambridge

Wilbur RB, Petitto LA (1981) How to know a conversation when you see one: discourse structure in American Sign Language conversations. J Natl Stud Speech-Lang-Hear Assoc 9:66–81

Wilcox MJ, Webster EJ (1980) Early discourse behavior: an analysis of children's responses to listener feedback. Child Dev 51:1120–1125

Wootton AJ (1994) Object transfer, intersubjectivity and third position repair: early developmental observations of one child. Child Lang 21:543–564. doi:10.1017/S0305000900009454

How Primate Mothers and Infants Communicate: Characterizing Interaction in Mother–Infant Studies

Maria Botero

Abstract All methodologies used to characterize mother–infant interaction in non-human primates include mother, infant, and other social factors. The chief difference is their understanding of how this interaction takes place. Using chimpanzees as a model, I will compare the different methodologies used to describe mother–infant interaction and show how implicit notions of communication and social interaction shape descriptions of this kind of interaction. I will examine the limitations and advantages of different approaches used in mother–infant studies, and I will sketch an alternative approach to studying mother–infant interaction in non-human primates that adopts Bruner's developmental studies on human infant communication.

Keywords Mother-infant interaction • Methods • Primate development • Chimpanzees

In creating the so-called "monster mothers"—mechanical models made of wire and covered in a soft cloth designed to resemble a monkey—Harlow and Zimmermann (1959) offered one of the most powerful images of how essential the mother is for an infant. Harlow described how an infant monkey will wait patiently in a corner for its "monster mother" to pull back its spikes so it can climb back for comfort. This conveys in a single image how essential it is for a primate infant to have a mother or at least a mother figure. Since then the challenge has become how to study the mother–infant interaction to be able to capture this essential relation common to all primates.

All methodologies used to characterize mother–infant interaction in non-human primates include mother, infant, and other social factors. The chief difference is their understanding of how this interaction takes place; that is, each methodology

M. Botero (✉)
Department of Psychology and Philosophy,
Sam Houston State University, Huntsville, TX, USA
e-mail: mbotero@shsu.edu

M. Pina and N. Gontier (eds.), *The Evolution of Social Communication in Primates*, Interdisciplinary Evolution Research 1, DOI: 10.1007/978-3-319-02669-5_5,

selects certain elements of this interaction as relevant. Within each methodological design, researchers explicitly or implicitly answer several questions. First, what is interaction and which is the best way to describe the interaction among these units? Second, what are the mental contents (if any) of the individuals interacting. As a result, each methodology focuses on different presuppositions leading to specific questions and leaving others aside. The aim of this chapter is to show how definitions of communication and social interaction have guided the design of methodologies that describe interaction in non-human primates and the kinds of questions and answers that result from such methodologies. Using chimpanzees as a model, I will compare the different methodologies used to describe mother–infant interaction and show how descriptions of this kind of interaction are shaped by implicit notions of communication and social interaction. The first model I will examine is what I call the *Ecological-Linear* approach, and I will offer as an example one of the most influential methodologies available to study mother–infant interaction, namely that designed by Altmann (1974). This approach focuses on describing the mother–infant interaction in terms of rates of observable behaviors (e.g., contact, proximity) and placing special emphasis on who initiates the behavior and who receives it.

As an alternative approach, I will examine Bruner's (1990) description of the development of infant human communication. In this approach, the infant's acquisition of the ability to communicate with others can be explained from the context of the mother–infant interaction without having to address the issue of whether the infant is conscious of the intentions in the communicative process; that is, instead of focusing on what kind of content (if any) is delivered when the mother performs a behavior toward an infant, the focus is placed on how within the mother–infant interaction, we can observe a communicative function. Only through the mother–infant interaction does the infant become proficient in *what* and *how* gestures and utterances can become meaningful in a specific community. Thus, by using Bruner's approach, I hope to provide a more complex picture of the mother–infant interaction, one that focuses on the infant's development not only as an interplay between biological and environmental factors but also as one that requires that the mother–infant interaction be observed under the presupposition that the infant's social, cognitive, and communicative development are tied together. As a result of this analysis, I will show first how the way we think about communication and social interaction shapes the kinds of questions and answers in a research program. Second, I will advocate for an approach to observing the mother–infant not only as a dyad but also as a socially embedded dyad.

1 Ecological-Linear Approach

In the first studies I examine in this chapter, researchers chose mother, infant, and the surrounding social and physical environmental elements as the basic units of analysis. I call this the *Ecological-Linear* approach; researchers who adopt this approach do not use mental states (e.g., intentionality, beliefs, desires) to explain the interaction between these units. They[1] limit themselves to observable behaviors (e.g., contact, proximity) that take place between these basic units. They also place special emphasis on who initiates the behavior and who receives it; thus, they characterize the interaction between these units as a communicative one. They presuppose that communication is the exchange of behaviors between the different units of analysis. In such a communicative exchange, one unit is the sender and the other is the receiver of the behavior. Such methodologies also presuppose that when the sender provides this information, the information causally affects the receiver. In other words, this set of studies presupposes a model of communication that flows in a linear way from a sender to a receiver which causally affects the receiver. Linearity means that any change in the communication process follows incremental variations.

An example of this approach to communication can be found in Cheney and Seyfarth's (1990) study of calls in vervet monkeys. All of these calls have similar acoustic structures,[2] but, depending on the predator (eagle or leopard), a monkey will act as a sender and select a specific kind of call that will elicit a specific escape behavior response. Alternatively, if the call has already been made by another monkey, the monkey may not select the call at all. All the other monkeys act as receivers that decode the call and depending on the sender (i.e., another adult or an infant) will act accordingly (climb a tree or remain in the ground). In this approach to communication, human observers will describe the behavior of a monkey using preset units of information (i.e., predefined calls with a specific acoustic structure).[3] They describe one monkey as acting as a sender, choosing a unit and sending it (or not sending it). The observers will also describe how these units are decoded by a receiver(s) who acts accordingly.

---

[1] Among the examples of how this approach is used in the research of different species we can find the following: Baboons (Altmann 1980), Japanese macaques (Bardi and Huffman 2002; Schino et al. 1995), rhesus macaques (Maestripieri 1993, 1998, 1999; Maestripieri et al. 2006), bonobos and chimpanzees (De Lathowres and Eslacker 2004), chimpanzees (Bloomsmith et al. 2003), and in general reviews of vervets, rhesus monkeys, Japanese macaques, and baboons (Fairbanks 1993, 1996; Fleming et al. 2002; Hinde 1983, 1984; Maestripieri 1999).

[2] The structure varies in female and male monkeys but does not alter the idea that a call is a unit of information which shares the same structure among individuals.

[3] When I describe these units as pre-set, this does not entail that they are not defined through observation. I am describing how a researcher, after hours of observation, divides her observations into categories or units and later uses these units to describe behaviors. In this way, by the time she observes the behavior she is interested in, these units that are already pre-set, ready to be used.

1.1 How This Approach is Used in Most Mother–Infant Studies in Non-human Primates

To understand the presuppositions of communication and mind in the mother–infant interaction, I will focus one of the most influential methodologies available to study mother–infant interaction, namely that designed by Altmann (1974). Although she does not use calls as the units of behaviors, she uses behaviors that are transmitted from a sender to a receiver to describe how animals interact. Altmann claims: "(a) Focal Animal Sample on animal *i* provides a record of all acts in which *i* is *either* the *actor* or *receiver*" [4](1974, p. 243).

In other words, in Altmann's approach, to describe the interaction between individuals, it is necessary to describe how they exchange behaviors. These behaviors communicate specific information, and it is necessary to distinguish who initiates and who receives, she argues that: "Most such behavior (social behavior) is directed ("addressed"); I shall distinguish between the actor or sender, and the object or receiver of each social act" (1974, p. 243). The explanation of this exchange does not require attributing the participants' mental states. Description is limited to the observation of how behavior is exchanged. This methodology presupposes that two animals interact in a communicative way: one is the sender and the other is the receiver; a subject can only perform one role (i.e., actor or receiver) at a time in an interaction. In this communicative process, discrete units of behavior (e.g., call, gestures) are exchanged and are summarized in the ethogram designed by the researcher. These units convey the same information in every context, whether they are found in play or aggression. What changes is the frequency with which these units are exchanged in different contexts. Summarizing the interaction between two or more individuals as the sum of these frequencies presupposes that the changes that take place in this communicative exchange can only change in an incremental way.

Altmann's (1980) methodology has been applied to different studies in primates where the common characteristic is that, because these units of behavior exchanged do not change in different contexts, all of these studies reduce the mother–infant interaction to the rates of exchanged behaviors. For example, one way in which units of behavior have been used to describe the mother–infant interaction is by describing the mother–infant interaction as parenting styles or variations in the dimension of *protectiveness* (i.e., variation in the degree to which the mother restrains infant exploration, initiates proximity and contact, and provides nurturing behaviors such as grooming) and *rejection* (i.e., degree to which the mother limits the timing and duration of suckling, carrying, and contact).

One of the advantages of adopting this linear model of communication is that using parenting styles allows researchers to make predictions of specific mother–infant interaction exchanges. Because the units of behavior or information are known, whenever it is observed that the sender sends this information in a specific

[4] Emphasis is not in the original.

context, it is possible to predict the receiver's behavior. For example, if the mother is low-ranking and protective (i.e., higher rate of restriction and initiation of contact), it is possible to predict that the infant will have a low exploratory behavior. Another advantage is that treating interaction as an exchange of fixed informational units allows, according to Altmann's (1974), an unbiased description of the behaviors of individuals. Altmann's methodology guarantees that the resulting data are not a record of only extraordinary events that the observer may choose based on what he/she considers important; rather, the data become a statistically significant description of the everyday behaviors of the mother and infant. Because of these two characteristics, other models have adopted the ecological measurements that presuppose Altmann's linear model of communication. These parenting styles are used to describe the mother–infant interaction in monkeys and apes in the field and the laboratory.

1.2 Notion of Communication

This model seems to follow a mathematical approach to information. In 1948, Claude Shannon suggested that communication can be described as a transmission process that can be treated as a matter of "encoding" the "information" that is contained in the "message" that is being sent. In this context, "information" is something which can be broken down into constituent elements and quantified. This presupposition of information can also be found in Saussure's *Course of General Linguistics*. In Saussure's famous "speech-circuit" diagram, the two speakers using a language are "encoding" and "decoding" the information they wish to convey. That is, linguistic interaction can be described as a code.

In this model of communication, information cannot be confused with meaning; according to Shannon and Weaver, "the semantic aspects of communication are irrelevant to the engineering aspects" (1978, p. 8). This does not mean that meaning is irrelevant. Rather than emphasizing *what* is said, the approach emphasizes what the sender *could* have said: "Information is a measure of one's freedom of choice when one selects a message." (Shannon and Weaver 1978, p. 9). For that reason, the emphasis in this approach to communication is on the logarithm of the number of available choices in a communicative process and how this choice is coded and sent to a receiver who decodes it. According to Ellis and Beattie (1986), when this model of communication is adopted, the criterion of when communication takes place is: "when one organism (the transmitter) encodes information into a signal which passes to another organism (the receiver) which decodes the signal and is capable of responding appropriately" (Ellis and Beattie 1986, p. 3). That is, if all relevant elements (i.e., sender, receiver, encoded units of information) can be identified, communication takes place. Within this linear model of communication, the roles of sender and receiver are fixed and can only be exchanged once the message has been delivered. The information transmitted can be divided and analyzed in units. Going back to the Cheney and Seyfarth example, the monkey selects from a repertoire of specific acoustic signals the alarm call for eagle instead of the alarm call for leopard, and a communicative interaction can be measured by the rate of exchange of these units.

2 Limitations of This Model

When this model of communication is used to design a methodology to capture the mother–infant interaction, the interaction has to be reduced to the preset elements. As a result, the methodology will miss much of the complexity present in the mother–infant interaction. It does not consider how the meaning of the units of behavior exchanged may vary depending on how this interaction takes place; that is, how the meaning of each behavior may change depending on the past history of the participants and how the partners respond to each other's behaviors. Why do we need to go into such detail? To answer this question let's go back to Harlow's experiments. He (1959) originally demonstrated how a soft, warm mother was essential for the infant's development. Later in his work with Harlow and Suomi (1971), he showed that the interactions with peers may compensate for the absence of the mother and absence of early social interaction. This is an example of why we need a model that provides a more textured account of how all the variables (i.e., mothers, peers, and social group) affect the infant's development. I will show this in more detail using chimpanzees as a model organism and showing the limitations of the Ecological-Linear approach.

Since the 1940s, it has been clear that the absence of the mother plays a significant role in infant chimpanzee social and cognitive development.[5] Recent studies have shown that early rearing conditions affect the performance of chimpanzees in cognitive tasks, such as the cognitive capacities that underlie gestural communication. For example, subjects exposed to early rearing conditions that include human social–linguistic tasks perform better in *object choice task.* In this task, a human experimenter points to one of the containers that were previously baited with food (out of sight of the ape), and then, the ape can point to request food (Lyn et al. 2010). Apes with enriched rearing conditions also performed better in *directional pointing,* a task in which the apes have to direct humans to a hidden goal in the woods (Menzel and Menzel 2012). Thus, we need to know the elements in this early interaction that have such a profound effect on the infant's cognitive and communicative development.

Second, non-human primates exhibit cultural/social conventions. Research in chimpanzees, done by Whiten et al. (1999), has emphasized that communication is sensitive to the different practices of different communities within the same

[5] For chimpanzees, the effects of a mother's absence have been observed in chimpanzees raised in isolation (for reviews see Yerkes 1943; Menzel 1964; King and Mellen 1994) and under captivity in enriched environments (for reviews see Clarke et al. (1982), Brent et al. (1991), Baker (1996) and van Ijzendoorn et al. (2009)).

species. For example, Whiten et al. show how the behavioral signal used to request grooming in chimpanzees from the Mahale and Kasakela communities (two communities that are not far apart) varies from hand-clasp to branch-clasp; there are no ecological explanations for this difference. A more complex notion of communication would be able to incorporate how the community-specific type of communication acquired by the infant through the interaction with its mother.

It may be argued that these questions are better answered in the context of experimental conditions rather than in the wild. There have been wonderful studies that described these conditions in controlled settings (see, for example, Matzusawa 2006); however, it is important to consider how these conditions may vary from captivity to social interactions in the wild. Moreover, if we are already observing differences in the outcomes produced by rearing conditions in captivity, it would seem important to know how this same mechanism works in the wild.

Let me illustrate the limitations of the *Ecological-Linear* approach with an example. As part of a larger study, Botero et al. (2013) examined the anxiety levels and social interactions of two orphan and four mother-reared adolescent chimpanzees (*Pan troglodytesschweinfurthii*) in the Kasekela community at Gombe National Park, Tanzania. The two orphan adolescent chimpanzees (Flirt and Titan) differed from other adolescents of a similar age, exhibiting higher levels of anxiety and lower levels of play. The findings of this study are at odds with observations done in laboratories and sanctuaries. Bloomsmith et al. (2005) found that laboratory infants orphaned after age 2 and then reared by peers behaved similarly to mother-reared infants as juveniles. Studies of wild-born chimpanzees orphaned by the bush and pet trade have also shown that individuals can successfully adapt to life in the wild (Humle et al. 2011; Beck 2010) or show no long-term signs of stress when raised in sanctuaries (Wobber and Hare 2011). Meanwhile, even though the orphans observed by Botero et al. were orphaned much later than 2 years of age, they still exhibited behavioral differences as juveniles.

The effects of a mother's absence in the wild, where an infant chimpanzee has the opportunity to be adopted and/or to interact with other members of its community, have been studied much less, and some of the findings are also contradictory. Some have found that orphans exhibit negative behavioral consequences (Pusey 1983; Goodall 1986), but others have found that orphan chimpanzees do not exhibit any behavioral problems (Boesch and Boesch-Ackerman 2000; Boesch et al. 2010). Thus, it seems that not only does variation exist in captive/sanctuary conditions and wild conditions but also social variations across different communities may influence orphan survival and behavior. For example, Boesch et al. (2010) found higher levels of adoption among the group members of the Täi Forest community than in other chimpanzee communities. Boesch et al. hypothesized that this behavior may be the result of unique within-group solidarity exhibited as a response to the threat posed by the population of leopards that coexists with the group.

Finally, even within the same community (e.g., Kasekela community), Botero et al. found that after losing their mothers at different ages, the subjects in this community developed somewhat different patterns of abnormal behavior. There may be a relation between the age at which a chimpanzee is orphaned and the behavior that follows; such a relation between age of being orphan and behavioral outcomes has also been found in observations of the rehabilitation process of wild-born animals (Carter 2003). However, it has also been found that different kinds of interaction between the infant and its mother seem to correlate with the different ways in which the infant behaves after its mothers' death (see Botero et al. in preparation). This is in accordance with previous studies that show how the kind of interaction the infant has with its mother affects the way the infant reacts to being separated from her (for a review, see Fairbanks, 1996.)

These results are intriguing and warrant further investigation to establish more fully whether age of orphaning, social characteristics, or the kind of interaction with its mother is a determining factor in the later development of abnormal behaviors. A linear account of these variations will be able to provide a description of ways in which infants in captivity and infants in the wild differ in terms of rates of behavior and the ways in which different orphans in different communities differ in terms of rates of behaviors such as contact or grooming. However, two questions remain, how do the different ways in which the mother–infant interaction takes place in different communities or within the same community come into existence in the first place? How do these differences affect the behavioral outcomes observed in the infant's development? Thus, we need an explanation that acknowledges not only different kinds of mother–infant interactions (such as parenting styles) but also how mother–infant interactions are situated in the specific social practices of the community in which the infants are born and how this affects the infant's development of social, cognitive, and communicative abilities. To include these elements, we need an explanation that allows us to go beyond the summary of rates to explain how the mother–infant interaction may affect the infant's behavior. We need to adopt a different notion of communication, one that is closer to the idea of pragmatics.

It may be argued that the users of the linear model frequently add another layer to this view of communication, a layer that includes how these exchanges can be influenced by other variables such as audiences, features of utterance, responses, and the modes for the correlation. To include these variables, the users of the linear model adopt Grice's (1989) idea of *reflexive intention* in which the speaker's intention is prior to communication and is directed to the listeners. According to Grice, the speaker attempts to produce an effect on the listeners partly by making them recognize his intention to produce this communication. In Grice's words: "U utters x M-intending that A produce r" (Grice 1989, p. 105). Adopting Grice approach in animal communication is not problematic since, according to Grice utterances may include, not just sounds and marks but also gesture, grunts, and groans. Thus, non-human animals are able to signal M-intentions without the use of verbal language. This characteristic can be found in the Cheney and Seyfarth reports of audience effects, such as a case when one the group disregards a call if it is emitted by an infant or a case when a member of the group fails to emit the signal when another member has already given the call. Moreover, in a more recent account, Cheney and Seyfarth (2007) argue that baboons have a language of thought, a language-like representational medium that allows them to deal with their complex social environment. This last point presents a problem for the observation of animal communication. If semantics can be reduced to propositional attitudes, then it is necessary to include a broader discussion of mental states, such as beliefs, desires, and intentionality, and most authors are not comfortable attributing these mental faculties to non-human animals.

Instead of focusing on internal mental states, I will argue for an approach that focuses on external observable behaviors. To include the different signals that mother and infant use to communicate and the way the infant learns how these signals are used in its community, we need to adopt a notion of communication that extends the meaning of the signal to external elements, to the social practices where these communicative interactions take place.

3 An Alternative Approach

One of the distinctive characteristics of mother–infant studies is that they deal with communication within the context of the mother–infant interaction and how this interaction is part of the infant's development. Bruner offers an approach to communication that highlights how communication is a developmental process that takes place within the context of the mother–infant interaction. Even though his approach focuses mostly on the development of human infants, it is possible to extrapolate it to other primates that share similar characteristics with the human caregiver–infant interaction, mainly with species in which the infant has a period of dependency on its mother allowing for the kind of interaction necessary to help the infant develop its social, cognitive, and communicative abilities.

Bruner's (1975) approach is useful to the observation of non-human primates in two ways. First, instead of describing only how complex linguistic abilities such as grammar develops, he focuses on what he calls the prespeech communicative acts. These acts are a set of complex, transferable skills that the child has to master to obtain a particular level of linguistic mastery. These skills are perceptual, motor, conceptual, social, and linguistic and allow the child to move from prespeech communication to the uses of language proper. The focus of this approach to communication concentrates on describing how the child exhibits a grasp of a basic form of understanding that becomes a prerequisite to a more complex utterance.

Second, in Bruner's (1990) approach, meaning is always a "culturally mediated phenomenon." For that reason, in development, "the child is not learning simply *what* to say but how, where, to whom, and under what circumstances" (1990, p. 71). In other words, for Bruner language is always used in a social behavior and as such, communication depends upon a mastery of cultural conventions.

According to Bruner, the mother–infant interaction is the context where social understanding becomes a form of practice. In each of these interactions, the child becomes a protagonist—an agent, a victim, an accomplice—and learns what is permissible and what leads to outcomes through action. These interactions usually take place in a conventionalized way, for example in a game of peekaboo. In these kinds of interactions, the mother attributes intention to the child's behavior, and the child learns how her efforts evoke a response. The child also learns that he/she can modify her responses to achieve a desired outcome. In these conventionalized interactions, the mother shows the child the ritual conventions in which the infant's gestures and utterances can help her achieve the child's desired outcomes. These different conventions vary between mothering styles and in socio-economic settings. Through this practice, the child understands: first, the interchangeability of roles between communicator and recipient and second, the link between the gesture and utterance in these conventions. This link makes the child's gesture or utterance meaningful.

Bruner claims that it is not necessary to focus on the question of whether and when a child is conscious. Instead, as researchers, we should focus on how communicative functions are shaped and how they are fulfilled. Even with human primates, according to Bruner (1975), the observer has to focus on the continuity between functionally equivalent forms of communication before and after the onset of speech and how these occur through the different ways in which the infant continually experiences and acts in the world, using language in different contexts. I will argue that to achieve a more detailed level of observation, following Bruner's ideas, we need to focus on the functional aspects of communication that can be observed in the way that the infant non-human primate interacts with its mother.

4 Application of this Definitions of Communication to Non-Human Primates Studies

Adopting this approach to the study of communication in mother–infant studies has several consequences for the design of such studies. First, it offers a different conception of the link between mind and communication. Traditionally, one of the main arguments against the ascription of mental states, such as intentionality, is that since animals do not have linguistic behavior, there is no way for the mental states to be manifest. In other words, thought and language go together, and the absence of the later makes it impossible to understand the mind (Davidson 1982). As stated, an approach to communication inspired by Grice, where the intentions of the communicator are fundamental to the communicative process, requires ascribing intentionality to the mother and infant even though there is no linguistic exchange.

If we adopt Bruner's approach to communication, the infant's acquisition of the ability to communicate with others can be explained in the context of mother–infant interaction without having to address the issue of whether the infant is conscious of intentions in the communicative process. That is, instead of focusing on what kind of content (if any) is delivered when the mother performs a behavior toward an infant, we will focus on how, within the mother–infant interaction, we can observe a communicative function. It is only through the mother–infant interaction that the infant becomes proficient in *what* and *how* her gestures and utterances can become meaningful in a specific community. Following Bruner, our emphasis would not be whether or not the infant has a conscious understanding of the intention; rather, we would focus on the function of communicative interactions.

The second and most important consequence of adopting Bruner's approach is that it forces us to understand how the child works at becoming a member of her linguistic community. To become part of the community, the child has to understand the social means to interact with others in her social environment. Bruner explains this through two fundamental concepts. First, the origins of language can be traced back to the child's earliest communicative and affective interactions with her mother. Bruner (1975) argues that the infant starts in demand modes, communicative routines where the infant uses different patterns to express discomfort, hunger, and demand for social interaction or fatigue. If these modes are attended by the mother, they create an expectancy of response. Once this expectancy is established, the next stage is the exchange mode. According to Bruner, this exchange mode in humans starts as early as two weeks of age, as infants will imitate facial gestures (Moore and Meltzoff 1977). Finally, the mother–infant interaction will be transformed into a reciprocal mode where the two participants enter a task with reciprocal non-identical modes where the roles between communicator and receiver are exchanged. Even though this model is designed for human primates, it is possible to extrapolate to other non-human primates based on the similarities found across primates in the way mothers interact with their infants. For example, early imitation of facial features as described by Moore and Meltzoff has also been found in chimpanzees (Takeshita et al. 2006) and monkeys (Ferrari and Fogassi 2012).

The idea of emphasizing the mother–infant interaction as a unit of observation is not new and has been illustrated among others in the attachment paradigm described by Bowlby (1958). The caregiver's face provides the infant a secure base that is used when exploring the world. The infant signals to let the caregiver know that she needs her, and the caregiver responds to these signals. The facial signals of the caregiver and the infant become synchronized. These ideas apply to both human and non-human primates. According to Suomi et al. (1995), Robert Hinde introduced Bowlby to the work of Harlow at the time that Bowlby was writing "The nature of the child's ties to his mother." In that paper, Bowlby included a footnote of Harlow's not-yet-published study of surrogated mothers. According to Bowlby, Harlow's work confirmed his hypothesis that it is "proximity and close bodily contact with a mother figure that cements the infant's attachment rather than the provision of food" (1991, p. 5). These observations have been made not only of monkeys but also of chimpanzees as well. From the beginning of attachment studies, chimpanzees have been used as examples of how non-human primates exhibit attachment behaviors. For example, Bowlby (1958), citing work by Yerkes and Tomilin, shows how behaviors necessary for the attachment bond, such as sucking, clinging, and crying, also take place in chimpanzees. Work by Masaki Tomonaga (2006) shows that one-month-old infant chimpanzees have the ability to discriminate their mothers' face from others.[6] Okamoto-Barth et al. (2007) describe how gaze is used as way of increasing proximity and how the infant chimpanzee uses its mother as a secure base.

However, the way in which interaction takes place in the mother–infant dyad can be examined in different ways. Vicedo (2013) shows that, when Harlow's experimental work is analyzed as a whole, it becomes clear that experimenters, experiments, and rhesus monkeys influenced each other. Many societal pressures, such as journalists, mothers, woman's clubs, and psychoanalysist's, influenced the interpretation of Harlow's findings. The monkeys overtime revealed a degree of plasticity and resilience that forced Harlow to reinterpret his description of primate developmental needs. Vicedo also shows that these changes in Harlow's conclusions were ignored by Bowlby when Bowlby was developing his attachment theory.

Thus, it becomes clear that first, to account for Harlow's research, it is necessary to use a model that offers a more complex description of social elements. Second, the presuppositions held by researchers on choosing the relevant elements when observing the mother–infant interaction have a profound effect in their theories; for Harlow outcomes in infant development depend on different social variables, as opposed to Bowlby who focused on instinctual interactions in the mother–infant dyad.

I claim that the Ecological-Linear approach also limits the mother–infant interaction to a reductionist view that describes the mother–infant dyad in terms of adaptive instinctual responses. Moreover, the Ecological-Linear approach attempts to describe how certain initial conditions of the mother–infant pair correlate with some aspects of the infant's outcome behavior but do not provide an explanation of *how* the mother–infant interaction affects the infant's development. As a consequence, researchers are unable to explain how the infant is affected by the mother–infant interaction. For example, the birth of a new sibling may be correlated with a decrease in contact between mother and infant and with more exploratory behavior on the part of the infant. This correlation may be explained in terms of cost and benefit for both mother and infant, how it creates or disrupts the attachment bond or how adaptive these behaviors are. But it does not explain how the decrease in contact causes the infant's exploratory behavior or why the birth of a new sibling may not cause a decrease in contact in other individual mother–infant pairs facing the same environmental conditions and with the same rank, or why this decreases in contact may not trigger exploratory behavior in other infants. These questions remain unanswered if the explanation is limited to a report of the correlations between rates of different behaviors observed.

What makes Bruner's approach an alternative to these previous descriptions is what he considers the second fundamental element in the acquisition of language, that is, language as the vehicle for the child's socialization. As Shanker and Talbot

---

[6] This behavior is important for forming an attachment bond because, as Bowlby (1958) notes, the infant chooses only one object of attachment. Thus, it needs to be able to discriminate among faces to be able to form its attachment bond.

(2001) claim, the central message of Bruner's approach to language development is that language and culture cannot be separated from one another. As part of her process of socialization, the child becomes a skilled participant in the culture-specific forms of communicative behavior used in the community where she is born. According to Shanker and Talbot, the acquisition of language is the gradual development of a set of practical, interactional techniques to engage with those in her social environment.

Following Bruner, Shanker and Talbot claim that when developing language, a child learns how to do things with words; thus, she learns not a simple behavior but a cultural technique that has meaning within that specific culture. For example, when a child truly learns the meaning of "I am sorry" she has learned the appropriate circumstances in which to express this sentence. For example, the apology has to be related to something she did rather than something that occurred in a movie. The child has also learned how to behave when expressing this sentence, for example, she learns that she needs to behave in a way that conveys that she really means it. The child also learns what she is doing by expressing this sentence. Finally, she has to understand that an apology has to be directed to the right person, and she has to be able to recognize when an apology is directed to her.

How does this translate to the study of mother–infant interaction in chimpanzees? Bruner's approach can be applied to a series of methodological decisions. Adopting this approach has consequences for the kinds of units we would be observing when studying the mother–infant interaction. Instead of observing mother, infant and rates of behavior exchanged, we would look at the interaction itself and how it unfolds in the specific cultural practices of the community observed. This approach will start where the linear method stops. That is, we can use as our starting point the different kinds of conventionalized mother–infant interaction as described by the parenting styles of non-human primates (Altman) such as protective (high in protectiveness and low in rejection), controlling (high in both protectiveness and rejection), rejecting (low in protectiveness and high in rejection), and laissez-faire. However, we will take a step further and describe in more detail how the infant learns within these conventionalized forms of interaction and how his/her gestures/utterances have meaning in the community where she lives.

That is, when understanding how an infant acquires a set of behaviors, we will not be limited to describe the rates of different behaviors, but we will emphasize how the infant acquires a meaningful set of behaviors that belongs to the community where he/she is born. These meaningful behaviors include any kind of behavior that is necessary for the social functioning of the community, such as aggressive displays, consolation, and gestures for requesting grooming or food. In other words, we will focus on any behavior that requires social partners to negotiate an outcome. We will also include expressions of emotional states among this set of meaningful behaviors. As Botero (2012) argues, variations in communication can be extended to the expression of emotions: the way emotions are expressed and how others understand these expressions are going to be specific to the groups where these emotional expressions were developed.

All of these different behaviors may be unique to the community, as described by Whiten et al. (1999), or more subtle behaviors that reflect the unique way in which the group interacts. This new emphasis requires that the researcher who is designing a study of mother–infant interaction in primates has to become familiar not only with species-typical behaviors but also with behaviors that are typical for the community being observed as well as the social hierarchies and patterns of interaction found in that community. These must be introduced in the design of his/her study.

Moreover, this approach requires that the researcher includes a description of the particular social interactions that the infant is exposed to in his/her community, especially those that involve a negotiation among two or more partners that will depend on who is present in the interaction and the context of the interaction. This is an important point because being part of a primate community means that communicative practices in a community may differ depending on the context, who is present, and the specific behavior being performed. For example, Slocombe and Zuberbühler (2005) show that chimpanzee vocalizations change depending on the information chimpanzees have about the specific social relationships of victims and aggressors and the severity of an attack; the call of the victim who is attacked may change if there is an equal or higher hierarchy subject than the aggressor in the audience. Therefore, to understand communication, it is not enough to describe it in terms of frequency of exchange of units of information. It is necessary to describe what kind of activity goes on in a community, how the communicative exchange happens, and how it is developed within a community. For a full description of a social event unfolding in the moment, it is necessary as King and Fox (2002) suggest to understand both particular interactions and community interactions.

As part of her development, the infant chimpanzee needs to understand all of the social variables in his/her community. She needs to learn how to become part of this community and to interact with the members of this community in meaningful ways. This learning/developmental process is done within the context of the interaction with her mother. Thus, following Bruner, a methodological design that attempts to follow the nuances of the mother–infant interaction and can account for variations across individuals and community will base its units of observation on how the infant gradually acquires these interactional techniques to engage with those in her social environment. This approach will provide us with a more complex picture of how the mother–infant interaction unfolds in a community and has an effect on the infant's cognitive, social, and communicative development. This picture will offer a more complex map of how individual differences in mother–infant interactions in a community and different mother–infant pairs in different communities result in different developmental outcomes for the infant.

5 Conclusion

The way we define communication and social interaction has a deep impact on the way we study human and non-human primates. In this chapter, I presented an analysis of different methodologies used in the study of mother–infant interaction in primates, and using chimpanzees as an example, I have shown how changing our understanding of communication and social interaction can have an effect on the design of the methods, the kinds of questions asked and the kinds of answers accepted when studying primates. I have also shown how a change in the definition of communication and social interaction, such as the one offered by Bruner, can lead to a more textured description of the mother–infant interaction. Adopting Bruner's approach to mother–infant studies entails focusing on the pragmatics of communication, how the infant acquires the capacity of understanding, how gestures, calls, or other forms of communicative expressions are meaningful in the community where the infants are born. This analysis starts with the findings of what I have called the *Ecological-Linear* method and takes them to a more detailed level. Even though this new level of detail requires more labor and cannot produce large samples of ecological measures, such as the ones produced by parenting styles, it allows us to understand in more detail how the infant learns to communicate through the interaction with its mother and how it becomes sensitive to the unique aspects of the interaction with its mother in its community.

Finally, I would like to suggest that even though this chapter has focused on mother–infant interaction in chimpanzees, it is plausible to think that these same conclusions can be extrapolated to other kinds of research in primates, that is, that the choice of definitions of communication and social interaction held in any research project that involves social interactions in a community of human or nonhuman primates will have similar effects in the way these primates are described.

References

Altmann J (1974) Observational study of behaviour: sampling methods. Behaviour 49:227–267

Altmann J (1980) Baboon mothers and infants. Harvard University Press, Cambridge

Baker KC (1996) Chimpanzees in single cages and small social groups: effects of housing on behavior. Contemp Top Lab Anim Sci 35:71–74

Bardi M, Huffman M (2002) Effects of maternal style on infant behavior in Japanese macaques (macaca fuscata). Dev Psychobiol Rev 41:364–372

Beck B (2010) Chimpanzee orphans: sanctuaries, reintroduction and cognition. In: Lonsdorf E, Ross SR, Matsuzawa T, Goodall J (eds) The mind of the chimpanzee: ecological and experimental perspectives. University of Chicago Press, Chicago

Bloomsmith M, Baker K, Ross S, Lambeth S (2005) Early rearing conditions and captive chimpanzee behavior: some surprising findings. In: Sackett G, Ruppenthal G (eds) Nursery rearing of nonhuman primates in the 21st century. Kluver, New York

Bloomsmith MA, Kuhar C, Baker K, Lambeth S, Brent L, Ross SR, Fritz J (2003) Primiparous chimpanzee mothers: behavior and success in a short-term assessment of infant rearing. Appl Anim Behav Sci 84:235–250

Boesch C, Boesch-Achermann H (2000) The chimpanzees of the Taï forest: behavioural ecology and evolution. Oxford University Press, New York

Boesch C, Bolé C, Eckhardt N, Boesch H (2010) Altruism in forest chimpanzees: the case of adoption. PLoS ONE 5(1):e8901

Botero M (2012) Reconstructing with more situated social interactions. Emot Rev 4:246

Botero M, MacDonal S, Miller R (2013) Anxiety-related behavior of orphan chimpanzees (*pan troglodytes schweinfurthii*) at Gombe National Park, Tanzania. Primates 54:21–26

Bowlby J (1958) The nature of the child's tie to his mother. Int J Psychoanal 39:350–373

Brent L, Lee DR, Eichberg JW (1991) Evaluation of chimpanzee enrichment enclosure. J Med Primatol 20:29–34

Bruner JS (1975) The ontogenesis of speech acts. J Child Lang 2:1–19

Bruner J (1990) Acts of meaning. Harvard University Press, Cambridge

Carter J (2003) Orphan chimpanzees in West Africa: experiences and prospects for viability in chimpanzee rehabilitation. In: Kormos C, Boesch C, Bakarr MI, Butynski T (eds) West African chimpanzees: status survey and conservation action plan. Primate Specialist Group

Cheney D, Seyfarth R (1990) How monkeys see the world. University of Chicago Press, Chicago

Cheney DL, Seyfarth RM (2007) Baboon metaphysics: the evolution of a social mind. University of Chicago Press, Chicago

Clarke S, Juno C, Maple T (1982) Behavioral effects of a change in the physical environment: a pilot study of captive chimpanzees. Zoo Biol 1:371–380

Davidson D (1982) Rational animals. Dialectica 36:317–328

De Lathowres M, Eslacker LV (2004) Comparing maternal styles in bonobos (pan paniscus) and chimpanzees (pan troglodytes). Am J Primatol 64:411–423

Ellis AW, Beattie G (1986) The psychology of language and communication. Weidenfeld & Nicolson, London

Fairbanks L (1993) What is a good mother? Adaptive variation in maternal behavior of primates. Curr Dir Psychol Sci 2:179–183

Fairbanks L (1996) Individual differences in maternal style. Adv Study Behav 25:579–611

Ferrari PF, Fogassi L (2012) The mirror neuron system in monkeys and its implications for social cognitive functions. In: De Waal FB, Ferrari PF (eds) The primate mind: built to connect with other minds. Harvard University Press, Cambridge

Fleming A, Kraemer G, Gonzalez A, Lovic V, Rees S, Melo A (2002) Mothering begets mothering: the transmission of behavior and its neurobiology across generations. Pharmacol Biochem Behav 73:61–75

Grice P (1989) Studies in the way of words. Harvard University Press, Cambridge

Harlow HF, Suomi SJ (1971) Social recovery by isolation-reared monkeys. Proc Natl Acad Sci USA 68(7):1534–1538

Harlow HF, Zimmermann RR (1959) Affectional responses in the infant monkey. Science 130:421–432

Hinde R (1983) Description of social behavior. In: Primate social relationships, an integrated approach. Blackwell Scientific Publications, Oxford

Humle T, Colin C, Laurans M, Raballand E (2011) Group release of sanctuary chimpanzees (*pan troglodytes*) in the Haut Niger National Park, Guinea, West Africa: ranging patterns and lessons so far. Int J Primatol 32:456–473

King B, Fox R (2002) On patterned interactions and culture in great apes. In: Anthropology beyond culture. Oxford, New York

King N, Mellen J (1994) The effects of early experience on adult copulatory behavior in zoo-born chimpanzees (*pan troglodytes*). Zoo Biol 13:51–59

Lyn H, Russell JL, Hopkins WD (2010) The impact of environment on the comprehension of declarative communication in apes. Psychol Sci 21:360–365

Maestripieri D (1993) Maternal anxiety in rhesus macaques (macaca mulatta). Measurement of anxiety and identification of anxiety-eliciting situations. Ethol [Zeitschrift Fur Tierpsychologie] 95:19–31

Maestripieri D (1998) Parenting styles of abusive mothers in group-living rhesus macaques. Anim Behav 55:1–11

Maestripieri D (1999) The biology of human parenting: insights from nonhuman primates. Neurosci Biobehav Rev 23:411–422

Maestripieri D, McCormack K, Lindell S, Highley JD, Sanchez M (2006) Influence on parenting style on the offspring's behavior and CSF monoamine metabolite levels in crossfostered and noncrossfostered female rhesus macaques. Behav Brain Res 175:90–95

Matsuzawa T (2006) Sociocognitive development in chimpanzees: a synthesis of laboratory work and fieldwork. In: Matsuzawa T, Tomonaga M, Tanaka M (eds) Cognitive development in chimpanzees. Springer, Tokyo

Meltzoff AN, Moore MK (1977) Imitation of facial and manual gestures by human neonates. Science 198:75–78

Menzel E (1964) Patterns of responsiveness in chimpanzees reared through infancy under conditions of environmental restriction. Psychol Forsch 27:337–365

Menzel CR, Menzel EW Jr (2012) Enquiries concerning chimpanzee understanding. In: De Waal F, Francesco-Ferrari P (eds) The primate mind. Harvard University Press, Cambridge

Okamoto-Barth S, Tanaka M, Kawai N, Tomonaga M (2007) Looking compensates for the distance between mother and infant chimpanzee. Dev Sci 10:172–182

Schino G, D'Amato F, Troisi A (1995) Mother-Infant relationship in Japanese macaques: sources of interindividual variation. Anim Behav 49:151–158

Shanker S, Taylor T (2001) The house that Bruner built. In: Language, culture, self: the philosophical psychology of Jerome Bruner. Sage, London

Shannon C, Weaver W (1949/1978) The mathematical theory of communication. University of Illinois at Urbana–Champaign, Urbana

Slocombe KE, Zuberbühler K (2005) Agonistic screams in wild chimpanzees vary as a function of social role. J Comp Psychol 119:67–77

Suomi S, Goldberg S, Muir R, Kerr J (1995) Influence of attachment theory on ethological studies of biobehavioral development in nonhuman primates. In: Muir G, Kerr J (eds) Attachment theory: social, developmental and clinical perspectives. Analytic, Hillsdale

Takeshita H, Myowa-Yamakoshi M, Hirata S (2006) A new comparative perspective in prenatal motor behaviours: preliminary research with four dimensional ultrasonography. In: Matsuzawa T, Tomonaga M, Tanaka M (eds) Cognitive development in chimpanzees. Springer, Tokyo

Tomonaga M, Matsuzawa M, Tomonaga M, Tanaka M (2006) Development of chimpanzee social cognition in the first 2 years of life. In: Cognitive development in chimpanzees. Springer, Tokyo

van Ijzendoorn MH, Bard KA, Bakermans-Kranenburg MJ, Ivan K (2009) Enhancement of attachment and cognitive development of young nursery-reared chimpanzees in responsive versus standard care. Dev Psychobiol 51:173–185

Vicedo M (2013) The nature and nurture of love: from imprinting to attachment in cold war America. University of Chicago Press, Chicago

Whiten A et al (1999) Cultures in chimpanzees. Nature 399:682–685

Wobber T, Hare B (2011) Psychological health of orphan bonobos and chimpanzees in African sanctuaries. PLoS One 6(6):e17147

Yerkes R (1943) Chimpanzees: a laboratory colony. Yale University Press, New Haven

On Prototypical Facial Expressions Versus Variation in Facial Behavior: What Have We Learned on the "Visibility" of Emotions from Measuring Facial Actions in Humans and Apes

Augusta Gaspar, Francisco Esteves and Patrícia Arriaga

Abstract It has long been recognized that behavior evolves as do other traits and that it may have great impact on evolution. It tends to be conservative when survival and fast responding are at stake, and because of that, similar patterns can be found across populations or species, typical in their form and intensity, and often also typical in context and consequence. Such fixed stereotypic patterns that evolved to communicate are known as displays, and their phylogenies can virtually be traced. In this chapter, we contrast and discuss two coexisting trends in the study of the meaning and origins of human facial expression: one, with a tradition of exploring cross-cultural commonalities in the recognition of facial expression, that may indicate species-specific displays of emotion (prototypical facial expressions) and another that builds upon the growing evidence that such expressive prototypes are outnumbered by a diversity of facial compositions that, even in emotional situations, vary in relation to culture, context, group, maturation, and individual factors. We present behavioral studies that look at links between basic emotion and facial actions in both human and non-human primates and discuss the role of multiple factors in facial action production and interpretation.

Keywords Behavior coding • Chimpanzee and bonobo expressive behavior • Development of facial expression in children • Emotional development • Evolution of facial expression • Facial expression • Perception of facial expression

A. Gaspar (✉) · F. Esteves · P. Arriaga
Centro de Investigação e Intervenção Social, ISCTE—Instituto Universitário de Lisboa (ISCTE-IUL), Lisboa, Portugal
e-mail: augusta.gaspar@iscte.pt

F. Esteves
Department of Social Sciences, Mid Sweden University, Sundsvall, Sweden

M. Pina and N. Gontier (eds.), *The Evolution of Social Communication in Primates*, Interdisciplinary Evolution Research 1, DOI: 10.1007/978-3-319-02669-5_6,

1 Studying Expressions: From Darwin Over the Differential Emotions Theory to the Facial Affect Program

> Individuals are prepared in some measure during their early development for the particular adult world they will enter.
>
> Patrick Bateson and Paul Martin (2000: 17)

Despite the fact that by now, the product of more than 40 consecutive years of research in facial expression has been amassed, researchers are still debating whether emotion translates in facial expression or not, and what it is exactly that facial expressions convey. This may sound rather disappointing, especially when we take popular sayings into account such as "the face is the mirror to emotions" and "the eyes are the windows to the soul." Even counselors and researchers who do not study facial expression directly are eager to acquire and use instruments that will allow them to measure facial emotion. Products for this purpose are actually sold for substantial amounts of money, paradoxically so, while the very foundations of such products are still under debate. The aforementioned instruments are based on one of the most widespread beliefs in Psychology and in society today: that we have a short number of universal facial expressions that communicate our innermost basic emotions.

Darwin's (1872/1965) seminal work "The expression of Emotions in Man and Animals" launched the idea that expressive movements had undergone evolutionary processes and that one could identify some archetypes of expression across species. Darwin's collection of human facial expressions came from photos of induced facial muscle contraction with electric current taken by Duchenne (1862/1990). Dr. Duchenne was a genius anatomist who pioneered facial EMG and who also, at the very dawn of photography, produced high-quality photos of faces showing the contractions of various muscles. As Darwin (1872/1965, p 8) describes it:

> (…) Dr. Duchenne galvanized, as we have already seen, certain muscles in the face of an old man, whose skin was little sensitive, and thus produced various expressions which were photographed on a large scale. It fortunately occurred to me to show several of the best plates, without a word of explanation, to above twenty educated persons of various ages and both sexes, asking them, in each case, by what emotion or feeling the old man was supposed to be agitated; and I recorded their answers in the words which they used. Several of the expressions were instantly recognized by almost everyone (…)

The latent reasoning is that *convergent appraisals must mean that the correct meaning has been found.*

Much later, in the 1960s–1970s, the subject of the evolution of facial expressions was picked up again, now with great impetus toward the evolutionary trends of primate facial expression (e.g., Andrew 1963; Chevalier-Skolnikoff 1982; van Hooff 1962, 1967). These studies raised the interest in the facial behavior of chimpanzees (e.g., Berdecio and Nash 1981; Gaspar 2006; Goodall 1986; Pollick and de Waal 2007; Preuschoft 1992; van Hooff 1972, 1973), which was generally conceived as comprised of archetypal facial displays.

Simultaneously, American psychologists revisited the evolution of human expression and would produce enormous quantities of data in support of a set of basic universal facial expressions of underlying basic emotions. This was achieved by submitting photographs of facial expressions to appraisal. This appraisal–recognition paradigm was picked up after Darwin as a means to decode the meaning of expressions by supporters of a view of emotions as discrete entities, with distinct motivations and physiological patterns of activation, distinct functions, and thus distinct action tendencies, known as the differential emotions theory (Izard and Malatesta 1987; Izard 1991). This perspective is in line with a view of facial expressions as evolution-driven programs of coding–decoding, unambiguous, prototypical, and, most of all, directly translating respective discrete emotions, which became to be known as the *Facial affects program* (Ekman 1972, 1973, Ekman et al. 1969; Izard 1971). The later has pushed forward much research and countless journal papers. Cross-cultural studies on the identification of posed facial expressions have led to the conclusion that most people decode the expressions in a similar way (Elfenbein and Ambady 2002) and that there are 6–7 universal facial expressions of emotion—for joy, fear, anger, sadness, surprise, disgust, and contempt (e.g., Ekman 1984, 1992, 1999).

Thus, central to this line of thinking was the identification of a few facial expressions which were proposed to be universal signals for certain basic emotions. The facial affect program proposes that when certain emotional states are triggered, there is a universal automatic neuromuscular activation of specific facial muscles that are an intrinsic component of the very emotional experience. The assumption underpinning the recognition paradigm is that recognition of emotional facial expressions across cultures demonstrates biological universality. Thus, exposure to similar learning opportunities is not a requirement to recognize the meaning of the expression, and physiognomic features to which one has or has not been exposed are not an obstacle either. The fact that a member of an isolated Papuan tribe recognized a happy expression in a photograph depicting a Western face and the reverse, a westerner recognizing an identical emotional expression in a Papuasian face, are taken as evidence of such universality. The outcome of that work was that over the years the association between a small number of facial action configurations and seven emotions became familiar and gained great acceptance, and the configurations (prototypes) became known as the human universal facial expressions of emotion (Ekman 1984, 1999).

2 Problems with the Facial Affect Program

Ekman's facial affect program and the universal facial expression configurations derived from common recognition became one of the bedrocks of emotion psychology. Nonetheless, today this view receives considerable criticism.

A first, common objection to this *rationale* is the method: These convergences result from participants having to choose between 6 and 10 emotional labels after observing a photographed facial expression, using a forced-choice questionnaire. Not only are they led toward identifying emotion, but the narrow set of response options is also largely conducive to labels. Moreover, these paradigms do not present dynamic information that is usually expressed by the facial muscular movements and which can be relevant for recognizing more subtle emotional displays (Bould et al. 2008). In addition, participants are identifying posed expressions rather than spontaneous ones.

A second criticism arises in regard to how discrete, uniform, and universal the facial prototypes are. Emotions can either be seen as discrete entities (the perspective of the facial affect program) or as in the fluctuations of a *core affect* system, comprised of experiential dimensions such as arousal and valence (Barrett 2006; Lindquist and Barrett 2008; Russel 1980, 2006). This later view has been supported by a myriad of behavioral, cognitive, physiological, and neurochemical studies (e.g., Adolphs et al. 1999; Carroll and Russell 1996; Grammer and Oberzaucher 2006; Lang et al. 1998; Öhman et al. 2001; Russell 1980; Russell and Fernandez-Dols 1997; Widen and Russell 2008). Even affective neuroscience (Panksepp 1998), which stems from a view of emotion as evolving in conservative "packages" or patterns of neural activity and predictable influences on the body and behavior that trigger adaptive responses, shows that there is considerable overlap in neurophysiological systems underlying *basic affective systems* (such as between the one driving maternal behavior and the one driving playful social interactions, which share structures and chemical modulators) rather than a clear-cut separation between them. If emotions are a flow of variations across dimensions of experience, does it make sense to expect expressive behavior to be made up of discrete prototypes?

A third criticism arises in regard to how facial expressions are associated with emotions. Many studies taking methodological approaches based on actual behavior observation have failed to demonstrate that the facial expression prototypes are elicited during the assumed trigger emotions (e.g., Camras 1992, 2002; Carroll and Russell 1996; Fernandez-Dols and Ruiz-Belda 1997; Gaspar and Esteves 2012; Mehu et al. 2012; Oster 2005; Rosenstein and Oster 1997; Scherer and Ellgring 2007; Russell and Fernandez-Dols 1997; Widen and Russell 2008). Part of those studies was naturalistic (based on the observation of natural emotional events) and part was emotion induction experiments, whereby the situations were created based on the definitions of specific emotions). The experimental studies on expectancy violation by Camras et al. (2002) and by Scherer et al. (2004) serve as an example. These scholars conducted an experiment that was set out to provide subjects (babies and children 5–12 years old, respectively) with a violation of their expectancy in order to observe expressions associated with surprise, which is considered to be a prototypical expression. However, only a small fraction of infants displayed the prototypical surprise expression. But as they differed in body behavior (stilling), gazing, and facial sobering, naive raters used those cues to identify the expressions as a violation of expectancy.

Another example comes from our own research with 3-year-old toddlers (Gaspar and Esteves 2012); we used an emotional context-coding scheme and the facial action descriptor system and facial action coding system (*FACS*), developed by Ekman and Friesen (1978) in order to objectively code behavior of the face and head. Trained coders were asked to analyze video samples using rigorous reliability procedures, but we found very few full facial configurations matching the universal facial emotion prototypes. The highest match was 27 % for joy/playful contexts. Individual and combined action units that are part of the fear and surprise adult prototypical facial expressions associated significantly with the expected contexts. Because a diversity of other facial configurations appeared in these and other emotional events, and the associations occurred with only a few action units rather than the full-face prototype, we proposed that the signal value is more plausible at the component level and only for a few emotional states—which we coined "visible emotions"—meaning those with the distinctive quality of eminent interactive nature, as in the case of joy/playful or anger states (sensus Panksepp 1998). Actually the latter have been shown to link to facial signals in a particular way. Facial muscle action units comprising the prototypical facial expressions of "happiness" and "anger" tend to occur at an intensity that is invisible to the human eye, but detectable by facial electromyography whenever subjects are experiencing increased positive or negative affect, respectively (e.g., Dimberg 1988, 1997).

Yet another important source of information on the facial expression and emotion relation is the naturalistic studies with non-human primates, where the most conspicuous outcome is not the display of prototypical facial expression, but the prevalence of diversity. Chimpanzees and bonobos are not less expressive than humans (Bard et al. 2011; Gaspar 2001, 2006) and, as in the human children example, not all emotions are equally visible: The full facial configuration corresponding to the human prototypical "happy face" stands out as the most predictably associated to the expected context—joy/playful (e.g., Bard et al. 2011; Gaspar 2001, 2006; Goodall 1986; Parr et al. 2005; Preuschoft and van Hooff 1997).

It is noteworthy that even the use of methods adopted by the proponents of the facial affect program has failed to provide support to the most facial expression of emotion prototypes. For example, Grammer and Oberzaucher (2006) departing from Ekman and Friesen's *FACS* introduce a system that randomly generates single-action units in an avatar face. The generated pictures, rated by hundreds of people, were submitted to a factor analysis, which indicated that appraisal was mostly explained by the components of emotion arousal and valence. Values of these two dimensions were able to generate all known facial expressions, including the universal facial prototypes. But parameter scores generating for example the gestalt of a prototypical "surprise face" were a combination of high arousal and high positive valence, which differs from the original assumptions of the "pure surprise." The same occurred for other expressions, such as the "sadness" brow, which was generated by positive valence, not negative as one would expect based on the facial affect program.

What remains uncontested is the adaptive nature of emotions and the existence of cross-species homologous emotional responses crucial to survival. It is also consensually recognized that there is some level of continuity in facial and bodily communication such as expressive withdrawal or approach in relation to aversive vs positive encounters (e.g., Plutchick 1980a, b), enabling the recognition of threat or friendliness, and that certain actions are more likely associated with positive affect than negative and vice versa.

So, currently, diverse lines of research and theoretical standpoints converge toward an increasingly supported view—that emotional phenomena are not discrete and are better explained by dimensional components and present extensive overlapping between their neurochemical and behavioral systems (e.g., Grammer and Oberzaucher 2006; Krumhuber and Schere 2011; Lang et al. 1998; Russell 1980).

3 What Do We Actually Know About the Human Facial Behavior of Emotion? Spontaneous Facial Behavior from Naturalistic and Experimental Studies

Freed from dogma, we can begin to ask questions and understand why there is much variation in expressive behavior, and particularly in facial expression, and why group differences in expression and individual idiosyncrasy make sense from an evolutionary perspective. Yet, prototypical expression does occur of course—there is laughter, more than one spontaneous type of smile, there is cry, anger faces, disgust faces, and so on—but they are less stereotyped than once thought, and their meaning is also more complex.

And it is precisely in the examples of play, anger, or cry facial behavior that we find the most consistent data between visible and "invisible" facial behavior and face perception studies, providing evidence to build a new model for understanding why different emotional systems behave distinctly in regard to their behavior interface and to their interindividual variation in expressiveness.

There were always good reasons to believe that we should openly look for diversity in human and non-human primate facial behavior and set out exploratory studies. All, albeit few, anatomical studies had been since Darwin, pointing in one single direction—the extraordinary facial muscle differentiation of primates, with nearly total overlap between human and chimpanzee, allowing for the production of countless combinations of facial actions (Huber 1930, 1931; Sonntag 1924; Swindler and Wood 1973; Pellatt 1979).

Studying people and other animals in their natural habitats is the key to finding *ecologically valid* cues to emotion, which may or may not have evolved as specialized communication signals. Some, if universal, should reveal themselves as prototypical displays; others, if culturally or interindividually diverse, should hold important information on the meaning of certain emotional experiences within cultures or translate trait information crucial for interpersonal interactions (within the rules, values, and expectations of a group), but with no generalizable meaning for humanity, or the entirety of another species.

The naturalistic approach to facial expression was never a popular approach, but some researchers have undertaken this laborious task, recording and coding children in their daily interactions. Some of the first studies were very descriptive and provided facial ethograms without actually shedding much light on the contexts and context consistency (e.g., McGrew 1969, 1972; Young and Décarie 1977). But Camras et al. (1991) examined whether prototypical facial expressions in children occurred in the expected emotional contexts and were surprised to find that the expressions seemed detached from the presumed emotion—e.g., prototypical faces of sadness were observed in anger or distress situations.

To effectively ponder on what we really know on prototypical emotional facial behavior in the sections bellow, we list findings resulting from actually observing facial behavior.

3.1 Eyebrow Flashes and Patterns of Approach Withdrawal

In the 1970s, the human ethologist Eibl-Eibesfeldt came up with a most creative idea—filming people *with* their consent but *without* their knowledge, so as to overcome posing and self-monitoring—as a false lens in a large movie camera pointed toward someone else; a true lens was actually recording their nonverbal behavior. With this resource and traveling throughout the world to record populations that had a minimum or no contact with Western culture, Eibl-Eibesfeldt (1989) documented in mother–child interactions, and sometimes other friendly interactions, a rapid eyebrow raising movement, accompanying a head lifting movement and a rapid smile, occurring immediately upon visual contact, and quickly followed by lowering the eyebrows and nodding. This extremely rapid eyebrow raising pattern became known as the eyebrow flash. In other contexts, such as the case of playing attention or being surprised, the eyebrow raising movement is slower and is followed by other actions, such as mouth opening or fixed visual contact. *Therefore, these facial action patterns are not only characterized by the presence or absence of certain facial units of movement, but also characterized by temporal properties* associated with context. Temporal properties seem to play an underestimated role in emotion communication (Grammer and Oberzaucher 2006).

Patterns of "approach–withdrawal" in expression have been described in ambiguous interactions, such as flirting (Eibl-Eibesfeldt 1989) or first stages of acquaintance and courtship behavior (e.g., Grammer and Eibl-Eibesfeldt 1990). They mostly comprise ambiguous signals, as the intermittent onset and offset of facial actions, such as opening the mouth or smiling, opening and closing eyelids, and approach–withdrawal of the head.

3.2 The Almost Undocumented Faces of Fear

Fear exists, undeniably, and as the ancient core affect that it is (e.g., Panksepp 1998; LeDoux 1996), it triggers archetypical physiological and behavioral responses, widely described, and doubtlessly adaptive. Brain circuits underlying fear are known and shared among vertebrates. Mammals, which are for the most part, social animals, have both advantages and disadvantages in displaying unequivocal signals of fear. Thus, it comes as no surprise that there are contradictory reports on the facial display of fear.

In intraspecific interactions, fear-triggering situations are threats and conflicts; low ranking or animals in an otherwise disadvantageous position tend to emit ritualized facial displays of submission, with common features across taxa, such as lip corner withdrawal (in humans, this corresponds to AU20), lips parting (AU25) and teeth showing, lowered head posture, and when applicable, lowered ears. These displays can be efficient in appeasing the opponent and fit within the definition of communication, as signal and decoding coevolved benefiting the survival of the signaler. The human universal facial expression described within the facial affect program incorporates (AU20+AU25) and adds wide open eyelids (AU5). But it is not that simple to find a universal/prototypical expression: Individuals can hide and control such "fear expressions," especially if they are high-ranking individuals or are on their way up the hierarchy ladder, as some anecdotal reports of non-human primate behavior testify (e.g., de Waal 1992, 1996; Tanner and Byrne 1993).

In interspecific interactions, displaying fear only plays a role in conspecifics; but it is not the sign of fear *per se* that is useful to other group members, it is the sign of alarm, the more informative the better. Accordingly, social animals, especially those in communities of highly related individuals such as Gunnison's prairie dogs (*Cynomys gunnisoni*) and suricatas (*Suricata suricata*), but also vervet monkeys (*Cercopithecus aethiops*) or chimpanzees take great risk sending out alarm calls that are specific in their sound properties to what type of danger is approaching (Manser et al. 2002; Slobodchikoffet al. 1991)—thus maybe called labels.

3.3 Pain Versus Pleasure in the Face

Doubtlessly pain, like fear, has an enormous adaptive value, as it prompts avoidance from sources of injury and promotes activities that aid healing. Pain is also an affective experience—as it shares circuits in the brain with the system that regulates attachment and panic of loss (Panksepp 1998) and attention (Crombez et al. 1998; Chapman 1995). Physical and psychological pains have been reported to largely overlap in regard to active neuronal circuits in the brain (Eisenberger and Lieberman 2005; Eisenberger et al. 2003; Panksepp and Watt 2011).

Williams (2002) reviews and describes studies attempting to identify facial actions that may communicate pain. Here, the wide range of actions, differing from one study to another, leave us puzzled—actions may include components of smile, such as pulling lip corners upward (AU12), cheek raising (AU6), or components of negative, aversive faces, such as nose wrinkling (AU9) or upper lip lifting (AU10), eye closing (AU43), and horizontal lip stretching (AU20), among others. So, no prototype emerges; instead, extensive overlapping occurs with actions described as components of the fear, sadness, and anger prototypical expressions. In babies, induction experiments reveal AU20 and other components of crying (for a review see Oster 2005).

Guinsburg (1999), reporting on pain and its expression in newborns, describes most of the above actions (but not AU12) and adds a chin tremor. She refers that different configurations incorporating these AUs can occur, but those distinct configurations do not differentiate the type or the intensity of pain.

In Fernandez-Dols et al. (2011) study of facial behavior during sexual excitement, the main pattern of facial behavior observed across the four stages of the sexual cycle (excitement, plateau, orgasm and resolution) and the prevalent throughout the first three stages was a combination of closed eyes and jaw drop (AU43+AU26). This behavior has similarly been observed by Gaspar (2001) in male bonobos. But had pain or orgasm's facial prototypes been selected for communication purposes, they should be easily decoded and distinguished. However, Fernandéz-Dols et al. (2011) reported that in the absence of contextual information, observers cannot distinguish orgasm faces from pain faces. If orgasm is pleasure, as most people would agree, and indeed it releases opioids in the brain pleasure centers, then this overlap is another demonstration of the uselessness of the term "facial expression of emotion."

Bring "triumph" or "happy" to the witness stand and the case for prototypes comes even closer to its end. If winning a gold medal brings pleasurable affective states, facial actions are not as consensual; these are tremendously varied (Fernandéz-Dols and Ruiz-Belda 1997) as we shall discuss in Sect. 3.5.

3.4 Disgust

Disgusting is a label that adults often apply to many different targets: from food and smell, to concepts, on the appearance or moral behavior of people. This is also an experience that involves increased parasympathetic activity and movements that protect the body from contamination, either by avoiding contact with the contaminant or by expelling it—if it was not possible to avoid at all (see Rozin et al. 2008).

The current view of disgust as an emotion, triggered by repulse, with an associated prototypical facial expression, originated in Darwin (1872/1965), who actually proposed a form of mental revolt and a form of physical repulse as both being disgust.

Although a cross-cultural facial expression of disgust has been described from naturalistic observation and experimental studies by independent research teams, there is no quantitative data on its triggers and formal description in natural contexts (but see Peleg et al. 2009). Its hallmark (prototype) has been described as the combined configuration of nose wrinkling with a marked deepening of the nasolabial fold and upper lip lifting (AU9+AU10), which are effective defensive responses against chemical intrusion into the respiratory system (Eibl-Eibesfeldt 1989). These characteristic features and other actions which are commonly associated with negative valence (AU15 or AU16+AU20) are, along with physiological markers, displayed by newborn babies with mere hours of life in response to typically aversive odors (butyric acid), indicating an overall aversive reaction (Soussignan and Schaal 2005). It remains debatable whether people are experiencing in all situations labeled as disgust experiences an emotion or a reflex.

3.5 Happy Faces, Playful Faces, and Laughter

For all that is worth, smiles and laughter faces are the most predictable patterns of emotional facial behavior, both in form and in context, and this consistency emerges from independent observational studies with adults and children (e.g., Dickson et al. 1997; Fogel et al. 2000).

Laughter and smile faces overlap in form and context, as van Hooff (1972) noticed and further attempted to explain by reconstructing their phylogenies in primates. He proposed that these facial expressions converged in form during human evolution, but had quite different origins: Whereas the smile derived from *bared-teeth-mouth faces,* associated with submission and appeasement seeking, laughter emerged in association with relaxed playful contexts in which the jaw tends to drop and a characteristic respiratory pattern (van Hooff 1972; Preushoft and van Hooff 1997). Since in both humans and chimpanzees there are open mouth smiles associated with positive affect, which converge in form and consequence with the onset (still silent) of laughter, the two can be said to have had recent convergent evolution. But this example also serves the purpose of illustrating the artificiality of assorting a face to an emotion. Patterns of facial actions either have predictable antecedents or predictable outcomes, in which case they can be said to be *visible facial expressions,* or they do not, in which case they are not to be called facial expressions of emotion, because the emotion remains invisible in the face or translates into variable, not stereotypical, actions. These variable actions may hold valuable information contingent upon familiarity with the individual or the culture.

Gaspar and Esteves (2012), reporting on the spontaneous facial behavior of 3-year-old children in kindergarten, stressed that for the particular contexts of positive friendly and playful interaction, children exhibit a close match to actions recognized by older children and adults as signs of positive affect: *cheek raising* and *lip corner upward pulling*, approaching, thus, the template of "adult prototypical facial expression of happiness." Because in the majority of events of this study the remaining actions of full adult happy face (lips parting and jaw drop—AU25+AU26) were absent, we can still speculate that the intensity of the emotion or of the interaction (possibly the level of physiological arousal) is modulating the expression, or otherwise, that experience may be playing a role, with one facial action triggering the receivers' response and a correct appraisal, and the other AUs just typically associate with it, which was originally proposed by Carroll and Russell (1996).

The scope of variation in this expression we are referring to, with or without mouth opening and jaw drop, but also with or without cheek raise and wrinkling in the eye corners, can be observed in chimpanzees and bonobos (see Fig. 1a–g), and within the same contexts (Gaspar 2001; Bard et al. 2011; Gaspar and Bard, unpublished manuscript).

This is an expression that does not always occur in adults as one would expect—in the moments of glory, triumph, and happiness—in fact, actions that are part of the prototypical facial expression of sadness occur in these situations and can be actually interpreted as the happiest (Fernández-Dolls and Ruiz Belda

Fig. 1 **a–f** Some joy–playful facial expression variations *clockwise from top*: chimpanzee AU12; bonobo AU12+25; human child AU6+12+25; chimpanzee AU6+12+25+26+droop lip; human child AU6+12+25+26; bonobo AU6+10+12+25+12. Photographs are still frames from video: **b–f** by A. Gaspar and **a** by C. Carvalho

1997). It is also worth mentioning that it is unreported in negative situations. As predicted by the behavior ecology model, the visibility of this expression associates with fostering an affinitive interaction. It is unambiguous in that regard. If that is so, it should also be the easiest to identify. And it is. Perception will be resumed in a forthcoming section.

3.6 Anger, Threat Facial Displays, and Attention Faces

With negative emotion, many more forms of facial behavior set foot on stage.

Charles Darwin, in "The Expression of Emotions in Man and Animals" (1872/1965), described and used compelling images of threat displays in humans and other animals, picking up the similarities not only in body posture and movement, but also in the common frown, with mouth opening and upper lip lifting, exposing canine teeth in the case of mammals. These expressive patterns seem to derive from intentional movements, and in incomplete form, they became highly stereotypical and adapted to communication. These patterns also seem to emerge during a limited set of contexts, characterized by explicit conflict, aggression, or by the imminent possibility of their occurrence (i.e., when the person or animal perceives that something is wrong and has to be corrected, he/she becomes angry and threatening, such as when one's resources or rights have been taken, or are at risk of being taken by someone else).

Inasmuch as threat displays associate with such an internal dispositional change that we call anger or fury, they seem to be part, in accordance with Darwin's proposal, of an ancient neuroethologic system (Panksepp 2005), resulting from prototypical contexts that have remained stable throughout millions of years of mammal evolution.

The expression of this anger/threat system has been reported to arise as early as at 4 months old (Lemerise and Dodge 1993). To be in line with a behavioral ecology view of facial expressions, the facial actions accompanying this affect should be highly visible; there should be no ambiguity in their signaling and interpretation. The facial actions pointed by Darwin, included in the proposed human prototypical facial expression, and reported from observational studies are, for the most part, convergent in that regard. Indeed, they include frowning (AU4) and a special type of mouth opening (AU27) in both humans (e.g., Shaver et al. 1987; Gaspar and Esteves 2012) and chimpanzees (Bard et al. 2011; Gaspar 2001; Goodall 1986). But they also comprise a variety of other mouth actions, including AU23, AU20, AU25, AU26, and even the combination AU1+AU2 (Gaspar and Esteves 2012)—which is thought to be typical of surprise—so there is great variability. To complicate things further, some of the convergent actions (the frown) are also associated with attention and protection from excessive light. It should come as no surprise then that the activity of the *Corrugator supercilii* muscle facial region is usually detected to increase upon the perception of new stimuli (roughly in the first 400–500 ms after stimulus onset), as part of what seems to be an attentional process, as well as upon the perception of negative stimuli (e.g., Dimberg 1997; Dimberg and Thunberg 1998) (See the perception section of this chapter). It is also likely that experience (e.g., Pollak and Kistler 2002) and individual traits are playing a role in the modulation of the prototypical anger expression (Bard et al. 2011), as individual traits also may affect the perception of angry faces (e.g., Öhman, Lundqvist and Esteves 2001) and emotional reactions to both angry and happy faces (Dimberg 1982; Dimberg and Thunberg 1998, 2012; Gaspar and Esteves in preparation).

4 Factors Modulating the Conveyed Meaning of Expressive Behavior Associated with Emotion

4.1 Attention Processes and Physiological Responses

As stated above, the way our body reacts to an emotionally relevant signal may be very informative of what the signal means, even when we are not aware of that reaction. The behavior ecology model of facial expression, proposed by Fridlund

(1994), emphasizes the coevolution of signal and detection abilities. It also calls, for the first time, to the importance of a number of factors that affect signal detection and interpretation. For example, the meaning of the signal is not independent of who the sender is, the emotional context of sender–receiver interaction, the presence of others, potential dangers, nor the current emotional state of the receiver, his (her) sensitivity to displays, and other internal traits of the receiver. We will discuss in the next section these factors in terms of conscious perception, but here we will address attention processes and physiological responses to emotion stimuli, as they are the first layer of evidence on the biological meaning of signals.

The idea that facial expressions conveying some kind or level of threat would be easier than others to detect in social contexts—due to the immediate survival value of this attentional bias in triggering prompt adaptive responses—was first tested by Hansen and Hansen (1988) using a visual search paradigm. This methodology requires participants to look for a "different" characteristic (the target) among several distracter stimuli. Thus, participants have to answer, for each trial, if the target is present or not. Comparing the time needed to detect the target, being it a happy or an angry face, Hansen and Hansen (1988) concluded that angry faces presented in a background of friendly faces were faster detected than happy faces among threatening ones. Although this "anger superiority effect" has not been replicated when photographs of real faces are used (e.g., Juth et al. 2005), data from experiments using schematic faces have quite consistently shown the advantage of threatening facial features in the capture of attention (e.g., Calvo et al. 2006; Öhman et al. 2001). Thus, it seems that threatening signals, even when reduced to simple features like the shape or position of the eyebrows, hold an important signal value that calls for priority when processing information from our environment. Furthermore, a recent study with photographs has revealed an interesting target–gender effect: Using redundant distracters (i.e., small stimulus sets), the authors found a faster detection of a deviant threatening photo when compared to a friendly face, but only when the target was a male face. With female targets, the facial happy expressions were detected faster. This result could be interpreted as the outcome of the prevalent dominant role of males in aggressive behavior throughout biological evolution (Öhman et al. 2012).

Another way of testing the relevance of threatening facial expressions is to present them in conditions of degraded perceptual information, e.g., with very short exposure times and backwardly masked in order to prevent conscious perception, and analyze the elicited physiological responses. Esteves et al. (1994a) have shown that angry faces could elicit larger skin conductance responses after a differential classical conditioning procedure than happy faces (i.e., more resistance to the extinction of the conditioned response), whereas this effect was not obtained when happy faces were used as the conditioned stimulus. Furthermore, the associative learning of a fear response to subliminally presented human faces has been obtained with threatening, angry facial expressions but not with friendly ones (Esteves et al. 1994b). This priority in the processing of threat cues presented at the preattentional level was also observed at a central level. For example, Morris et al. (1998), using a backward masking procedure to prevent conscious awareness of the stimuli, obtained activation of the right amygdala to threatening faces, thus showing that the activation of a fear response to the threat face does not require conscious perception.

4.2 Appraisal and Conscious Perception

Additional sources of information on the meaning and on the flexibility of the meaning of facial expressions are the perception and appraisal studies. We organize those sources in the forthcoming subsections. First in Sect. 4.2.1, we provide evidence of the relation between phylogenetically stable and more stereotyped facial behavior and the early timing not only of the onset but also of the correct interpretation of that behavior. It has long been assumed that the relation between phylogeny and ontogeny predicts that nature does not leave to chance the acquisition by a newborn or small child of the necessary repertoire to recruit parental care or ensure basic survival, but that the basic actions are genetically encoded and prompted, developing in conformity with the adequacy of environmental stimulation.

In the Sect. 4.2.2, we report studies that show the role of contextual cues in the interpretation of the content of formerly considered prototypical facial expressions of emotion.

4.2.1 The Earlier and the Easier Identifications

Campos et al. (1989) report that children respond with expected appropriate behavior to the facial expressions of other people as early as 5 months old, and this is even more salient for the expression of joy and anger, which made the authors suggest that these are the earliest to acquire meaningful content.

The "prototypical expression of happiness," which includes the lip corner raising, the cheek raising, and the wrinkling around the eyes, has been documented among the first expressions in newborns, in relation to hedonic experiences (e.g., Soussignan and Schaal 2005) and reported as the easiest to recognize by both adults and children (e.g., Hager and Ekman 1979), even when these facial expressions are presented at low intensity (Gao and Maurer 2009), upside down, or outside the focus of overt visual attention (Calvo et al. 2010). From a wide array of observational studies of the infant–parent interaction, this configuration stands out as unambiguously associated with positive emotion and graded in intensity in relation to the intensity of the positive interaction (for a review see Messinger et al. 2014). Izard (1971) reported that 2-year-old children were able to recognize the universal facial expressions of happiness, sadness, anger, fear, and surprise; however amazing, this result might have been inflated in its method, which involved presenting 3 answer options per each photograph shown to children. He also obtained better results for happy and anger faces, by that order. Widen and Russell (2003) also found this to be the order by which children start labeling emotions correctly, and at about 3 years of age that can do so quite successfully for happy and joy expressions, but Widen (2013) also clarifies that children begin by making only distinctions between positive and negative emotions from the appraisal of facial expression.

In addition, appraisal of facial expressions in terms of similarity/dissimilarity, by 2–4-year-old children, fits a Valence X Arousal dimensional space but not a discrete emotions model (Russell and Bullock 1986) and, like adults, 4-year-old children classify facial expressions according to a third dimension, which authors tentatively named as "assertiveness versus taken aback."

There are many more developmental studies that could be quoted here, but they are convergent in regard to one point, which is the message we want to convey with these examples: All in all, there seems to be a prevalence of preprogramming for *reacting to* and *recognizing the* expressions of the *basic affective states* "joy/playful" and "anger" (as in Panksepp 1998), at least at an early stage in ontogeny. With context entering later, and the accumulating experience on other emotional situations and the diversity of facial expression, recognition of other expressions almost levels off later with maturity and this salience of "joy/playful" and "anger" faces dilutes.

4.2.2 The Role of Contextual Cues

Recently, researchers became aware of the role of context in the interpretation of the alleged "universal facial expressions of emotion."

As previously stated, humans are usually capable of extracting emotional information from perceptual cues within brief moments, but especially when stimuli are full-face configurations making up the prototypical emotional expressions (e.g., Elfenbein and Ambady 2002). However, most studies used paradigms in which perceptual visual cues of nonverbal language were presented in isolation, thus providing only information on the contribution of each visual cue in isolation and not in combinations of body and face or face and background context. This information might be relevant when we consider interactions that rely on only one type of cue. But we usually interact and communicate using distinct perceptual channels at the same time, and therefore, a more ecological perspective in emotional recognition should take into account a holistic view of human interaction communication, by investigating how different individual units are processed when in combination (Aviezer et al. 2012). Besides, effective non-ambiguous communication, in humans and other animals, is redundant in more than one way—it signals in more than one channel (visual, auditory, mechanical, and chemical), and the signal is repeated time and time again. That is the hallmark of displays. And indeed, several other physical cues can influence the way we judge other individuals' facial expressions, including body postures (Aviezer et al. 2008, 2012) and voices (de Gelder et al. 1999). Body posture, for example, is relevant to communicate emotional information. Aviezer et al. (2008) demonstrated that emotional body context (e.g., posture) and few other contextual cues can dramatically alter the recognition of emotions from prototypical facial expressions. They combined Ekman's pictures of facial affect with different body postures and other contextual elements, like a hand pointing a gun. For example, the prototypical facial expression of sadness, recognized as sad by 74 % of participants when it had been presented in isolation by Ekman and Friesen (1976), was recognized in Aviezer et al. (2008) study as fearful by more than 70 % of participants when a fearful context (body posture and objects) was embedded, and as said by less than 20 % of the participants. In addition, the authors found that context also affects the facial regions to which individuals attend, which in turn is very relevant for the identification of the emotional displays in the face. Recently, Hassin et al. (2013) published a review of studies showing the crucial role of context in the interpretation of facial expressions, proposing that without context, most expressions are inherently ambiguous and that their categorization as "angry" or "fearful" depends on the contingencies of the facial configuration and the situation in which it is displayed. The authors also challenge diagnoses of neuropsychological conditions that are made based on the assumption that there are "correct" and "wrong" categorizations of pure facial expressions (i.e., without context).

In addition to multichannel communication, there are contextual features, such as scenes and words, which have also been shown to affect the way we perceive emotion in another human face (e.g., Barrett and Kensinger 2010; Righart and De Gelder 2008; Lindquist et al. 2006). Recognizing the value of the context is actually an important developmental task as Gnepp (1983) noted in a study in which pictures containing expression alone or a combination of expression and context cues were presented to 4-, 7-, and 12-year-old children, the younger children relied solely on facial cues to identify emotion, whereas the older ones tended to progressively use more the contextual cues as well.

It is beyond the scope and limits of this chapter to review all contextual factors that may influence the interpretation of facial expressions, so for a detailed review on this topic, see Barrett et al. (2011). Our major point is that the weight of context and channel redundancy increases throughout ontogeny, narrowing the reliance on facial expression to identify emotion in the human face.

5 Evolutionary Significance: Why Should We Expect to See Prototypical Facial Expressions and to What Extent?

Konrad Lorenz (1967/1986) proposed mechanisms for the evolution of behavior, demonstrating that behavior evolves similar to other phenotypic traits and it evolves in a rather conservative way, so that one can virtually trace phylogenies of behavior. *Ritualization* is one of the processes by which behavior evolves. It comprises changes in a behavioral pattern so as to specifically serve a communication function. Often, this involves the action becoming incomplete and typical in intensity, rhythm, and other properties—i.e., it becomes stereotyped. Many threat displays are thought to have evolved this way, as the comparison of related species shows that incomplete actions of attack became signals of threat. It is as if aggression is simulated in a ritual form. Actual aggression may be avoided in species that have unambiguous threat signals that can be well detected and understood by the receiver. Human populations have many cultural rituals that meet that function such as simulated ritual combats of all sorts, many of which are in the form of dance (e.g., Brown 1991; Eibl-Eibesfeldt 1989).

In the process of ritualization, the action may change function as well. For example, in several bird species, courtship behavior often includes modified infantile begging behavior patterns. Ritualized behavior can also derive from conflicting motivations, in which case the actions themselves seem quite ambivalent, incorporating ritualized elements from one and the other action. Patterns of approach–withdrawal, and threat combined with avoidance, are commonly described in animal behavior (for further details see McFarland 1987). Repetition of the signal and exaggerated intensity are also key ingredients often added to these ritualized patterns, as they result in less ambiguity, an important feature if a deadly combat can be avoided.

This "economy" of evolution constantly preserving actions and their neuromuscular structure in the building of signals should lead us to expect very stable signals (great similarity in form) across *taxa*, and an even greater stability—prototypical forms—at the species level. And, some degree of overlap in repertoires across species is to be expected as a result of the aforementioned conservative nature of evolution's mechanisms to generate signals, but also as a result of evolutionarily stable contexts, particularly when surviving depends on speedy reaction (which is when behavior is the most conservative). This seems to apply extensively to the repertoires of fishes, reptiles, and many other groups. However, many animals, including humans, non-human primates, whales, and songbirds, while preserving some stereotyped signals, diversify their communication repertoires throughout their life spans.

Before 1997, when the first author began compiling data on the repertoire of bonobo facial expressions for her doctoral dissertation (Gaspar 2001), she could not imagine that bonobos, and later, chimpanzees, had such diversified repertoires of facial expressions as they actually do. When comparing chimpanzees' and bonobos' facial expression repertoire, Gaspar (2001) reports more variation in facial behavior due to colony belonging than to species belonging. And although the study set out to seek for cross-species stereotypical facial expressions, over 60 pages of the final 385 page dissertation are devoted to findings of idiosyncratic facial behavior. Because chimpanzees, bonobos, and humans share common ancestry and homologous facial structures (muscles and nerves), we can expect to observe similar facial configurations. Our research, however, showed that in addition to a few identical combinations of facial configuration and function, many similar configurations have distinct functions (Gaspar 2001; Bard et al. 2011). Another finding was that greater expressive similarities can occur between communities of different species than between communities of the same species (Gaspar 2001). Both indicate that aspects of the signal repertoire are acquired during the life span.

In humans, the search for prototypical behavior targeted mostly babies and small infants. This is based on the rationale that newborns must be endowed with some genetically programmed unambiguous signals, so as to not leave communication crucial to survival vulnerable to the contingencies of environmental stimulation. But so far, infant and children emotional developmental studies have failed to show discrete spontaneous facial expressions of emotion. Elegant experiments designed to elicit distinct negative experiences in babies and infants were unsuccessful in finding any clear differentiation in facial behavior. Notwithstanding, some common denominators to negative tone were found in the frequency of some facial action units (not full facial configurations) such as *lip withdrawal* or *lowering of the corners of the lower lip* (for a review see Oster 2005). Elicited positive emotion led to the production of varied facial actions, but these were more convergent than in negative valenced emotions: They basically consisted of smiles (Bennett et al. 2002) in predictable varieties, according to differences in activity level, continuity, and interactor (mother/father), suggesting that arousal and pleasure, as well as expectations based on past experience modulate programs of facial behavior in babies and infants (Dickson et al. 1997).

Regarding distant and proximal causes of behavior, it makes sense that stereotyped facial displays coexist with more flexible facial compositions, suiting both the requirements of stable and flexible contexts, the latter more common in complex societies and in species dwelling in a wide range of habitats, such as humans or chimpanzees. Realistically, stable contexts are probably few, such as the early stages of mother–infant interaction and bonding, threat, or a general playful mood. Yet some facial displays may preserve their formal properties while gradually acquiring new functions. For example, the human laughter face kept its stereotypicallity, but only partially, as it blended smoothly into wide open smiles and laughter became associated with a number of different positive meanings (van Hooff 1972). Social contexts demand innovation, and they are often unpredictable, so they exert great pressure toward behavior flexibility (Bateson 2004), but even if plasticity were all that was favored in communication within a population, the breakup with old codes generated by entire new signals emerging de novo would bring communication misunderstandings and subsequent harmful consequences. So in conditions where signal plasticity is favored, some extent of continuity seems fundamental. This could be achieved by preserving entire compositions of behavior units, with their duration and repetition patterns (displays/facial expressions), or only some of the composition elements, such as one action unit (e.g., brow furrowing instead of an anger full face).

Whichever of the above paths expression evolution might have taken, we still have to ponder on what could have been the evolutionary advantages of having preprogrammed involuntary, stereotyped expressions for each and every emotional state. Expressing one's willingness to engage in some form of interaction with others during an emotional episode, exuding information on one's motivational state, is entirely different from expressing behaviorally *all of one's basic affective states*. Not only are there reasons for basic affective states not be expected to be discrete, but the states are associated with very distinct biological quests, so even if they were discrete, not all should be expected to translate in signaling to another animal with whom one is interacting.

One would have to *always* derive some kind of survival/reproductive benefit from such ubiquitous signaling—that is, effective decoding by observers would reliably promote the sender's survival. Always showing core negative affect, such as fear or sadness, would only be beneficial to the sender if there were no risk that the observer would use this information on the sender's vulnerability to put him at even greater danger, and altruistic or at least cooperative interactions were the only possible outcome. But exploitation and cheating are always around the corner in social groups, and only in protected contexts can one afford to openly show certain emotions, or even exaggerate them, either to one's own benefit or to the benefit of the group, such as in the case of alarm signals. Exaggerated signals of distress are displayed by infants toward their mothers, as part of their et-epimeletic repertoire, and this is obviously adaptive and deep seated in animal species with long developmental periods. And, cheating, by simulating or exaggerating a signal of negative affect, such as distress, only occurs in certain contexts, and in more cooperative, less aggressive groups. An example is the temper tantrums thrown by adult female bonobos in frustration (Bard et al. 2011; Gaspar 2001), which probably would not be tolerated in the highly formal chimpanzee communities, but which foster attention in the egalitarian less aggressive male bonobo communities (Preuschoft and van Hoof 1997).

Although social groups have protective mechanisms against exploitative interactions and cheating with signals, based on the memory of past events and on interindividual recognition (Gaspar 2006), we would not assume that what observers are detecting in those situations are "lies." A practical and parsimonious approach is rather to examine the actual behavior *per* context and determine which future action is most likely to take place from both sender and receiver, that is, the consequence of the behavior. From thereon, we may be in a better position to ascertain whether the action is related to emotion.

The logical reasoning in evolutionary terms is to address the consequences of a rapidly widening and changing repertoire of facial behavior in non-human primate and human evolution. We could expect some prototypical and ritualized facial expression in our universal repertoire, but much more non-prototypical diversity in expressive behavior. This diversity is not just made of intensity control, cheating attempts, cultural display rules, and emotion blending over composite facial expressions, as once proposed (Ekman 1972, 1984; Ekman and Friesen 1976; Ekman et al. 1976, 1984) but of a multiplicity of functions and modulatory effects of context and individual traits of sender that give content to unitary facial actions.

6 Conclusions

From this review, three main conclusions emerge: (1) That there is insufficient supporting evidence for the proposed emotional content of some human facial configurations known as universal facial expressions of basic emotions; (2) that the diversity of facial behavior, rather than prototypicality, stands out from direct behavior observation; and (3) that we begin to see a landscape where some affective states have visible expressive landmarks, whereas others do not—for example "joy/playful" is more accurately identified by observers because it entails more than other states—predictable actions in the face that emerge early in ontogenetic development and remain stable throughout life, a feature common across ape species. Happy/joyful faces and anger faces are, by all standards, the closest match to the definition of prototypical facial expressions in humans and non-human primates. However more predictable than other complex signals related to affect, and reportedly the easiest and earliest in life to react to and recognize, they seem to move through ontogeny toward modulation by other variables, both in the sender and in the observer.

The chapter stresses that actual behavior studies of emotional facial expression have been neglected in the past few decades and their contribution to understanding the content and function of facial behavior has been largely ignored. Here, we highlight the need to devote more attention to descriptive studies of affective behavior and to combine them with experimental approaches in order to understand the links between emotion and expression. The expressive diversity of humans as well as that of non-human primates has been also largely downplayed, and we think that this topic will call for a great deal more attention in future studies.

Taken together, the evidence from the studies we have reviewed in this chapter strongly suggest that we are born with an action–recognition template for a set of facial actions that are crucial to our first interactions with caretakers, and as our brain matures and experience invades us with new information, new features unfold over that original program. Still, the program never loses some of its properties—for example, some facial action units tend to remain in positive or negatively valenced affects and the timing of onset and offset of certain actions is never the same when posed. So these perennial features that predictably emerge in the face (and tend to be adequately interpreted by observers) in utterly positive states, such as the facial actions and dynamic properties of the Duchenne smile, or in the blow of fury, such as the threat brow, are what creates the differential "visibility" of some affective states: Other affective states are simply not as predictable or detectable by observers and do not carry typical interactive intentions. Although actions associated with these two affective states also expand on variability, they do so to a lesser degree than other emotional states, including some considered basic, such as fear or sadness.

From a conceptual standpoint, this differential "visibility" of affective states is partly compatible with a view of emotions as discrete categories—or at least natural kinds (sensus Panksepp 1998)—and with a constructionist view, by which basic affect components, such as arousal and valence, vary and flow as part of a *core affect* that is never discontinue (sensus Russel 1980, 2003). But the diversity of expressive behavior is generally best explained by a componential approach. Throughout ontogeny, and beyond, facial behavior develops integrating personality, contextual variation, the meaning of a situation to a specific person, and cultural rules. This multi-componential and customized program generates immense expressive diversity that we are only beginning to realize and understand. These wide, complex, and flexible repertoires of emotion expression in humans and great apes summon to an arms race for sender and receiver, especially regarding deceit and eavesdropping of intentions and of affective states. In humans and other social primates, wide expressive diversity should come as no surprise, because behavior tends to change more rapidly and to exert greater influence on evolution (i.e., the destiny of its bearers) when it is responding to demands of the complex social environments (Bateson 2004). So this unfolding diversity and flexibility, departing from a very few early facial affect signals that, as part of a wider et-epimeletic repertoire, ensures the survival of a baby must be around for many millions of years.

A major and useful challenge lies ahead in understanding how humans and great apes resolve the conundrums of decoding, amid such great expressive diversity, crucial information to regulate their social interactions.

Acknowledgements Funding for research on facial expression and perception of facial expression has been provided by the Portuguese national funding agency for science, research, and technology—FCT, through the following grants to A. Gaspar PTDC/PSI-PCO/104170/2008; POCTI/PSI/47547/2002;SFRH/BPD/26387/2005; PRAXIS XXI BD/9406/96); and to F. Esteves (PRAXIS/FCSH/C/PSI/90/96; POCTI/PSI/14118/2001). The authors wish to express their gratitude to José-Miguel Fernández-Dols, anonymous reviewers, and the editors for their comments on an earlier version of the manuscript.

References

Adolphs R, Tranel D, Hamann S, Young AW, Calder AJ, Phelps EA, Anderson A, Lee GP, Damasio AR (1999) Recognition of facial emotion in nine individuals with bilateral amygdala damage. Neuropsychologia 37:1111–1117

Andrew RJ (1963) The origin and evolution of the calls and facial expressions of the primates. Behaviour 20:1–109

Aviezer H, Hassin RR, Ryan J, Grady C, Susskind J, Anderson A, Moscovitch M, Bentin S (2008) Angry, disgusted, or afraid? Studies on the malleability of emotion perception. Psychol Sci 19(7):724–732. doi:10.1111/j.1467-9280.2008.02148.x

Aviezer H, Trope Y, Todorov A (2012) Holistic person processing: faces with bodies tell the whole story. J Pers Soc Psychol 103(1):20–37. doi:10.1037/a0027411

Bard K, Gaspar A, Vick SJ (2011) Chimpanzee faces under the magnifying glass: emerging methods reveal cross-species similarities and individuality. In: Weiss A, King J, Murray L (eds) Personality and temperament in nonhuman primates. Springer, New York, pp 193–231

Barrett LF (2006) Are emotions natural kinds? Perspect Psychol Sci 1:28–58

Barrett LF, Kensinger EA (2010) Context is routinely encoded during emotion perception. Psychol Sci 21(4):595–599

Barrett LF, Mesquita B, Gendron M (2011) Context in emotion perception. Curr Dir Psychol Sci 20(5):286–290

Bateson P (2004) The active role of behaviour in evolution. Biol Philos 19(2):283-298. doi:10.1023/B:Biph.0000024468.12161.83 (Book review Evolution and learning: the Baldwin effect reconsidered)

Bateson P, Martin P (2000) Design for a life: how behavior and personality develop. Simon and Schuster, New York

Bennett DS, Bendersky M, Lewis M (2002) Expressivity at 4 months: a context by expression analysis. Infancy 3(1):97–113

Berdecio S, Nash LT (1981) Chimpanzee visual communication. Facial, gestural and postural expressive movement in young, captive Chimpanzees (*Pan troglodytes*). Anthropological Research Papers n.26. Arizona state University, Tempe, AZ

Bould E, Morris N, Wink B (2008) Recognising subtle emotional expressions: the role of facial movements. Cogn Emot 22(8):1569–1587

Brown DE (1991) Human universals. McGraw-Hill, New York

Calvo MG, Avero P, Lundqvist D (2006) Facilitated detection of angry faces: initial orienting and processing efficiency. Cogn Emot 20:785–811

Calvo MG, Nummenmaa L, Avero P (2010) Recognition advantage of happy faces in extrafoveal vision: featural and affective processing. Vis Cogn 18(9):1274–1297. doi:10.1080/13506285.2010.481867

Campos JJ, Campos RG, Barrett KC (1989) Emergent themes in the study of emotional development and emotion regulation. Dev Psychol 25:394–402

Camras LA (1992) Expressive development and basic emotion. Cogn Emot 6(3–4):269–283. doi:10.1080/02699939208411072

Camras LA, Malatesta C, Izard CE (1991) The development of facial expressions in infancy. In: R. Feldman & B. Rimé (eds.) Fundamentals of nonverbal behavior. Cambridge University Press, Cambridge, pp. 73–105

Camras LA, Meng Z, Ujiie T, Dharamsi S, Miyake K, Oster H, Wang L, Cruz J, Murdoch A, Campos J (2002) Observing emotion in infants: facial expression, body behavior, and rater judgments of responses to an expectancy-violating event. Emotion 2(2):179–193

Carroll JM, Russell JA (1996) Do facial expressions signal specific emotions? Judging emotion from the face in context. J Pers Soc Psychol 70(2):205–218. doi:10.1037/0022-3514.70.2.205

Chapman CR (1995) The affective dimension of pain: a model. In: Bromm B, Desmedt JE (eds) Pain and the brain: from nociception to cognition. Advances in pain research and therapy, vol 22. Raven Press, New York, pp 283–301

Chevalier-Skolnikoff S (1982) A cognitive analysis of facial behavior in Old World monkeys, apes, and human beings. In: Snowdon CT, Brown CH, Petersen MR (eds) A cognitive analysis of facial behavior in Old World monkeys, apes, and human beings. Cambridge University Press, Cambridge, pp 303–368

Crombez G, Eccleston C, Baeyens F, Eelen P (1998) When somatic information threatens, catastrophic thinking enhances attentional interference. Pain 75(2–3):187–198

Darwin C (1872/1965) The expression of the emotions in man and animals. University of Chicago Press, Chicago

de Gelder B, Böcker KB, Tuomainen J, Hensen M, Vroomen J (1999) The combined perception of emotion from voice and face: Early interaction revealed by human electric brain responses. Neurosci Lett 260:133–136

de Waal F (1992) Chimpanzee politics: power and sex among apes. Johns Hopkins University Press, Baltimore

de Waal F (1996) Good natured: the origins of right and wrong in humans and other animals. Harvard University Press, Cambridge, MA

Dickson KL, Walker H, Fogel A (1997) The relationship between smile type and play type during parent-infant play. Dev Psychol 33(6):925–933

Dimberg U (1982) Facial reactions to facial expressions. Psychophysiology 19(6):643–647

Dimberg U (1988) Facial expressions and emotional reactions: a Psychobiological analysis of human social behavior. In: Wagner H (ed) Social psychophysiology: theory and clinical practice. John Wiley, New York, pp.131–150

Dimberg U (1997) Facial reactions: rapidly evoked emotional responses. J Psychophysiol 11(2):115–123

Dimberg U, Thunberg M (1998) Rapid facial reactions to emotional facial expressions. Scand J Psychol 39:39–45. doi:10.1111/1467-9450.00054

Dimberg U, Thunberg M (2012) Empathy, emotional contagion, and rapid facial reactions to angry and happy facial expressions. PsyCh Journal 1:118–127. doi:10-1002/pchj.4

Duchenne GB (1862/1990) The mechanism of human facial expression. In: Ed. and translated by Cuthbertson A. Cambridge University Press, Cambridge

Eibl-Eibesfeldt I (1989) Human ethology. Aldine de Gruyter, New York

Eisenberger NI, Lieberman MD (2005) Why it hurts to be left out: the neurocognitive overlap between physical and social pain. In: Williams KD, Forgas JP, Hippel Wv (eds) The social outcast: ostracism, social exclusion, rejection, and bullying. Psychology Press, New York, pp 109–130

Eisenberger NI, Lieberman MD, Williams KD (2003) Does rejection hurt? An fMRI study of social exclusion. Science 302(5643):290–292. doi:10.1126/science.1089134

Ekman, P (1972) Universal and cultural differences in facial expressions of emotions. In: Cole JK (ed) Nebraska symposium on motivation 1971. University of Nebraska Press, Lincoln, pp 207–238

Ekman P (1973) Cross cultural studies of facial expression. In: Ekman P (ed) Darwin and facial expression: a century of research in review. Academic Press, New York, pp 169–222

Ekman P (1984) Expression and the nature of emotion. In: Scherer K, Ekman P (eds). Erlbaum, Hillsdale, pp 319–344

Ekman P (1992) Facial expressions of emotion: an old controversy and new findings. Phil.Trans. R. Soc. Lond. B 335:63–69

Ekman P (1999) Facial expressions handbook of cognition and emotion. In: Dalgleish T, Power M (eds) Handbook of cognition and emotion. Wiley, New York, pp 301–320

Ekman P, Friesen WV (1976) Pictures of facial affect. Consulting Psychologists Press, Palo Alto

Ekman P, Friesen WV (1978) Facial action coding system (FACS). Consulting Psychologists Press, Palo Alto, CA

Ekman P, Sorenson ER, Friesen W (1969) Pan-cultural elements in the facial displays of emotion. Science 164:86–88

Elfenbein HA, Ambady N (2002) On the universality and cultural specificity of emotion recognition: a meta-analysis. Psychol Bull 128(2):203–235

Esteves F, Dimberg U, Öhman A (1994a) Automatically elicited fear: conditioned skin conductance responses to masked facial expressions. Cogn Emot 8:393–413

Esteves F, Parra C, Dimberg U, Öhman A (1994b) Nonconscious associative learning: pavlovian conditioning of skin conductance responses to masked fear-relevant facial stimuli. Psychophysiology 31:375–385

Fernandez-Dols JM, Carrera P, Crivelli C (2011) Facial behavior while experiencing sexual excitement. J Nonverbal Behav 35(1):63–71. doi:10.1007/s10919-010-0097-7

Fernandez-Dols JM, Ruiz-Belda MA (1997) Spontaneous facial behaviour during intense emotional episodes: artistic truth and optical truth. In: Russell JA, Fernandez-Dols JM (eds) The psychology of facial expression. Cambridge University Press, New York, pp 255–274

Fogel A, Nelson-Goens C, Hsu H (2000) Do different infant smiles reflect different positive emotions? Soc Dev 9:497–520

Fridlund AJ (1994) Human facial expression: an evolutionary view. Academic Press, San Diego

Gao X, Maurer D (2009) Influence of intensity on children's sensitivity to happy, sad, and fearful facial expressions. J Exp Child Psychol 102(4):503–521

Gaspar A (2001) Comportamento facial em Pan e Homo. Contribuição para o estudo evolutivo das expressões faciais [Facial behavior in Pan and Homo. Contribution to the evolutionary study of facial expressions]. Doctoral thesis, Universidade Nova de Lisboa, Lisboa

Gaspar A (2006) Universals and individuality in facial behavior-past and future of an evolutionary perspective. Acta Ethol 9(1):1–14. doi:10.1007/s10211-006-0010-x

Gaspar A, Esteves FG (2012) Preschooler's faces in spontaneous emotional contexts-how well do they match adult facial expression prototypes? Int J Behav Dev 36(5):348–357. doi:10.1177/0165025412441762

Gaspar A, Esteves, FG, Dimberg (in prep) Decoding facial expression and emotional responses in adolescents and adults—how does it relate to emotional empathy?

Goodall J (1986) The chimpanzees of Gombe: patterns of behavior. Harvard University Press, Cambridge

Gnepp J (1983) Children's social sensitivity: inferring emotions from conflicting cues. Develop Psychol 25:277–288

Grammer K, Eibl-Eibesfeldt I (1990) The ritualization of laughter. In: Koch W (ed) Naturlichkeit der sprache und der kultur: Acta colloquii. Brockmeyer, Bochum, pp 192–214

Grammer K, Oberzaucher E (2006) The reconstruction of facial expressions in embodied systems: New approaches to an old problem. ZiF Mitteilungen 2:14–31

Guinsburg R (1999) Avaliação e tratamento da dor no recém-nascido. Jornal Pediatria 75(3):149–160

Hassin RR, Aviezer H, Bentin S (2013) Inherently ambiguous: facial expressions of emotions in context. Emotion Rev 5(1):60–65

Hager JC, Ekman P (1979) Long distance transmission of facial affect signals. Ethol Sociobiol 1:77–82

Huber E (1930/1972) Evolution of facial musculature and facial expression. Reprinted in evolution of facial expression. Two accounts. Arno Press, New York

Huber E (1931) Evolution of facial musculature and facial expression. The John Oxford University Press, Baltimore (Hopkins Press: Humphrey Milford: London)

Izard C (1971) The face of emotion. Appleton-Century Crofts, New York

Izard CE (1991) The psychology of emotions. Plenum Press, New York

Izard CE, Malatesta CZ (1987) Perspectives on emotional development: differential emotions theory of early emotional development. In: Osofsky JD (ed) Handbook of infant development, 2nd edn. Wiley Interscience, New York, pp 494–554

Krumhuber EG, Schere KR (2011) Affect bursts: dynamic patterns of facial expression. Emotion 11(4):825–841

Lang PJ, Bradley MM, Cuthbert BN (1998) Emotion, motivation, and anxiety: brain mechanisms and psychophysiology. Biol Psychiat 44(12):1248–1263

LeDoux JE (1996) The emotional brain. Simon & Schuster, New York

Lemerise EA, Dodge KA (1993) The development of anger and hostile interactions. In: Lewis M, Haviland JM (eds) Handbook of emotions. Guilford Press, New York, pp 537–546

Lindquist KA, Barrett LF (2008) Emotional complexity. In Lewis M, Haviland-Jones JM, Barrett LF (eds) Handook of emotions, 3rd ed. Guilford Press, New York, pp 513–532

Lindquist KA, Barrett LF, Bliss-Moreau E, Russell JA (2006) Language and the perception of emotion. Emotion 6(1):125–138

Lorenz K (1967/1986) Evolution and modification of behavior. 2nd edn. University of Chicago Press, Chicago

Manser MB, Seyfarth RM, Cheney DL (2002) Suricate alarm calls signal predator class and urgency. Trends Cogn Sci 6(2):56–58

McFarland D (1987) The Oxford companion to animal behaviour. Oxford University Press, Oxford

McGrew WC (1969) An ethological study of agonistic behaviour in preschool children. In Proceedings of the 2nd congress of primatology (Atlanta, Georgia, 1968) Karger, Basel, New York, pp 149–159

McGrew WC (1972) An ethological study of children´s behavior. Academic Press, New York

Mehu M, Mortillaro M, Banziger T, Scherer KL (2012) Reliable facial muscle activation enhances recognizability and credibility of emotional expression. Emotion 12(4):701–715

Messinger D, Mahoor MM, Chow S-M, Haltigan JD, Cadavid S, Cohn J (2014) Early emotional communication: novel approaches to interaction. In: Gratch J, Marsella S (eds) Social emotions in nature and artifact. Oxford University Press, Oxford, pp 162–212

Morris JS, Öhman A, Dolan RJ (1998) Conscious and uncouscious emotional learning in the human amygdala. Nature 393:467–470

Öhman A, Lundqvist D, Esteves F (2001) The face in the crowd revisited: a threat advantage with schematic stimuli. J Pers Soc Psychol 80(3):381–396. doi:10.1037/0022-3514.80.3.381

Öhman A, Soares S, Juth P, Lindström B, Esteves F (2012) Evolutionary derived modulations of attention to two common fear stimuli: serpents and hostile humans. J Cogn Psychol 24:17–32

Oster H (2005) The repertoire of infant facial expressions: an ontogenetic perspective. In: Nadel J, Muir D (eds) Emotional development. Oxford University Press, Oxford, UK, pp 261–292

Panksepp J (1998) Affective neuroscience: the foundations of human and animal emotions. Oxford University Press, New York

Panksepp J (2005) Affective consciousness: core emotional feelings in animals and humans. Conscious Cogn 14(1):30–80. doi:10.1016/j.concog.2004.10.004

Panksepp J, Watt J (2011) Why does depression hurt? ancestral primary-process separation-distress (PANIC/GRIEF) and diminished brain reward (SEEKING) processes in the genesis of depressive affect. Psychiatry 74(1):5–13

Parr LA, Cohen M, de Waal F (2005) Influence of social context on the use of blended and graded facial displays in chimpanzees. Int J Primatol 26(1):73–103. doi:10.1007/s10764-005-0724-z

Pellatt A (1979) The facial muscles of three African primates contrasted with those of Papio ursinus. South Afr J Sci 75:436–440

Peleg J, Katzir G, Peleg O, Kamara M, Brodsky L, Hel-Or H, Keren D, Nevo A (2009) Facial expressions in various emotional states in congenitally blind and sighted subjects. Israel J Ecol Evol 55(1):11–30. doi:10.1560/IJEE.55.1.11

Plutchick R (1980a) Emotion: a psychoevolutionary synthesis. Harper & Row, New York

Plutchick R (1980b) Emotion: theory, research and experience, vol 1, theories of emotion. Academic Press, New York

Pollak SD, Kistler DJ (2002) Early experience is associated with the development of categorical representations for facial expressions of emotion. P Natl Acad Sci USA 99(13):9072–9076

Pollick AS, de Waal FB (2007) Ape gestures and language evolution. Proc Natl Acad Sci U S A 104(19):8184–8189. doi:10.1073/pnas.0702624104

Preuschoft S (1992) "Laughter" and "smile" in Barbary Macaques (*Macaca sylvanus*). Ethology 91(3):220–236. doi:10.1111/j.1439-0310.1992.tb00864.x

Preuschoft S, van Hooff JARAM (1997) The social function of "smile" and "laughter". Variations across primate species and societies. In: Segestrale UC, Molnár P (eds) Nonverbal communication: where nature meets culture. Lawrence Earlbaum Associates, Hillsdale, pp 171–190

Righart R, De Gelder B (2008) Recognition of facial expressions is influenced by emotional scene gist. Cogn Affect Behav Neurosci 8(3):264–272

Rosenstein D, Oster H (1997) Differential facial responses to four basic tastes in newborns. In: Ekman P, Rosenberg E (eds) What the face reveals: basic and applied studies of spontaneous expression using the facial action coding system. Oxford University Press, New York, pp 302–319

Rozin P, Haidt J, McCauley CR (2008) Digust. In: Lewis M, Haviland-Jones JM (eds) Handbook of emotions, 3rd edn. Guilford Press, New York, pp 757–776

Russell JA (1980) A circumplex model of affect. J Pers Soc Psychol 39(6):1161–1178. doi:10.1037/h0077714

Russel JA (2003) Core affect and the psychological construction of emotion. Psych Rev 110(1):145–172

Russell JA (2006) Emotions are not modules. Can J Philos 32:53–72

Russell J, Bullock M (1986) On the dimensions preschoolers use to interpret facial expression of emotion. Dev Psychol 22(1):97–102

Russell JA, Fernandez-Dols JM (1997) What does a facial expression mean? In: Russell JA, Fernandez-Dols JM (eds) The psychology of facial expression. Cambridge University Press, New York, pp 3–30

Shaver P, Schwartz J, Kirson D, Oconnor C (1987) Emotion knowledge: further exploration of a prototype approach. J Pers Soc Psychol 52(6):1061–1086. doi:10.1037/0022-3514.52.6.1061

Scherer KR, Zentner MR, Stern D (2004) Beyond surprise: the puzzle of infants' expressive reactions to expectancy violation. Emotion 4(4):389–402

Scherer KR, Elgring H (2007) Multimodal expression of emotion: affect programs or multimodal appraisal programms. Emotion 7(1):158–1711

Slobodchikoff CN, Kiriazis J, Fischer C, Creef E (1991) Semantic information distinguishing individual predators in the alarm calls of Gunnison's prairie dogs. Anim Behav 42(5):713–719

Sonntag CF (1924) The morphology and evolution of apes and man. Sons and Danielsson Ltd, London

Soussignant R, Schaal B (2005) Emotional processes in human newborns: a functionalist perspective. In Nadel J, Muir D (eds) Emotional development. Oxford University Press, Oxford, pp 127–159

Swindler DR, Wood CD (1973) An atlas of primate gross anatomy baboon, chimpanzee and man. University of Washington Press, Seattle

Tanner JE, Byrne WB (1993) Concealing facial evidence of mood: perspective taking in a captive gorilla. Primates 34:451–457

van Hooff J (1962) Facial expressions in higher primates. Symp Zool Soc L 8:97–125

van Hooff J (1967) The facial displays of the Catarrhine monkeys and apes. In: Morris D (ed) Primate ethology. Aldine, Chicago, pp 7–68

van Hooff J (1972) A comparative approach to the phylogeny of laughter and smiling. In: Hinde RA (ed) Non-verbal communication. Cambridge University Press, Cambridge, pp 209–241

van Hooff JARAM (1973) A structural analysis of the social behavior of a semi-captive group of chimpanzees. In: M. Von Cranach M, Vine I (eds) Social communication and movement. European Monographs in Social Psychology. Vol 4 Academic Press, London, pp 75–162

Widen SC, Russell JA (2008) Children acquire emotion categories gradually. Cognitive Dev 23(2):291–312. doi:10.1016/j.cogdev.2008.01.002

Widen S (2013) Children's interpretation of facial expressions: the long path from valence-based to specific discrete categories. Emot Rev 5(1):72–77

Williams ACD (2002) Facial expression of pain: an evolutionary account. Behav Brain Scis 25(4):439–488

Young G, Decarie TG (1977) An ethology-based catalogue of facial/vocal behaviors in infancy. Anim Behav 25(1):95–107

The Evolution of Joint Attention: A Review and Critique

Timothy P. Racine, Tyler J. Wereha, Olga Vasileva, Donna Tafreshi and Joseph J. Thompson

Abstract Joint attention can be defined as the ability to intentionally coordinate an attentional focus on some object or state of affairs with another. This capacity is believed by most theorists to be logically, developmentally, and evolutionarily prior to language and further forms of social cognition tied up with human social communication. However, although there has been a good deal of empirical and theoretical work on joint attention, there has been less attention paid to the evolution of joint attention in its own right. There has also been sustained debate concerning whether other primates can be said to engage in joint attention, which in turn conditions the evolutionary theories that are offered. In this chapter, we define and describe joint attention, discuss the skills it involves, and the extent to which we share these with other animals. Next, we review work that has been done on the evolution of joint attention and related capacities and classify it as a function of its mode of explanation. We then discuss the aforementioned forms of evolutionary explanation in the light of recent evolutionary theories and findings that question adaptationist thinking, and consider the potential relevance of non-adaptationist thinking for theoretical work on the evolution of joint attention.

Keywords Joint attention • Aadaptationist stance • Evolutionary psychology • Core knowledge • Shared intentionality • Non-human primates

Perhaps unsurprisingly, providing a satisfactory and relatively complete account of the evolution of social communicative abilities in human and non-human primates has proven to be a difficult task. Although it is important to understand the origins and properties of any animal communication system as an end in itself, interest

T. P. Racine (✉) · T. J. Wereha · O. Vasileva · D. Tafreshi · J. J. Thompson
Department of Psychology, Simon Fraser University, 8888 University Drive,
Burnaby, BC V5A 1S6, Canada
e-mail: tracine@sfu.ca

M. Pina and N. Gontier (eds.), *The Evolution of Social Communication in Primates*, Interdisciplinary Evolution Research 1, DOI: 10.1007/978-3-319-02669-5_7,

in evolutionary analyses of communicative systems that appear similar to those in humans is almost universal. That is, whereas a non-human organism does not need a full-fledged language, and the biological, sensorimotor, and cognitive abilities that human languages seem to require, in order for "social communication" to be possible, such communication exhibits near universal interest to the extent that it can reveal something about the common descent, or perhaps convergent evolution, of the skills under investigation.

Part of the reason for the difficulty of this task is that there might be, pace Darwin, qualitative rather than quantitative differences in linguistic skill and repertoire between humans and our nearest living relative, the common chimpanzee, Pan troglodytes, or indeed the chimpanzee's nearest living relative, the bonobo, Pan paniscus (see Tallis 2011 for a recent forceful expression of one such view, see also e.g. Penn et al. 2008). Another reason is that natural selection itself might not be the right or at least sole theoretical tool required to understand the evolutionary origins of this particular phenomenon for the simple reason that "social communication" might not arise from a simple adaptation or set of adaptations (Gould and Lewontin 1979; Lickliter and Honeycutt 2013; Racine 2013). In this chapter, we first briefly introduce the evolution of social communication in general terms (Sect. 1) to justify focussing on a particular capacity called "joint attention" that most agree is required for language and intentional social communication in general. In Sect. 2, we define and describe joint attention further, discuss the skills it involves, and the extent to which we share these with other animals. In Sect. 3, we review work that has been done on the evolution of joint attention and related capacities and classify it with respect to its degree of reliance on adaptationist and innatist thinking. In Sect. 4, we conclude by discussing the aforementioned forms of evolutionary explanation in the light of recent evolutionary theories and findings that question adaptationist thinking, and consider the potential relevance of non-adaptationist thinking for theoretical work on the evolution of joint attention. In particular, we discuss the relation between development and evolution and their potential reintegration.

1 Introduction: Language, Social Communication, and Joint Attention

Even if, as Darwin would have it, there is continuity rather than discontinuity in the social communication systems of humans and other primates, it has proven difficult, at least if understood in selectionistic terms, to explain the evolution of language through a series of gradual small steps. Darwin's (1879) own view, laid out in The Descent of Man, emphasizes changes in cognition and general intelligence followed by sexual selection acting on vocal control. Through this process, a "musilanguage", as Brown (1999) calls it, is created that adds meaning to these voicings, which in turn is the result of, and leads to, changes in general intelligence. Although few, including Darwin, argue for direct selection on a language organ or instinct (cf. Pinker 1994), this has proven to be an area of evolutionary theory that is quite resistant to significant progress or at least consensus. Part of the reason for this is probably that the concept "language" itself is thorny and used in a variety of ways, each of which potentially requires slightly different skills. There is also a large and somewhat unwieldy literature in the philosophy of language, philosophy of mind, and philosophy of biology, that is, relevant for this discussion, but is difficult to incorporate in a sustained manner.

This is not to say though that there have not been many attempts. Although several others could be listed, for comparative purposes, let us consider three very different recent works in this vein. Hauser et al. (2002), Place (2000), and Whiten and Erdal (2012) account for language evolution chiefly by emphasizing the evolution of a particular mental-combinatorial skill (recursion), manual and representational skills (including pointing), and sociocultural skills (mostly mindreading abilities), respectively. Given that there is little agreement in explanatory scope or focus in these sorts of "grand theory" attempts, it may make some sense to focus on a thorough explanation of a particular aspect of language that make it possible. This would mean resisting the urge to single out one factor that one theorizes to be chiefly responsible for human language, such as recursion in the case of Hauser et al. (2002). For this reason, we are more in sympathy with Whiten and Erdal's (2012) attempt at identifying a potential suite of such factors. However, the fact that Whiten and Erdal (2012: 2127) added a footnote at the proof stage in response to an article on the "cultural niche" by Boyd et al. (2011) suggests that Whiten and Erdal realize their notion of a "(socio-)cognitive niche" might be associated with the Evolutionary Psychology theorizing of Pinker (2010), an association that Whiten and Erdal are clearly attempting to avoid. Furthermore, from our perspective and probably Boyd et al. (2011) as well, Whiten and Erdal (2012) might be assuming too much adaptive value in "mindreading" skills in an ancestral context.

We think we can make some headway on these issues by focussing on a capacity that most agree is required for language and intentional social communication in general. This capacity is joint attention, which Leaven and Racine (2009: 240) define as "the ability to intentionally co-orient towards a common focus". However, as with language and social communication more broadly, joint attention may not be a single capacity, and depending on the researcher, can involve gaze following of various forms, social referencing, gesture including different forms of pointing gesture, and instrumental imitation (for a review, see Racine and Carpendale 2007). Joint attention is also used in the field in an inclusive manner to refer to this entire suite of behaviours (or some subset of them), or an exclusive manner to mean literal episodes of joint (typically visual) attention. However, if joint attention is necessary for language, it is therefore logically and evolutionarily prior to it, and is therefore an ability that needs to be explained in a satisfactory evolutionary account of social communication.

One can also easily see why such a skill would be implicated in language and intentional social communication: it would seem that minimally an individual has to appreciate that another is referring to something in order to learn the meaning of words or to intentionally inform another of some state of affairs (but see Akhtar and Gernsbacher 2007).[1] Although this will not be the focus of our chapter, intentional social communication would likely necessarily involve joint attention as well, although there might be classes of social communication (e.g. alarm calls) that may not be intentional, at least not in the same sense and therefore do not qualify as involving joint attention in the way that it is used in the research community. In either case, although joint attention is a relatively straightforward ability, we need to briefly acknowledge the complexity of the joint attention conceptual domain before proceeding. This is because this complexity has implications for theories concerning the evolution of joint attention, and also explains why in our review we focus on a variety of accounts, some of which are broader classes of theory.

1.1 Joint Attention and a Motley Crew of Related Social Cognitive Concepts

Trevarthen (1979) and Trevarthen and Hubley (1978) classified two important developmental transitions in human infants in the first year of life that he termed primary and secondary intersubjectivity, respectively. The first phase refers to a shared awareness shown in social games, turn taking, and emotional exchanges with a caregiver that is evident in the social smiles that begin around 2 months. This early period is dyadic—back and forth between infant and caregiver, but does not include an additional common object of focus. In a semantic idiom, the interaction lacks extension. The absence of a common focus precludes there being referent for the interaction and, in this sense, the interaction is not about anything. This is in contrast to secondary intersubjectivity that involves a common focus on some object or state of affairs. What psychologists, philosophers, linguists, and cognitive and brain scientists call "joint attention" is in many respects the same phenomenon that Trevarthen terms secondary intersubjectivity. Other than reflecting or perhaps enabling secondary intersubjectivity, joint attention is also viewed as an early developing form of a general "mindreading" ability, and discussed as an aspect of metacognition, metarepresentation, and/or "theory of mind" (Brinck and Liljenfors 2013; Call and Tomasello 2008; Carruthers 2009). Although these are all somewhat related social cognitive domains, part of the difficulty in explaining the evolutionary origins of joint attention is that this concept is tied up with a variety of other social cognitive ones.

[1] Although Akhtar and Gernsbacher (2007) argue for robust effects of "overhearing" in typical language development, the concern in the present paper is with the evolutionary origins of social communication. In this context, it is difficult to imagine a scenario where language or protolanguage could evolve independently of joint attention. However, if Akhtar and Gernsbacher (2007) are reacting to the often highly cognitive descriptions of language that abound in the developmental literature, we share their concern.

It might be tempting to unite these various social cognitive concepts by speaking of early versus later developing social cognitive skills, and therefore of joint attention in contradistinction to the understanding of more complex states, such as beliefs, generally believed to be typical of preschoolers. However, as we will see, theorists such as Baillargeon et al. (2010) have used looking time studies to argue for belief understanding in infants. Although others such as Charles and Rivera (2009) argue that infant looking time procedures are more appropriately interpreted as measures of infant perception and object-oriented behaviours rather than measures of infant knowledge, Luo (2011), for example, adopts the latter interpretation and claim that belief understanding, and therefore metarepresentation, is present as early as 10 months of age. This is the same time period as Trevarthen's secondary intersubjectivity, and the rudiments of joint attentional skills like pointing and gaze following. Thus, although we focus on joint attention in what follows, we need to return to these more general concepts when discussing evolutionary theories of joint attention. This is in part because there is little evolutionary work on discrete joint attention behaviours (e.g. pointing and gaze following), and also because its role in further social cognitive and communicative development is often situated in these broader social cognitive accounts.

2 Joint Attention: Continuity or Discontinuity?

Joint attention would seem to be a capacity, that is, evolutionarily speaking, quite ancient.[2] After all, it is the ability to coordinate one's another focus with that of a conspecific ("to jointly attend"). It also should be obvious when two individuals of a given species are or are not engaged in an act of joint attention. However, a complication is that two individuals might be looking at the same thing by happenstance without an awareness of the attentional focus of the other jointly attending individual. Perhaps though this should not necessarily mean that it does not count in some sense as joint attention, for example Butterworth (1998: 171) defines joint attention as "simply…looking where someone else is looking". Similarly, Bakeman and Adamson (1984) allow for a category they term "passive" joint attention. However, the way the concept is typically used in the field denotes a mutual awareness of each other's attentional state. To put this differently, the way "joint attention" is typically understood within the field, both parties must intend to jointly attend to some shared object or state of affairs, and be aware that the other's attention is also focussed on the same object or state of affairs (Leavens and

[2] Given that the original discovery of mirror neurons some 20 years ago was in macaques, it would seem that a capacity for joint attention might have its roots as far back as old world monkeys. However, it is not clear what mirror neurons really do, or the extent to which they imply "mindreading" abilities like joint attention (Racine et al. 2012). For one thing, although all species of great ape pass mirror self-recognition tests, monkeys do not.

Racine 2009). That is, they must engage in what is often called triadic interaction (e.g. Tomasello 1995). As long as one keeps in mind that this just means that the attribution of joint attention to two agents involved in the right sort of activity (e.g. gaze following, pointing, and so forth) requires that their coordinated mutual attentional focus be non-accidental ("that they intend to attend"), then this is a reasonable terminological stipulation. And, on the face of it, one would still expect joint attention to be a capacity shared with some other primate species through common descent, and probably other highly social species through convergent evolution, as has been suggested, for example, in work on the corvid, Corvus corax, the common raven (Pika and Bugnyar 2011).

It might be surprising, then, for the reader without a previous background in this research area to learn that there is a considerable amount of pessimism concerning the mentality of non-human animals when it comes to explaining basic joint attentional capacities. This is not the place to review the causes and consequences of this state of affairs, but rather to see how they play out in evolutionary work on joint attention. However, a summary of some of the key concerns is required in order to proceed (for more detail, see, e.g. Racine 2012a, b).

Scepticism regarding non-human joint attentional capacities often involves two steps of argument, the first of which is conceptualizing the triad in "triadic interaction" not as "two agents coordinating their attention to an object", but rather as two "coordinated subject–object mental relations". That is, the activity of joint attention has been redescribed as a shared second-order representational state.[3] As long as it is understood that this is just to say there is a mutual awareness of an intention to attend, this is not inherently problematic. However, the second-order mental states in question are not typically understood in this manner, but rather causally. That is, the mental states are theorized to give rise to the behaviour of interest rather than the joint attentional behaviour being grounds for the attribution of the mental state. The underlying reason for conceiving as joint attention as mental representations of subject–object relations—and causal ones at that—is that joint attention is typically understood from within the lens of what is often called in the philosophy and cognitive sciences, the representational theory of mind (RTM) (see Racine 2012b; Slaney and Racine 2011). The next step, of course, is to discount activities that can be given ostensibly "less mental" interpretations. And the pessimism about animal minds comes full circle.

A highly influential group of such pessimists is Tomasello and colleagues who have consistently asserted that shared attention involving great apes such as chimpanzees should not qualify as joint attention in the way used in the field (e.g. Tomasello et al. 2005, 2007; Tomasello 2008, 2014). The evolutionary force

[3] See Andrews (2012) and Hutto and Myin (2013) for discussion and critiques of the more general position that folk psychology implies the uncovering of propositional attitudes. Alternative approaches maintain that "the folk" understand one another as entire organisms with histories, embedded in particular contexts, with moods, temperaments, and so on, not typically as bearers of propositional attitudes.

of their argument will be considered in more detail below (see also Racine 2012a, 2013; Wereha and Racine 2012), but the most relevant point, for them, is that apes' motivations for sharing attention are mostly instrumental, whereas humans do the same activities for non-instrumental purposes (to simply share intentions as an end in itself). This follows a classical distinction between prelinguistic imperative and declarative acts (so-called protoimperatives and protodeclaratives) introduced by Bates et al. (1975). But, unlike Bates et al., Tomasello et al. conflate the cause of the behaviour with the meaning of the behaviour. This fundamental difference plays out in Tomasello's evolutionary work on joint attention and related work by others.

Consider, by parity, the following. Assume that in an act of shared attention one organism is highly motivated to attend because of their interest in the interaction, but the other organism is only doing so begrudgingly, or even for a completely different reason (e.g. they were paid, or otherwise reinforced, for doing so), but they were still both intentionally jointly attending. Would we be justified in saying that therefore only one of the organisms is really sharing attention (or really pointing cf. Tomasello 2006)? We might say one was more motivated than the other, and it might well be that the more motivated one was more likely to initiate or continue episodes of joint attention, but it is an obvious tautology that this would not and could not mean that only one of the two was jointly attending. Therefore, to claim that only one organism is really jointly attending, as stated, is to conflate the cause and meaning of the activity. To claim that both are attending because of some common underlying neural causes is, of course, equally problematic. The typical person lacks knowledge of the neural causes of behaviour, yet still has sensible grounds for describing the activity correctly.

We assume that many readers will agree that substituting a motivational construct for an intentional–attentional one is scientifically very confusing and presumably ill-advised. However, it is this very move that Tomasello (2008, 2014) performs in his "shared intentionality hypothesis", which is more an illustrative use of a body of work in the philosophy of action than a scientific hypothesis (see Racine 2012b). By contrast, Leavens and Racine (2009) concluded that while there is joint attentional variation among apes that is attributable to differential rearing histories, apes engage in all behaviours considered to show joint attention in humans.[4] Although we will not repeat this point again in the present chapter or include it in our classification of theories, Leavens and Racine (2009) argue that therefore there is no clear evidence for uniquely human cognitive adaptation(s) for joint attention.

Even if Leavens and Racine are wrong, it is important to bear in mind that claiming that chimpanzees are aware of the attention of others and accordingly intentionally engage in joint attention when they follow gaze or gesture is not

[4] In terms of relevant similarities, it seems noteworthy that all great ape species can pass any test of gaze following ability that a human child can, including being aware that individuals cannot see though obstructions, and also that great apes use manual gestures, including pointing gestures, particularly in captive environments (Brauer et al. 2005; Leavens and Racine 2009; Pika 2008).

tantamount to claiming that a chimpanzee would have to be aware of everything about another's attentional focus. Or, that an ape would need to understand all the subtle ways that attention can be deployed (Wilkins 2003). Or, even that the identical genetic, neural, sensorimotor, cognitive, or behavioural substrates would need to be responsible for joint attention in humans and other apes.[5] After all, the same would apply if comparing a prelinguistic human to a human child or adult. But this is not the question we asked to begin with. We simply asked whether an ape (or a prelinguistic infant as the case may be) intentionally shares in the attentional focus of a conspecific around some object of mutual attention and is aware that his or her interlocutor is doing the same. To forget this would again invite potential confusion of definitional issues of what counts as joint attention and empirical (and sometimes causal) issues concerning how joint attention occurs (Racine 2012b).

Although many researchers seem to assume that these matters can be resolved empirically or by creating more adequate methodologies, the foregoing should suggest these are not simple empirical or methodological matters.[6] That is, one cannot stipulate that other animals do not have a capacity for joint attention just because one believes that their joint attentional behavioural can be explained without the attribution of second-order representational states—particularly when there is no clear evidence that human infants require second-order representational states to engage in joint attention (Leavens 2012; Leavens and Racine 2009; Racine 2012b; Racine et al. 2012). After these considerations, we are now, we believe, in a better position to review and discuss the evolution of joint attention.

3 Evolutionary Theories of Joint Attention

Evolutionary claims about joint attention follow the tendency in the field to conceive of a joint attention in an inclusive or exclusive manner. It is also common to focus on a particular capacity, such as pointing or gaze following. Pointing in particular is thought to have played an important role in human evolution, being a human specific adaptation in that it is considered to be a human universal (Butterworth 2003; Povinelli et al. 2003). And, it is argued to be absent among wild populations of great apes (Tomasello 2006, 2008). The pointing gesture itself has also been considered to be an adaptation of human physiology (Butterworth

[5] Although this might seem bad news for an attempted evolutionary analysis of joint attention, it might on the other hand be diagnostic of the limitations of adaptationist and selectionist thinking. An important change in thinking in the past 20 years or so is to question the opposition between evolution and development (e.g., Gottlieb 2002; Jablonka and Lamb 2005; Lickliter and Honeycutt 2013; Oyama 2000).

[6] This is not to say though that empirical matters are independent of these sorts of conceptual concerns (Glock 2013). The point is that improved methods or additional empirical work cannot avoid conceptual problems that result for the application of the RTM framework when the framework itself is built into the interpretation of the new methods or findings.

2003) and Povinelli and Davis (1994) have documented differences in the resting state of the index finger in humans and chimpanzees, suggesting biological preparation in humans.

The focus on the evolution of pointing is not surprising because pointing is taken to be the least ambiguous indicator of shared attention skill in that it inherently requires another and some shared object or state of affairs. In a strictly behavioural sense, pointing is a very useful indicator because it is more overt than gaze following and the latter develops earlier than pointing when its function is less clear. However, gaze following often develops into a form that involves so-called visual checking, wherein one individual double-checks to ensure that the attended visual referent was the intended one (Morissette et al. 1995), which seems as cognitively demanding in a joint attentional sense. Also, as summarized earlier, there is quite a bit of debate about whether pointing requires second-order representation, and therefore, whether it should be seen as a "superior" indicator of joint attentional skill in a logical sense. Therefore, a behaviour-by-behaviour comparison of evolutionary joint attention claims seems ill-advised, and we will present the theories according to their form of evolutionary theory. As we will see, the three approaches we compare are all explicitly adaptationist. Try as we might, we were also not able to find other forms of explanation used for joint attention.

We will describe the theories in broad strokes and essentially outline what are common but potentially problematic ways to think about the evolution of joint attention. In this sense, we are not concerned with giving the high-resolution details of a given account, and suggest, if desired, that interested readers consult the accounts in question for more detail. We turn to possible remedies in the following section. We begin with an orienting discussion of Trevarthen's work on secondary intersubjectivity to show the difficulty in making an evolutionary explanation of an innatist claim. We then proceed to review three main ways of thinking about the evolution of joint attention beginning with what is the most extremely adaptationist, and to our mind, straightforwardly problematic account, which is that of evolutionary psychology. We then consider the core knowledge account of social cognitive abilities, which are argued to be present at the same time that joint attentional skills emerge. Here, the emphasis switches from adaptationism to innatism, but the common thread is domain-specificity. We conclude with Tomasello's shared intentionality theory, which is often understood to be a more moderate alternative to evolutionary psychology and core knowledge accounts, despite the fact that it contains many of the same problematic assumptions.

3.1 Intersubjectivity and the Perils of Innatist Explanation

Trevarthen's innatist explanation of secondary intersubjectivity anticipates in some respects accounts as diverse as core knowledge theory in the sense of the implicit Chomskyan "poverty of the stimulus" flavour to the argument, and Tomasello's shared intentionality theory. Trevarthen unabashedly claims that the capacity for intersubjectivity, and by extension, joint attention, is innate. Although we do not have the space to review his theory and the support for it in detail, he recently summarized his position, in what in many respects is a reflection on his distinguished career (Trevarthen 2011: 119), in the following manner: "We are born to generate shifting states of self-awareness, to show them to other persons, and to provoke interest and affectionate responses from them", and "cultural intelligence itself is motivated at every stage by the kind of powers of innate intersubjective sympathy that an alert infant can show shortly after birth". Beyond Trevarthen's landmark infant development studies which demonstrated what he called protoconversations between infant and caregivers and coupling between infant gestures and prespeech, Trevarthen's theory is supported to some extent by Meltzoff's work on neonatal imitation (but see e.g. Jones 2009), and the discovery of mirror neurons which have been argued by some to underlie intersubjectivity (e.g. Gallese 2001, but see e.g. Racine et al. 2012). However, the chief impetus for this theory, we think, is more intuitive.

Trevarthen often criticizes the information processing and overly cognitive manner is which infant life is—or at least was—often explained. Like Bruner (e.g. 1983), who cites Trevarthen's work as support for his own, the goal is to emphasize sociocultural processes that are presupposed in mainstream accounts, in Bruner's case behaviourism (Racine 2012b), and in Trevarthen's Piagetian accounts. In many respects, we are in sympathy with the motivations of Trevarthen and Bruner—and also Tomasello, another prominent theorist who emphasizes sociocultural factors; information processing and unqualified learning or constructivist theories are too simple at worst and redescriptions of familiar phenomena at best to be of much theoretical use in a general sense. However, claiming that something is innate or the product of natural selection (i.e. an adaptation) is not necessarily that much of an improvement. That is, although we do understand the limitations of the unqualified "socioculturalism" or "interactionism" that, for example, Tomasello has criticized in cultural theories (Wereha and Racine 2012), claiming that something is innate or the product of an adaptation tells us as much about the development of psychological capacities as cultural notions of appropriation and internalization do. That is, very little.

3.2 Evolution Psychology: Massively Modular and Massively Ambitious

Evolutionary psychologists, by which we mean to single out the so-called Santa Barbara school of Evolutionary Psychology (EP) associated with the work of Tooby and Cosmides, Buss, Pinker, and a few others, have also gained ascendance within psychology in the past 25 years for reasons that are ironically very similar to those of Trevarthen, Bruner, and Tomasello. Although EP takes this to fanciful levels by proposing that psychology itself should come home to EP as a core explanatory principle, in criticizing what EP has called the standard social science model, they are essentially taking the field to task for neglecting the biological substrate of cultural learning. Tooby and Cosmides (1992) and Pinker (1994) in particular have traded on the notion that psychologists too easily fall back on naive forms of learning theory to explain human capacities. This critique is well taken as far as it goes, but of course the remedy suggested by EP—namely that of adopting EP—has been thoroughly criticized on a number of fronts, including its fundamental conceptual and biological adequacy (see e.g. Heyes 2012; Lickliter and Honeycutt 2013; Racine 2013; Wereha and Racine 2012). As this topic has been covered very well in the literature, we will not repeat these critiques here, but will focus on what is more relevant for present concerns. The evolutionary foundations of core knowledge and shared intentionality theory have not received nearly the same level of scrutiny, and we therefore focus more attention on those sections.

According to EP, joint attention and other aspects of social cognition are adaptations in a classical sense. In particular, they are adaptations to selective pressures our hominid ancestors faced in the Pleistocene. As Tooby and Cosmides, the founders of the approach, write in the foreword of Baron-Cohen's (1995) book Mindblindness: An Essay on Autism and Theory of Mind, "our cognitive architecture resembles a confederation of hundreds or thousands of functionally dedicated computers (often called modules) designed to solve adaptive problems endemic to our hunter-gatherer ancestors" (p. xiii). Baron-Cohen's (1995) "human mindreading system" is his explicit contribution to the EP programme of research.[7] Within this system, he describes four major modules that process social information that he refers to as the intentionality detector (ID), the eye-direction detector (EDD), the shared attention mechanism (SAM), and the theory of mind mechanism (ToMM). These modules process information on agents' volition, perception, shared attention, and epistemic states, respectively. This model has been persuasive in regards in its proposed aetiology of autism as damage to the SAM or ToMM, according to the model, leads to the kind of social cognitive impairments that presents in autism spectrum disorder (ASD).[8] From this perspective, although Baron-Cohen claims the ID and EDD are shared with other primates, a specific adaptation, the SAM, does the heavy lifting in explaining the existence of human joint attention capacities. True to their adaptationist programme, Tooby and Cosmides (1995: xvi) contend that, "natural selection is the only known natural process that builds functional organization into the species-typical designs of organisms". Be that as it may, this reasoning has failed to convince many in the field that EP's use of natural selection makes is sufficient as an explanation. Furthermore, the biological sciences have moved on since the 1990s and biologists

---

[7] Interestingly, although Baron-Cohen was once somewhat of a poster child for Evolutionary Psychology, Baron-Cohen himself has moved on to other aspects of autism research and theory that, while perhaps in some ways similar in form to EP-styled explanation, does not explicitly cast itself as EP.

[8] "Mindblindness" is a term that Baron-Cohen (1995) in fact coined for autism spectrum disorder. It is notable that social cognition is only one aspect of what is often compromised in ASD.

routinely speak of non-genetic and epigenetic channels of inheritance that are clearly relevant to functional organization (Jablonka and Lamb 2005), and which are processes that should presumably be taken seriously in any evolutionary account of human cognition.

3.3 Core Knowledge Theory: Massively Modular and Massively Innate

The so-called core knowledge explanation of human development postulates the existence of domain-specific ("core") competencies upon which later skills critically depend. Core knowledge theories have become a bit of a cottage industry in the past 20 years or so with core domains theorized to exist for number concepts, language concepts, physical concepts such as gravity and containment relations, folk biology, folk psychology, and most recently moral intuitions. Ironically, although EP-styled forms of explanation seem to be falling out of favour in psychology, core knowledge theories have become more common and influential. This is surprising because although core knowledge approaches are less explicitly "evolutionary", they are equally domain-specific, make similarly selectionist arguments—and cite the work of evolutionary psychologists for support. We surmise that the reason that core knowledge accounts seem to be thriving in our psychology when EP might be on the way out is the apparent empirical rigour of core knowledge accounts. However, as with Trevarthen, core knowledge accounts are explicitly innatist. And, as with Trevarthen, this aspect of their theorizing is quite out of step with recent work in the biological sciences (Lickliter and Honeycutt 2013).

However, the early social cognitive work of Baillargeon and colleagues has provoked quite a bit of recent discussion in psychology (for a review, see Baillargeon et al. 2010). Their account of false belief understanding has invigorated early social cognitive research since Onishi and Baillargeon's (2005) landmark study in which they concluded, using familiarization and looking time (so-called violation of expectation measures), that 15-month-old infants understand false beliefs. There has since been a study conducted using a modified version of the Baillargeon paradigm that fixed false belief understanding at 10 months (Luo 2011). The innovation—and limitation—of core knowledge theories are that they rely on presenting conceptually relevant stimuli (e.g. a situation depicting false belief) and then determine if infants look surprised when an actor violates this expectation. In this sense, they rely on a perceptual rather than a conversational paradigm where children can answer questions about false beliefs and make correct predictions (or not) about a protagonist.

Baillargeon et al. (2010: 111) "assume that infants are born with a psychological reasoning system that provides them with a skeletal causal framework for interpreting the actions of others". Like Trevarthen, they implicitly rely on a Chomskyan poverty of the stimulus form of argumentation, which draws its force in part from the claim that the skills in question develop very early and, consequently, suggest unlearned and domain-specific knowledge. However, the fact that something appears early does not, of course, mean it cannot in some sense be learned. But, in either case, this simple dichotomization of learning and acquired is also a form of nature–nurture reasoning that we doubt anyone would want to defend in public (Bateson and Mameli 2007). When one adds to this the fact that their conclusions only follow if core knowledge researchers are actually investigating an earlier developing form of a logically identical behaviour—and that there are no clear mapping rules from their perceptually based studies to the conceptually based ones performed in other paradigms—it gets even more puzzling. Explaining the false belief of a protagonist provides reasonable logical grounds for attributing at least a rudimentary, but possibly not adult-like, understanding of false beliefs to a child. However, it is not at all clear that the looking longer at a violated false belief situation entails anything of the sort (Müller and Racine 2010). Moreover, given that the associations assessed in this paradigm could well be learned, there is little empirical force to this argument anyway.

Although some core knowledge researchers make some appeals to evolutionary logic, for example Wynn (1998) pointing out that counting skills might have advantaged our hominid ancestors and adducing some comparative "counting" studies in other model species, Baillargeon and colleagues make no typical evolutionary arguments whatsoever. To the extent that they do so, it is through association with the work of evolutionary psychologists (Barrett et al. 2013). Therefore, we are left with the claim that infant social cognitive skills, which we would assume to include joint attention, must have evolved because they develop early and cross-culturally. This inference rests, however, on a notion of "innate" that Bateson and Mameli (2007) have taken pains to show is problematic: early and robust development, even when it involves little apparent learning, need not imply innateness in the sense of an inner biological constraint.

3.4 The Shared Intentionality Hypothesis: Explicitly Adaptationist and Implicitly Innatist

By contrast, Tomasello and colleagues rely heavily upon comparative research in human infants and chimpanzees for their evolutionary account of joint attention, and make the case for the existence of many shared capacities between the two lineages as well as important differences that they contend account for the unique psychological capacities of humans. Tomasello et al. argue that apes and humans share many social cognitive capacities, such as being able to gaze follow, knowing what an opponent in a competitive competition can and cannot see, and grant that captive chimpanzees can come to point for their caretakers to food that they want. Thus, they acknowledge that chimpanzees, like humans, understand psychological states, however, they do not understand as many states as humans, nor do they understand them to the same extent that humans do (Tomasello 2008; Tomasello et al. 2005, 2007). The key difference between humans and apes, they contend, lies in an adaptation they call shared intentionality. Shared intentionality is a prosocial motivation to share attention with others, which they claim transforms capacities to share attention, exhibited both by humans and chimpanzees, into those psychological capacities unique to humans. Thus, shared intentionality "is a big part of what makes humans unique in the animal kingdom, serving as a psychological foundation for all things cultural" and "skills and motivations for shared intentionality are…direct expressions of the biological adaptation that enables children to participate in the cultural practices around them" (Tomasello and Carpenter 2007: 124).

The claim that a phenotype is a "direct expression" of an adaptation seems to militate against the sort of moderation and dismissal of unqualified innatism that Tomasello and colleagues argue against in other places. They claim, for example, (e.g. Tomasello et al. 2005: 688) that "to understand the origins of a human cognitive skill we must go beyond simply labelling it as 'innate'. Indeed, although we concur that understanding actions as goal directed is a biological adaptation, this says nothing about the ontogenetic process". We obviously agree. However, in the same way that "innate" says nothing about how a phenotype develops, it also tells us nothing about how the phenotype evolved.[9] The underlying problem is that despite appearances and occasional appeals to the contrary, innatist claims are actually not, as they stand, evolutionary claims. If innatist claims are understood in the typical selectionist and adaptationist manner in which they frequently are, minimally, some plausible account of differential reproductive success is required. Here, Tomasello (2008) does quite well with an account of how our evolutionary ancestors might have benefitted from their shared intentionality, but of course, these sorts of "just so stories" as Gould and Lewontin (1979) (and Kipling) called them, have their own issues, and what seems to pass evolutionary muster in psychology would probably not in the biological sciences (Racine 2013).

4 Beyond Innate and Adaptation: The Reintegration of Evolution and Development

We have argued that "innate" is simply a vacuous concept especially when used as an explanatory device (e.g. Bateson and Mameli 2007; Racine 2013), and is essentially an admission that one does not understand how a given phenotype develops or has evolved, and is at best a promissory note. Unqualified adaptationist claims run similar risks for reasons that were well documented by Gould and Lewontin; it is simply tricky to reverse engineer an adaptation from a presently adaptive behaviour. As Gould and Lewontin (1979) note, previous adaptations can be co-opted to take on new functions and presently adaptive behaviours

[9] Of course, this should not be taken to mean that we dispute that there are evolved adaptations, or that we believe that processes of adaptation are somehow unimportant in natural selection.

can have no corresponding selective regime. Gene frequencies also change as a result of a variety of well-understood processes such as genetic drift, gene migration, and assortative mating, which do in some cases have clear phenotypic effects. This is not even to consider newer work in the biological sciences that has begun to reintegrate evolution and development (e.g. evolutionary developmental biology or "evo-devo"). However, the neo-Darwinian integration of Darwinian natural selection and Mendelian genetics, in which many innatist claims are purportedly grounded, has been very successful in the biological sciences. Progress though often comes with a price.

The price in this case was that, in order for the population genetic model that underwrites the modern synthesis of Darwin and Mendel to work, reliable developmental processes were simply presupposed (Lickliter and Honeycutt 2013). This reflected the separation of the science of development, or embryology as it was then called, from the study of heredity that was occurring at the same time. Although there are many interesting details to this that we do not have the space to discuss, what is most relevant is that neo-Darwinian thinking, despite Darwin's own interest in development, came to exclude development from evolution. This is reflected in Mayr's (1961) separation of ultimate versus proximate causation, and Tinbergen's (1963) explicit addition of ontogeny to Huxley's three problems of biology, which added a distinction between phylogeny and ontogeny to the intellectual landscape. Although these are reasonable assumptions to make if one's goal is to characterize changes in allele frequencies in a population, the selectionist and adaptationist mode of explanation has been criticized for its lack of use in explaining the developmental emergence of phenotype, and for its potential genetic determinism and reductionism (Lickliter and Honeycutt 2013; Racine 2012a; Racine et al. 2012; Wereha and Racine 2012).

Although epigenetics has become an important topic in the biological and psychological sciences in the past decade, and psychologists such as Gottlieb (2002) have used the term "epigenesis" in a related but historical sense to indicate opposition to preformationist views of phenotype, and to emphasize the importance of developmental processes in the construction of phenotype, most of the shifts in thinking that characterize the current evolutionary theoretical landscape have barely made their way into psychology at all. It is with more than a touch of irony, then, that in recent years adaptationist thinking has become increasingly common in development and comparative psychology, and joint attention theory in particular. All three forms of evolutionary theorizing reviewed earlier are implicitly or explicitly predicated on a neo-Darwinian view whose limitations are more fully recognized outside of the discipline.

In essence, it is not just EP, but adaptationist approaches in general that focus on the "design features" of a given phenotype by reasoning about the particular problem the putative adaptation might have solved in an ancestral environment. Although the empirical foundation for much of the core knowledge and shared intentionality work is creative and empirically sound, if developmentalists who explicitly—or even implicitly—draw on evolutionary theory wish to avoid Gould and Lewontin's (1979) "just so story" characterization of adaptationist explanation, they need to be avoid unqualified uses of concepts like "innate", "adaptation", "adaptive", "instinctual", or "designed", and the adaptationist forms of thinking with which they are tied up.

The first place to start is realizing that calling a phenotype the product of an adaptation when attempting to explain developmental processes is to not only give no explanation, but is to use neo-Darwinian ideas for a purpose for which they were not constructed. Second, at least some of the theoretical tools found in developmental systems approaches that consider the tight relation between developmental and evolutionary processes are already in place, and it is clear that the pioneering work of Oyama, Gottlieb, Lickliter, and others is exerting an influence in psychology. Ironically though, developmental systems approaches are often understood to be ways of taking into account the complexity of developmental processes without appreciating that systems views have an implicit evolutionary underbelly. The negative evolutionary critique of DST is one that is critical of EP, core knowledge, and shared intentionality's predetermined epigenetic notions of adaptation (i.e. adaptations that arise from developmental information in sets of genes). A developmental systems perspective calls for more complete explanations and rigorous science. What this means in clear practical terms is conducting careful work that involves many levels of explanation—coupled with the realization that this is part of evolutionary explanation (Lickliter and Honeycutt 2013). In terms of social communication and particularly "high-level" abilities like language, we need to study more basic, but logically related, abilities like joint attention because such work is intrinsic to constructing plausible evolutionary explanations of the more complex abilities (Whiten and Erdal 2012).

In the rejection of the familiar nature–nurture and innate–acquired dichotomies which many psychologists, including those of EP, core knowledge, and shared intentionality stripes claim to agree on, the evolutionary and developmental landscape is changing in such a way that development–evolution becomes another problematic dichotomy that is impeding scientific progress. If so, the current generation of researchers and theoreticians will be the beneficiaries.

References

Akhtar N, Gernsbacher MA (2007) Joint attention and early vocabulary: a critical look. Lang Linguist Compass 1:195–207

Andrews K (2012) Do apes read minds? Toward a new folk psychology. MIT Press, Cambridge

Baillargeon R, Scott RM, He Z (2010) False-belief understanding in infants. Trends Cogn Sci 14:110–118

Bakeman R, Adamson LB (1984) Coordinating attention to people and objects in mother-infant and peer-infant interaction. Child Dev 55:1278–1289

Baron-Cohen S (1995) Mindblindness: an essay on autism and theory of mind. Cambridge University Press, Cambridge

Barrett HC, Broesch T, Scott R, He Z, Baillargeon R, Wu D, Bolz M, Henrich J, Setoh P, Wang J, Laurence S (2013) Early false-belief understanding in traditional non-western societies. Proc R Soc B Biol Sci 280:1–6

Bates E, Camaioni L, Volterra V (1975) The acquisition of performatives prior to speech. Merrill-Palmer Q 21:205–226

Bateson P, Mameli M (2007) The innate and the acquired: useful clusters or a residual distinction from folk biology? Dev Psychobiol 49:818–831

Boyd R, Richerson PJ, Henrich J (2011) The cultural niche: why social learning is essential for human adaptation. Proc Natl Acad Sci USA 108:10918–10925

Brauer J, Call J, Tomasello M (2005) All great ape species follow gaze to distant locations and around barriers. J Comp Psychol 119:145–154

Brinck I, Liljenfors R (2013) The developmental origin of metacognition. Infant Child Dev 22:85–101

Brown S (1999) The "musilanguage" model of music evolution. In: Wallin NL, Merker B, Brown S (eds) The origins of music. MIT Press, Cambridge, pp 271–301

Bruner JS (1983) Child's talk: learning to use language. Norton, New York

Butterworth G (1998) What is special about pointing in babies? In: Simion F, Butterworth G (eds) The development of sensory, motor and cognitive capacities in early infancy: from perception to cognition. Psychology Press, Hove, pp 29–40

Butterworth G (2003) Pointing is the royal road to language for babies. In: Kita S (ed) Pointing: where language, culture, and cognition meet. Erlbaum, Mahwah, pp 9–33

Call J, Tomasello M (2008) Does the chimpanzee have a theory of mind? 30 years later. Trends Cogn Sci 12:187–192

Carruthers P (2009) How we know our own minds: the relationship between mindreading and metacognition. Behav Brain Sci 32:121–138

Charles EP, Rivera SM (2009) Object permanence and method of disappearance: looking measures further contradict reaching measures. Dev Sci 12:991–1006

Tooby J, Cosmides, L (1995) Foreword. In: Baron-Cohen S (ed) Mindblindness: an essay on autism and theory of mind. Cambridge University Press, Cambridge, pp xi–xviii

Darwin C (1879) The descent of man, and selection in relation to sex, 2nd edn. John Murray, London

Gallese V (2001) The shared manifold hypothesis: from mirror neurons to empathy. J Conscious Stud 8:33–50

Glock H-J (2013) Animal minds: philosophical and scientific aspects. In: Racine TP, Slaney KL (eds) A Wittgensteinian perspective on the role and use of conceptual analysis in psychology. Palgrave Macmillan, Basingstoke, pp 130–152

Gottlieb G (2002) Individual development and evolution: the genesis of novel behaviour. Erlbaum, Mahwah

Gould SJ, Lewontin RC (1979) The spandrels of San Marco and the Panglossian paradigm: a critique of the adaptationist programme. Proc R Soc Lond B 205:581–598

Hauser MD, Chomsky N, Fitch WD (2002) The faculty of language: what is it, who has it, and how did it evolve? Science 298:1569–1579

Heyes C (2012) New thinking: the evolution of human cognition. Philos Trans R Soc B 367:2091–2096

Hutto DD, Myin E (2013) Radicalizing enactivism: basic minds without content. MIT Press, Cambridge

Jablonka E, Lamb MJ (2005) Evolution in four dimensions: genetic, epigenetic, behavioural and symbolic variation in the history of life. MIT Press, Cambridge

Jones SS (2009) The development of imitation in infancy. Philos Trans R Soc B 364:2325–2335

Leavens DA (2012) Pointing: contexts and instrumentality. In: Pika S, Liebal K (eds) Developments in primate gesture research. Benjamins, Amsterdam, pp 181–197

Leavens DA, Racine TP (2009) Joint attention in apes and humans: are humans unique? J Conscious Stud 16:240–267

Lickliter R, Honeycutt H (2013) A developmental evolutionary framework for psychology. Rev Gen Psychol 17:184–189

Luo Y (2011) Do 10-month-old infants understand others' false beliefs? Cognition 121:289–298

Mayr E (1961) Cause and effect in biology. Science 134:1501–1506
Morissette P, Ricard M, Décarie TG (1995) Joint visual attention and pointing in infancy: a longitudinal study of comprehension. Br J Dev Psychol 13:163–175
Müller U, Racine TP (2010) The development of representation and concepts. In: Overton WF, Lerner RM (eds) Handbook of lifespan human development, vol 1., Methods, biology, neuroscience and cognitive developmentWiley, Hoboken, pp 346–390
Onishi KH, Baillargeon R (2005) Do 15-month-old infants understand false beliefs? Sci 308:255–258
Oyama S (2000) The ontogeny of information: developmental systems and evolution, 2nd edn. Duke University Press, Durham
Penn DC, Holyoak KJ, Povinelli DJ (2008) Darwin's mistake: explaining the discontinuity between human and nonhuman minds. Behav Brain Sci 31:109–130
Pika S (2008) Evidence for intentional and referential communication in great apes? In: Zlatev J, Racine TP, Sinha C, Itkonen E (eds) The shared mind: perspectives on intersubjectivity. Benjamins, Amsterdam, pp 165–186
Pika S, Bugnyar T (2011) The use of referential gestures in ravens (Corvus corax) in the wild. Nature Commun 2:560
Pinker (1994) The language instinct: the new science of language and mind. Penguin, New York
Pinker S (2010) The cognitive niche: coevolution of intelligence, sociality and language. Proc Natl Acad Sci USA 107:8993–8999
Place UT (2000) The role of the hand in the evolution of language. Psycoloquy. http://www.cogsci.soton.ac.uk/psyc
Povinelli DJ, Davis DR (1994) Differences between chimpanzees (Pan troglodytes) and humans (Homo sapiens) in the resting state of the index finger: implications for pointing. J Comp Psychol 108:134–139
Povinelli DJ, Bering J, Giambrone S (2003) Chimpanzee 'pointing': another error of the argument by analogy? In: Kita S (ed) Pointing: where language culture and cognition meet. Erlbaum, Mahwah, pp 35–68
Racine TP (2012a) Cognitivism, adaptationism, and pointing. In: Pika S, Liebal K (eds) Developments in primate gesture research. Benjamins, Amsterdam, pp 163–178
Racine TP (2012b) Getting beyond rich and lean views of joint attention. In: Seemann A (ed) Joint attention: new developments in psychology, philosophy of mind, and social neuroscience. MIT Press, Cambridge, pp 21–42
Racine TP (2013) How useful are the concepts 'innate' and 'adaptation' for explaining human development? Hum Dev 56:141–146
Racine TP, Carpendale JIM (2007) The role of shared practice in joint attention. Br J Dev Psychol 25:3–25
Racine TP, Wereha TJ, Leavens DA (2012) To what extent nonhuman primates are intersubjective and why. In: Foolen A, Lüdtke U, Racine TP, Zlatev J (eds) Moving ourselves, moving others: motion and emotion in intersubjectivity, consciousness and language. Benjamins, Amsterdam, pp 221–242
Slaney KL, Racine TP (2011) On the ambiguity of concept use in psychology: Is the concept 'concept' a useful concept? J Theor Philos Psychol 31:73–89
Tallis RF (2011) Aping mankind: neuromania, Darwinitis, and the misrepresentation of humanity. Acumen, Durham
Tinbergen N (1963) On aims and methods of ethology. Zeitschirift für Tierpsychologie 20:410–413
Tomasello M (1995) Joint attention as social cognition. In: Moore C, Dunham P (eds) Joint attention: its origins and role in development. Erlbaum, Hillsdale, pp 103–130
Tomasello M (2006) Why don't apes point? In: Enfield NJ, Levinson SC (eds) Roots of human sociality: culture, cognition and interaction. Berg, Oxford, pp 506–524
Tomasello M (2008) Origins of human communication. MIT Press, Cambridge
Tomasello M (2014) A natural history of human thinking. Harvard University Press, Cambridge, MA
Tomasello M, Carpenter M (2007) Shared intentionality. Dev Sci 10:121–125

Tomasello M, Carpenter M, Call J, Behne T, Moll H (2005) Understanding and sharing intentions: the origins of cultural cognition. Behav Brain Sci 28:675–735

Tomasello M, Carpenter M, Liszkowski U (2007) A new look at infant pointing. Child Dev 78:705–722

Tooby J, Cosmides L (1992) The psychological foundations of culture. In: Barkow J, Cosmides L, Tooby J (eds) The adapted mind: evolutionary psychology and the generation of culture. Oxford University Press, New York, pp 19–136

Trevarthen C (1979) Communication and cooperation in early infancy: a description of primary intersubjectivity. In: Bullowa M (ed) Before speech: the beginning of interpersonal communication. Cambridge University Press, Cambridge, pp 321–347

Trevarthen C (2011) What is it like to be a person who knows nothing? Defining the active intersubjective mind of a newborn human being. Infant Child Dev 20:119–135

Trevarthen C, Hubley P (1978) Secondary intersubjectivity: confidence, confiding, and acts of meaning in the first year. In: Lock A (ed) Action, gesture, and symbol: the emergence of language. Academic Press, London, pp 183–229

Wereha TJ, Racine TP (2012) Evolution, development, and human social cognition. Rev Philos Psychol 3:559–579

Whiten A, Erdal D (2012) The human socio-cognitive niche and its evolutionary origins. Philos Trans R Soc B 367:2119–2129

Wilkins D (2003) Why pointing with the index finger is not a universal (in sociocultural and semiotic terms). In: Kita S (ed) Pointing: where language, culture, and cognition meet. Erlbaum, Mahwah, pp 171–216

Wynn K (1998) An evolved capacity for number. In: Cummins D, Allen C (eds) The evolution of mind. Oxford University Press, New York, pp 107–126

Describing Mental States: From Brain Science to a Science of Mind Reading

Shoji Nagataki

Abstract The rise of cognitive science is, without doubt, promoting research on the mind since the middle of the twentieth century. In particular, by the advancement of observation instruments that makes an elaborate scan of brain states possible, it contributes to explicate how the mind works. There remains, however, the difficult problem of specifying correspondence between the physical and mental states. What matters most in this context is that we seem to have only precarious ways to know the latter. Recently, some researchers have been trying to revive the apparently long-discarded method of introspection. In the present chapter, we discuss the difficulties of this method and suggest instead an alternative way, known as mind reading which has been acquired in the process of human evolution, to describe others' mental states. We consider what descriptions it can give of others' mental states, to what extent its reliability can be endorsed experimentally, and discuss its viability in the science of mind.

Keywords Mental states • Bodily behavior • Quasi-third-person perspective • Folk psychology

1 Introduction

More than one hundred years have passed since philosophy lost its privileged status in the study of mind; particularly of late, philosophers' influence on researchers in the field seems to have been decreasing. This is due mainly to the rise of a synthetic discipline, known as cognitive science, and especially to the recent development of brain science. Some brain scientists go so far as to argue that

S. Nagataki (✉)
Department of International Liberal Studies, Chukyo University, Nagoya, Japan
e-mail: nagataki@lets.chukyo-u.ac.jp

M. Pina and N. Gontier (eds.), *The Evolution of Social Communication in Primates*, Interdisciplinary Evolution Research 1, DOI: 10.1007/978-3-319-02669-5_8,

philosophical endeavor for a few thousand years has been utterly in vain. It is true that it would be a task of some interest to examine and rebut, in a direct way, that kind of charge against us, philosophers. We would like, however, to make an indirect move of answering it by searching for a cooperative relationship between philosophy and cognitive science, including brain science and psychology. If it is ever possible for philosophy to contribute to the science of mind, it would be by way of unraveling conceptual entanglements or proposing new methodological perspectives. In the present chapter, we will give a brief view of the history of the science of mind, identify some of the problems thereof, and re-examine the old method of introspection in order to suggest an alternative way in the study of mind.

2 Psychology as a Science: From the Inner to Behavior

It was just past the middle of the nineteenth century that psychology set its course toward a positivistic discipline, modeled after the modern natural sciences. The basis on which psychology as a science was built was furnished by prominent scholars, who have a background in physics and physiology, such as Gustav Fechner and Hermann von Helmholtz. Fechner made a detailed attempt to handle mental processes in a scientific way and developed an allegedly successful means with which to measure and quantify them. Helmholtz exploited the possibility of studying visionary and auditory processes by setting up and confirming hypotheses in a scientific manner. The foundation of a branch of modern psychology, known as experimental psychology, was laid in that very period (Titchener 1896).

Arguably, the biggest contributor to the establishment of experimental psychology was Wilhelm Wundt—though, in fact, he was a rather speculative scholar. According to him, consciousness is a mental complex, which can be decomposed into various mental elements. The task of psychologists is to explicate the combinatory laws between them. This position was called elementalism or constructivism, because what mattered for him was to clarify what those elements were like and how they were composed. The experimental method Wundt employed for that purpose was so-called introspection, which shares some similarities with those used by the subjectivistic tradition in modern philosophy. Typically, the subject in the laboratory was given stimuli from outside. She was required to observe introspectively what was going on as a response to them and to make a report to the experimenter. The strategy is well endorsed by our naïve intuition that no one is better versed in the states of mind than the one who bears it. In effect, psychology in its infancy tried to find its underpinning in the first-person descriptions provided by the subject's introspective reports (Giorgi 1970).

A serious suspicion arose; however, can the evidence gathered by introspection be considered as scientific data? Wundt tried to improve their reliability by training his subjects. Despite his efforts, the introspective reports from various laboratories came to be invalidated as not being sufficiently objective, because they produced different listings of the basic feelings and sensations that purportedly composed consciousness. People thought that the psychology based on introspection failed to attain the objectivity essential to a scientific discipline and thus ended in deadlock. Wundt became aware of it by himself (Wundt 1912: 3f.).

Prevented from employing introspective observations, scientifically oriented psychologists tried to construct an experimental paradigm that excluded as many subjective reports as possible. They denied "the validity of introspection as a tool" (Hebb 1980: 8) and looked for the support in the behavior of the subjects observable from outside, which promised to provide the evidential basis for psychology as a science. However, the amount of information gathered from observing behavior was pretty less than that from introspective reports; it was only a limited aspect of the human mind that psychological behaviorists could thus observe. Put another way, much of what we usually mean by mental phenomena and events fell out of the sphere of interest for them. The destined failure of behaviorism would have been unforeseeable to J. B. Watson, the prominent figure of this school, when he wrote: "Psychology as behavior will, after all, have to neglect but few of the really essential problems with which psychology as an introspective science now concerns itself" (Watson 1913: 177).

Though introspectionism and psychological behaviorism collapsed in their own ways, there is a sense in which each approach was taken over by the succeeding movement of cognitive science in the second half of the twentieth century; the study of unobservable mental processes was conducted anew using computer simulation, and much attention was focused on the brain states as a newly observable domain of the human behavior.

3 Brain as Scientists See It

Since the end of the last century, brain scientists have accumulated a lot of empirical data using powerful tools, which can probe the brain in greater detail. Supported by ample funds from the governments, brain science has been holding the limelight in the academic scene. There is a common belief among them that the study on brain activities will shed much light on the way the mind works. Has scanning the brain totally replaced the psychological study of mental activities? One might say that they are united by the assumption that the physicochemical states of the brain are identical to the mental states. Changes occurring in the mind are realized exhaustively in the brain. Some of them go so far as to say that psychology will be assimilated into brain science and thus walk silently away from the scene. Making an elaborate scan of the brain will totally replace the psychological study of mental activities, because "brain activity is a truly objective alternative" (Frith 2002: 374) for the latter.

Things cannot, however, be expected to run so smoothly. One of the main reasons is that a functional MRI, which plays a major role in the investigation of the encephalic activities, can at best detect the changes in the amount of cerebral vascular flow. Besides, its temporal resolution is not sufficient to pick up the changes of signals between small intervals.

From a philosophical point of view, what matters most is the relationship between the changes in the amount of vascular flow and the mental states. Can we explain all the properties of the latter by reducing them to the former? If we grasp some of the physical states of the brain, does it mean that we explain a distinctive feature of the mental states, known as qualia, belonging to the person who bears these states? Some argue that there is no difference between the brain and the machine in that both detect objects in the external world, encode, and process the information about them. Nevertheless, the machine does not seem to have "qualia," they say. This problem is known as "the explanatory gap" (Jackson 1982: 127–136). No doubt, serious thought should be given to whether the explanatory gap really exists, or whether it really means the existence of certain limitations to the brain studies.

Admittedly, one might well be accused of throwing the baby out with the bath water, if we bring up such a hard problem in order to suggest inadequacies of a newborn discipline. There remain, however, knotty problems within the research program of brain science, even if we focus on less foundational issues. As is widely known, while accumulating empirical data with regard to the brain states, we have enormous difficulty specifying the corresponding mental states, because there are various problems related to assembling the data: How can we translate the information about the subject's mental states into objective data? When the situation in which one gives instructions to and conduct relevant experiments on the subject has some sort of influences on her very mental states, how can we control them?

Let us think, for instance, of an experiment involving such instructions as "push the button when you feel pain in the index finger of your right hand" or "push the button when you see a red light." When instructions are as simple as these, one might think that the influences of the experimental setting on the subject's mental states are, if there are even any, relatively limited; the correspondence between the mental state of feeling a pain and the relevant brain state might be detected rather accurately. There would be virtually no brain scientists or psychologists who consider the act of pushing the button as involving an introspective report. Whether a certain mental state exists or not seems to be ascertainable by means of apparently objective procedures. However, what about the instruction that says "push the button when you feel anger?" Contrary to the case above, the experience of feeling anger may well be distorted by the procedure of introspection which immediately precedes the act of pushing the button. The intervention of introspection may render the feeling of anger quite different from the one we experience in the ordinary context.

What we have just said is pertinent to another problem, which is not to be underestimated both from a scientific and a philosophical viewpoint: It is very difficult to identify the intricate correspondence between the brain and mental states on the basis of empirical data. In order to establish the correspondence between two given series of events, one needs to specify in advance what events each series is composed of. This task is, of course, a prerequisite for interpreting the cerebral physical states. One might think that this is just what brain scientists are engaged in. The fact is that while various kinds of sophisticated equipment have been developed, we do not have reliable tools for specifying mental states yet. Most of the brain scientists are concentrating on the objective data from a third-person perspective, and they rarely incorporate into their data inventory what is admitted as legitimate from a first-person and a "quasi-third-person" perspective.[1] Even if a first-person report about mental states, such as pain, is employed, that is supposedly different from a subjective introspection. As a result, it cannot be said that they rightly treat the description of mental states which are amenable to observation and verification.

There does not seem to be any exact descriptions of subjects' (or others') mental states yet. However, if there is any way whatsoever of getting ones which can stand up to scientific purposes, it will find clues in the very mundane practice of understanding others, namely:

1. Having recourse to their first-person reports of thc "inncr" statcs.
2. Reading others' mental states in terms of their bodily behavior and facial expression.[2]

The latter is related to the quasi-third-person perspective. We will examine the possible applicability of these two practices in the following sections.

4 Reconstructing the Introspective Method: Neurophenomenology

Since we cannot observe immediately what the subject's mental states are like or what she is experiencing within herself, we are naturally inclined to think that we should have recourse to the first-person reports made by the subject in order to identify her mental states. Surely, the way scientists describe the brain states is far removed from our first-person experiences. As if endorsed by this fact, there have been appearing some proponents who argue that introspective reports on subjective experiences should be incorporated into cognitive science. This approach is most strikingly illustrated by neurophenomenology, which was pioneered by Francisco Varela and has been developed by many figures, such as Natalie Depraz, Evan Thompson, Antoine Lutz, and Shaun Gallagher.

[1] The term "third-person perspective" is often used in reference to the observation of brain states, which are themselves non-personal and objectively quantifiable. We will call the description of bodily behavior and facial expressions "quasi-third-person" because, though observable from outside, they are of a qualitative and personal character.

[2] One might suspect that there is no difference between our position and that of behaviorism if it means just an observation of the subject's behavior from the outside. However, what is meant here by the descriptions of "behavior" is not objective data gathered by behaviorists, such as that of rapidity of response to perceptual stimuli.

As is well known, there are some serious problems with the reliability of first-person reports, including the charges against the introspective method of Wundt.

1. Even given the same stimuli, there is variability in first-person reports by different subjects.
2. The very process of giving first-person reports may distort the subject's inner experiences.
3. There is a hard problem, referred to above as the explanatory gap, of how we should relate the first-person domain of subjectivity to the third-person domain, i.e., brain, body, and behavior.

According to neurophenomenologists, who propose to apply the methods of phenomenology to brain science, the first two problems can be coped with by training appropriately the subjects. One might argue against them by saying that they would follow the same fate as their predecessors, i.e., introspective psychologists, who tried to deal with the difficulties in a similar way. Neurophenomenologists respond to this objection by arguing that the subject would produce sophisticated data using the first-person methods that are derived from the continental phenomenology; the first-person methods would not just make slippery introspective descriptions stable, but also "facilitate the subjects becoming aware of previously inaccessible aspects of his/her mental life" (Lutz and Thompson 2003: 33). They consist of a systematic training of attention and the control of feelings, which purportedly enable the subject to amplify the sensitivity to moment-to-moment varying experiences and provide more accurate introspective reports. Particularly, pivotal to these methods is the procedure of epoché, which has, according to Depraz et al. (2000), three intertwining phases:

a. Suspension, which is to "bracket" our ordinary beliefs in order to give an unbiased description of inner experience.
b. Redirection of attention, which is to redirect the subject's attention from its habitual immersion in the experienced object toward the lived qualities of the experiencing process.
c. An attitude of receptivity, which is to let the experiencing process go without any active intervention from the subject herself.

By making a conscious effort to go through a cyclic process of these phases, one should be able to capture emerging categories or invariants within our experiences, they argue.

Granted, such a procedure may be helpful in making first-person reports more sophisticated. There is, however, a natural suspicion that it is, in fact, not a better method than Wundt's, in that it might distort the way inner experiences are really made. Neurophenomenologists reply that though phenomenological training can indeed modify experiences, this fact is not a limitation, but an advantage. Lutz and Thompson write: "Anyone who has acquired a new cognitive skill … can attest that experience is not fixed, but dynamic and plastic. First-person methods help to stabilize phenomenal aspects of this plasticity so that they can be translated into descriptive first-person reports" (Lutz and Thompson 2003: 39). Furthermore, they "help 'tune' experience, so that such translation and intersubjective corroboration can be made more precise and rigorous" (ibid.).

To the eyes of those who do not place a particular reliance on phenomenology, the alleged distinction between having the subject learn the phenomenological methods and training her in an appropriate way (whatever that way might be) may seem to be blurred; after all, what matters ultimately for both is that the subject reflects attentively on her own experience. Some argue that "with a few days of training and in the hands of a skilled questioner, most people can answer the DES question (i.e., give descriptions of inner experience) with ease and with substantial accuracy" (Hurlburt and Heavey 2001: 401).

There is another problem with neurophenomenology. They prohibit the subject from making reports in terms of folk psychology, because they think that its categories are unscientific (Gallagher and Sørensen 2006: 131). This prohibition would make their experimental design rather hard to implement, for not every subject is likely to have a good command of phenomenological terms and concepts.

In order to reconstruct the introspective method successfully, one needs, first of all, to secure the way to make the experimenter share the subject's experience. For that purpose, the experimenter has to inform the subject of her intentions correctly, be aware of the possible influence of her questions, and try to be sensitive to the subject's communicative intentions. This is what might be called a second-person perspective (Jack and Roepstorff 2002: 373) on the introspective method. Neurophenomenologists give such considerations as supportive of securing the process of introspective data collection. They say: "Without this reciprocal, empathetically grounded exchange, there is no refined first-person data to be had" (Lutz and Thompson 2003: 40). What is meant here is the importance of the subject's adequate understanding of the experiment, and the desirability of the ideal situation in which both the experimenter and the subject attain full comprehension of the other's intentions. In other words, the experimenter should more consciously regard her subject as a person, rather than just an owner of certain brain states.

When we typically approach the mental states of someone in her totality, there is a natural and primitive way to it, which is different from observing her brain states or relying on her first-person reports. Though similar to the second-person method in some ways, it is a kind of third-person approach in that the focus is centered on the subject's bodily behavior and facial expressions. As the observation of physicochemical states of the brain belong to an objective, scientific method, this approach can be called quasi-third-person one.

5 Folk Psychology Revisited

The ability to attribute mundane mental categories, such as belief and desire, to others—or even to ourselves—is known as "folk psychology." The body of such categories is also sometimes so called. It is a hard fact that such ability and concepts play a pivotal role in our mental life. We usually use a rather definite set of categories when ascribing certain mental states to others and thereby explain or predict their behavior—in this sense, folk psychology can be regarded as the ability of social communication. The ability of using them constitutes, as a product of human evolution,[3] one of our basic faculties to live on earth. Though, as mentioned above, neurophenomenologists disparage folk psychology as unscientific, not all theorists fall in with their view. Its status is a contentious issue since 1980s, in which three viewpoints are of special importance: the theory-theory, the rationality theory, and the simulation theory.

The theory–theory argues that folk psychology is a theory, or a body of general knowledge, which is supposed to be implicit in our mind. According to this view, mental states are essentially equivalent to physical ones in that both are explained and predicted by means of unobservable, theoretical entities (Carruthers and Smith 1996). Humans acquire, in some way or other, a theory of the mental realm, which is analogous to a causal conception of the physical world. They have, so the theory-theory argues, a theory of mind as an inferential system, which enables them to explain and predict others' behavior by taking a theoretical stance. There is a disagreement among its proponents about whether the theory is inherited genetically or developed during infancy.

The simulation theory insists that folk psychology is a practice of simulating others. Put simply, it presupposes that I am similar to others; given the same situations, others will have similar mental states and behavior as I do. In effect, we can understand and predict others' behavior by putting ourselves in their places. Those endorsing the simulation theory—among others, Gordon and Heal—introduced this idea as an alternative to the theory-theory in the mid-1980s (Gordon 1986). It is the latter that is dominant among psychologists.

Proponents of the rationality theory hold that humans use principles of rationality when they understand others' minds. According to this theory, they assume that others are rational in matters of belief, desire, and decision making. Dennett calls this kind of procedure "the intentional stance" (Dennett 1987). The intentional stance treats others whose behavior you want to predict as a rational agent with mental states.

---

[3] Much attention has been devoted to mind reading in philosophy of mind as well as in other areas such as psychology and primatology. Empirical studies show that non-human primates can also read others' minds in some degree. For example, apes like chimpanzees and gorillas "understand others as intentional" (Tomasello 2008: 49). However, communication of non-human primates differs from that of humans in a fundamental way. Only humans can participate in joint attentional interactions. Non-human primates can recognize intentionality in other individuals, but they—even great apes—cannot "participate in shared intentionality" (ibid. 331). While humans can detect others' intentions just from the moves or direction of their eyes, non-human primates do it from the direction of their bodies and heads (Emery 2000: 592). Furthermore, compared to non-human primates, humans have, in addition to the flexibility of the body, an elaborate system of facial musculature which realizes a high expressiveness. These physiological conditions also greatly contribute to an effective detection of others' intentions and to our participating in shared intentionality. Put another way, we have skills to understand the meaning of others' facial expressions and a variety of bodily behavior. It is these skills that provide a basis for social communication.

Of course, it is not so easy to sort out the problems facing those viewpoints, not least when considering that each designation has been used, more or less, in a variety of ways. However, we can point out some basic deficits. The first one is that, as Gallagher and Zahavi succinctly formulates it: "The theory–theory and the simulation theory both deny that it is possible to directly experience other minded creatures; this is supposedly why we need to rely on and employ either theoretical inferences or internal simulations. Both accounts consequently share the view that the minds of others are hidden" (Gallagher and Zahavi 2008: 183).

What about the rationality theory? According to Dennett, the intentional stance works like following: "first you decide to treat the object whose behavior is to be predicted as a rational agent; then you figure out what beliefs that agent ought to have, given its place in the world and its purpose" (Dennett 1987: 17). Dennett also does not seem to think that we can directly experience others' minds.

What is spotted and criticized here is, in essence, the Cartesian conception of others' minds, according to which a sharp dichotomy can be drawn between the mental and the physical. Genuine knowledge rests upon indubitable foundations within the mental realm, immediate access to which is possible only to its owner, the self. Others' minds are, therefore, beyond the scope of clear and distinctive comprehension.

The simulation theory is also challenged by an obvious, though not easy, problem. As Donald Davidson says: "If the mental states of others are known only through their behavioral and other outward manifestation, while this is not true of our own mental states, why should we think our own mental states are anything like those of others?" (Davidson 2001: 207) The point is that we cannot easily speak of similarity between two things, when they are known by different means, respectively.

In general, skeptical challenges, like that of Davidson, are hard to meet. If, however, it is knowledge by inference when he refers to that given "through behavioral and other outward manifestation," he may be neglecting another possible mode of understanding, i.e., intuition in folk psychological terms.

6 A Phenomenological Viewpoint

We read, in some way or other, others' minds from a quasi-third-person perspective. On the face of it, we can make clear others' intentions, thoughts, feelings, and dispositions to the extent necessary to ensure our smooth communication. This practice seems to be just an everyday experience we have, and indubitable unless people with some pathological conditions are concerned.

Part of this remarkable aspect of understanding others is clearly, though figuratively, suggested by Merleau-Ponty's discussion on the perception of faces. He pointed out a unique mode of understanding when we see human faces. According to him, "[a] face is a center of human expression, the transparent envelope of the attitudes and desires of others, the place of manifestation, the barely material support for a multitude of intentions" (Merleau-Ponty 2002: 167).

What he says of a face holds also true of the body, because we often use it in its entirety when expressing anger, joy, and the like. Others' bodies are not just physical objects, but intentional or meaningful ones, whose intentionality we can share with others with great immediateness.[4]

On the phenomenological conception, the mind comes to be visible in bodily expression; mental states are quasi-present in bodily behavior, thus intuitively given to us. So, there never exists a "ghost in the machine." As Merleau-Ponty puts it, "[w]e must abandon the fundamental prejudice according to which the psyche is that which is accessible only to myself and cannot be seen from outside" (Merleau-Ponty 1964a: 116). Merleau-Ponty also writes this. "Faced with an angry or threatening gestures, I have no need, in order to understand it, to recall the feelings which I myself experienced when I used these gestures on my own account. I know very little, from inside, of the mime of anger so that a decisive factor is missing for any association by resemblance or reasoning by analogy, and what is more, I do not see anger or a threatening attitude as a psychic fact hidden behind the gesture, I read anger in the gesture" (Merleau-Ponty 1962: 184).

In sum, we have the ability to "read" immediately the mental states of others in their bodily behavior, such as faces, gestures, and voices. The other's body is not just an automaton. Rather, it appears in front of us as a body with a mind: It has "a singular eloquence" (Merleau-Ponty 1964b: 169). He says: "What I perceive to begin with is a different 'sensibility' (Empfindbarkeit), and only subsequently a different man and different thought" (ibid. 168). In effect, what he is arguing for on the basis of his own interpretation of the later Husserl is that we can understand the other's mind intuitively. One might say that mental states are contagious to other people by means of the body (By way of illustration, we can name a phenomenon known as "affect contagion").

A few remarks should be made about the terminology of intuition. Some might find it misleading and divergent from its traditional usage. In the philosophical context, it usually means a mode of immediate understanding, or something immediately given to the conscious mind. Because of its alleged transparency, directness, and immediateness, this epistemological given sometimes assumes a privileged role in providing the foundations of empirical knowledge in the history of modern philosophy. As is well known, however, intuition as defined above has been attacked by many philosophers, especially by Charles S. Peirce.[5]

---

[4] Of course, Merleau-Ponty is not suggesting, which would be absurd, that what is going on in others' minds can be completely transparent. He is just referring to one aspect of multifaceted understanding of others.

[5] Peirce says that "[n]ow, it is plainly one thing to have an intuition and another to know intuitively that it is an intuition, and the question is whether these two things, distinguishable in thought, are, in fact, invariably connected, so that we can always intuitively distinguish between an intuition and a cognition determined by another. [...] There is no evidence that we have this faculty, except that we seem to feel that we have it" (Peirce 1992: 12). Bernstein succinctly reconstructed Peirce's argument against the traditional notion of intuition in his "Charles S. Peirce's Critique of Cartesianism" (Bernstein 2010).

Despite the fact that the concept seems no longer tenable as it was, we use this term only because of its connotation of epistemological directness. What we mean by intuition is a mode of understanding which seems to be given—consciously or unconsciously—immediately, though open to fallibility. It operates, at every stage of human development, as a basis of social interaction, and is always tested and revised through its own process. One might call it "detranscendentalized intuition."[6]

7 Intuition, Inference, and Learning

As mentioned above, there are, generally speaking, two modes of understanding others: intuitive and inferential. The intuitive one is illustrated by the way babies read off others' emotions. It is based on the ability to put oneself into the relationship of emotive resonance. The inferential, or discursive, mode is typically exhibited by the case in which we understand verbal behavior of others. Verbal behavior can be seen as an object of inferential understanding, because it involves an empirically acquired process in which one infers from a series of sounds or characters to its meaning.

These are not exclusive with each other; nor is there any definite demarcation line between them. Others' mental states are sometimes given in a mixed way. However, intuitively given, sympathetic resonance of embodiment provides the basis to the understanding of others' mental life and thus underpinning higher-level, inferential communication with them. Of course, it does not mean that the inscrutability or impenetrability of the mental states of others is completely dispelled. Other minds are neither completely observable nor unobservable. Sometimes we can understand them well, and sometimes we cannot. It is our day-to-day experiences. Then, reading ability is, in part, innately present and, in part, acquired by learning in a social and cultural context.

There are typically two reasons why we sometimes fail to understand inner states of others. The first one is the case in which her ability to read behavioral and other outward manifestation is, for some reason or other, impaired or not fully developed. For instance, it is widely known that autistics find it difficult to read emotions from the behavior of others, even though the etiology of the disorder is not well understood.[7] The second is because there is a realm of mind which does not tend to manifest itself in bodily behavior and facial expressions. This is best illustrated by thoughts expressed in the form of proposition. We can gain access to the thoughts of others almost exclusively by their linguistic behavior. Lack of the subtle ability to use language may lead to a serious failure in communication.

---

6 Some phenomenologists might feel aversion to that appellation.

7 To cope with that problem, some autistics are known to adopt an inferential strategy; they sometimes employ the method of cataloging typical patterns of others' behavior and mapping them onto the class of typical modes of emotion.

Table 1 Modified version of the state trait anger expression inventory (STAXI)

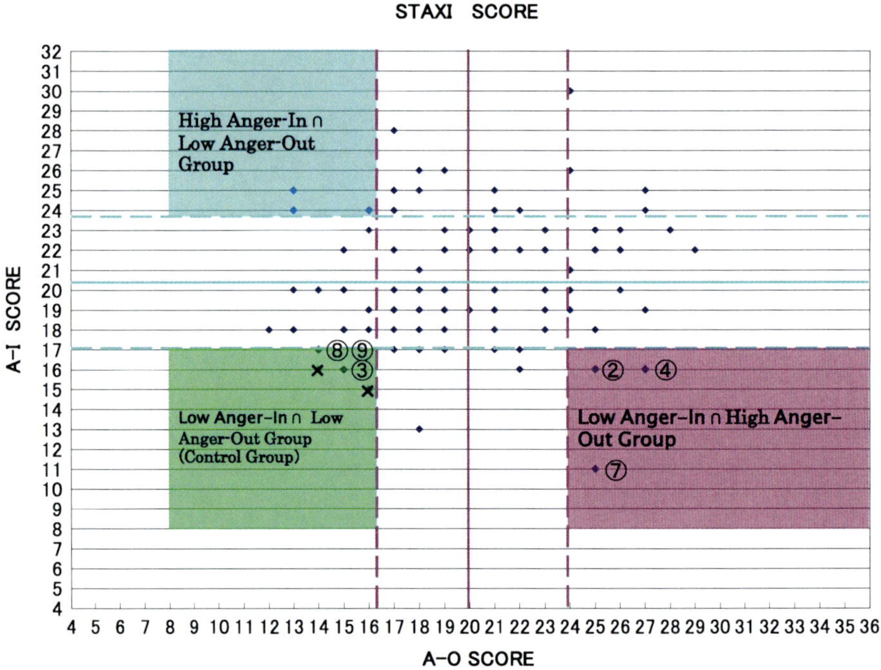

What is intuitively given by behavior and facial expression of others is typically instantiated by their present emotions and moods. However, we sometimes read off various meanings on those manifestations by way of inference, which ability can be honed and promoted by empirical learning. The experiment below shows that understanding the characteristic dispositions of others depends, to a certain extent, on such learning.

We evaluated the reading ability of occupational therapists (OTs) by an experiment in which OTs were required to identify persons with a particular type of character. We chose OTs in psychiatry as our subjects, because they are expected to be adept at understanding subtle mental condition of patients.

To begin with, we administered a personality test, a modified version of the State Trait Anger Expression Inventory (STAXI), to 104 mentally healthy people. Based on that result (See Table 1), we got two different groups by drawing three representative samples from each category:

1. "Low Anger-In and High Anger-Out" group
2. "Low Anger-In and Low Anger-Out" group

"Anger-Out" refers to the tendency of expressing anger toward other persons or objects in the environment, whereas "Anger-In" designates that of holding in or suppressing angry feelings. Thus, the latter group consists of those who do not have much anger or whose anger does not last long.

We asked six people of both groups to work on leather craft and took a video of those scenes. They had not had any experience of that kind before. Then, we showed that video to some 30 subjects who had been working as OTs for more than 5 years and asked them to guess who belonged to which group. The subjects were required to describe what aspects of bodily behavior and facial expressions they drew on in their determination. Particularly, when they marked someone as belonging to the group of Anger-Out type, they were also required to guess what her mental states were. As a control group, we asked some 60 undergraduate students to guess the same things as required for OTs.

The result was that the accuracy rate in the OT group was significantly higher than random chance. The value in the control group of students was significantly lower than that of the OT group. The experiment also showed that many of the subjects took notice of common and distinct elements of bodily behavior and facial expressions. In addition, they made highly similar guesses about the inner states of those working on leather craft when those elements were observed. Let us cite a few examples. When behavior, such as some strain in her movements, resting her elbows on the table, and pouting her lips, was exhibited, the OTs typically guessed that she felt irritable, or that she felt like it was too much like work. Based on that guess, they inferred that she had less emotional control and thus classified her correctly into "Anger-Out" group. The classification under it also happened when they felt a slapdash air about her performance. When asked to verbalize the slapdashness they felt, they gave descriptions after some thought like the following: A friction sound was heard when she whipstitched with a lace; she slightly hit the dye bottle with another one; she hit her hand to the desk when she took scissors. It should be noted that they were bodily mimicking the workers' behavior when describing what they felt.

When calm behavior was exhibited almost on a steady basis, the great majority of the subjects—almost all the students and about half the OTs—classified that person into "Low Anger-Out and Low Anger-In" group. The remaining OTs, however, identified him as a member of "Anger-Out" group, because they noticed something a little aggressive in his behavior of setting up the working environment. They interpreted this as a sign of wanting to do everything just as he pleased and thought that he might express anger over unfavorable things.

8 Concluding Remarks

Our experiment suggests that we can detect, to a certain extent, dispositions which are opaque or hidden in one's character and that the skills of experienced OTs typically include the ability to evaluate the personality of others only by limited aspects of their behavior. True, one might say that a cognition of this kind is uncertain and uninformative, thus not providing substantial knowledge of others' inner states. However, it gives not only to OTs, but to every of us, valuable clues to understand others in varying degrees. We all engage, consciously or unconsciously, ourselves in this practice.

It is noteworthy that experienced OTs decided on which category each worker belonged to, by intuitive rather than inferential understanding. They discerned the workers' dispositions rather immediately and intuitively, even though they could later point to an evidential basis discursively. As we already initiated, intuitive, and inferential understandings are not exclusive. They operate in a parallel and complementary way at every scene of our social life. Furthermore, no definite line of demarcation can be drawn between them. Even if a vague boundary apparently exists, it can be dislocated by empirical learning. We can extend the span of intuitive understanding as we accumulate experience in inferential one.

Also, remarkable is the fact that many of the OTs took notice of much the same features of behavioral and facial expressions and thereby cognized much the same inner states, which were pertinent to "Anger-Out," within them. It should come as no surprise that the way of manifestation as well as the content of one's character is formed in the social context. However, the very existence of such quasi-uniformity, which can be easily overlooked by those who are socially less experienced, suggests that information on the correspondence between distinctive manifestations and less-noticeable dispositions of character can be of great importance to us. As experienced OTs acquire such kind of skills, the experimenters, if trained, could reach a common understanding of some aspects of the subject's mental states even without recourse to her introspective report.

The alternative way to describe mental states will be convenient especially for the investigation of rather intricate mental processes, like those of emotions. Emotions are likely to be more susceptible to the influences of introspection than the usual experiences of perception.

One might suspect how such experiment could be implemented within the framework of brain science: What kind of scanning device should we use to carry out our alternative? Evidently, such present technologies as fMRI, MEG, or PET are not suitable for our purpose since the subject's head or body needs to be fixed for the successful measurement. While ERP (event-related brain potential) and near-infrared spectroscopy (NIRS) allow greater degree of freedom to the bodily movement of the subject, they have only limited measurability. At present, we have to admit that further development of measuring technology is needed in order to incorporate our program fully into brain science.

In the present chapter, we sorted out several problems which bedevil brain science, examined the new methods of introspection based on phenomenology, and suggested another way of experimenting from a quasi-third-person perspective. What we insist is not that we should abandon the first-person in favor of the quasi-third-person method. Rather, we should adopt suitable methods and approaches depending on the subject of the study of mind. It is necessary to enrich and refine our inventory of methods, and to connect it with scientific investigations in a prolific way. We would be glad if this little piece could contribute that way.

References

Bernstein RJ (2010) Charles S. Peirce's critique of cartesianism. In: The pragmatic turn. Polity Press, Cambridge

Carruthers P, Smith PK (eds) (1996) Theories of theories of mind. Cambridge University Press, Cambridge

Davidson D (2001) Subjective, intersubjective, objective. Clarendon Press, Oxford

Dennett DC (1987) The intentional stance. MIT Press, Cambridge

Depraz N, Francisco JV, Vermersch P (2000) The gesture of awareness: an account of its structural dynamics. In: Velmans M (ed) Investigating phenomenal consciousness. John Benjamins Publishing Company, Amsterdam

Emery NJ (2000) The eyes have it: the neuroethology, function and evolution of social gaze. Neurosci Biobehav Rev 24:581–604. doi:10.1016/S0149-7634(00)00025-7

Frith C (2002) How can we share experiences? Trends Cogn Sci 6(9):374

Gallagher S, Sørensen JB (2006) Experimenting with phenomenology. Conscious Cogn 15: 119–134. doi:10.1016/j.concog.2005.03.002

Gallagher S, Zahavi D (2008) The phenomenological mind: an introduction to philosophy of mind and cognitive science, 2nd edn. Routledge, London

Giorgi A (1970) Psychology as a human science. Harper & Row Publishers Inc, New York

Gordon RM (1986) Folk psychology as simulation. Mind Lang 1(2):158–171. doi:10.1111/j.1468-0017.1986.tb00324.x

Hebb DO (1980) Essay on mind. Lawrence Erlbaum Associates Publishers, Hillsdale

Hurlburt RT, Heavey CL (2001) Telling what we know: describing inner experience. Trends Cogn Sci 5(9):400–403. doi:10.1016/S1364-6613(00)01724-1

Jack SI, Roepstorff A (2002) The 'measurement problem' for experience: damaging flaw or intriguing puzzle? Trends Cogn Sci 6(9):372–374

Jackson F (1982) Epiphenomenal qualia. Phil Quart 32:127–136. doi:10.2307/2960077

Lutz A, Thompson E (2003) Neurophenomenology: integrating subjective experience and brain dynamics in the neuroscience of consciousness. J Conscious Stud 10(9–10):31–52

Merleau-Ponty M (1962) Phenomenology of perception (trans: Smith C). Routledge & Kegan Paul Ltd, London

Merleau-Ponty M (1964a) The child's relation with others. In: Edie JM (ed) The primacy of perception: and other essays on phenomenological psychology, the philosophy of art, history and politics (trans: Cobb W). North Western University Press, Evanston

Merleau-Ponty M (1964b) Signs (trans: McCleary RC). Northwestern University Press, Evanston

Merleau-Ponty M (2002) The structure of behavior (trans: Fisher AL). Duquesne University Press, Pittsburgh

Peirce CS (1992) The essential Peirce: selected philosophical writings, vol. 1: 1867–1893. In: Houser N, Kloesel C (ed). Indiana University Press, Bloomington

Titchener EB (1896) An outline of psychology. The Macmillan Company, New York

Tomasello M (2008) Origins of human communication. The MIT Press, Cambridge, London

Watson JB (1913) Psychology as the behaviorist views it. Psychol Rev 20(2):158–177. doi:10.1037/h0074428

Wundt W (1912) Elemente der Völkerpsychologie. Alfred Kroner Verlag, Stuttgart

Part III
Evolutionary Transitions from Social Communication Systems to Language

Bodily Mimesis and the Transition to Speech

Jordan Zlatev

Abstract The chapter reviews evidence for the bodily mimesis hypothesis, which states that the evolution of language was preceded by an adaptation for improved volitional control of the body, giving our ancestors advantages in the domains of imitation, empathy, and gestural communication. Much of this evidence is also shared by other gesture-first theories of language origins, but they face the problem of explaining the "switch" from a gestural (proto) language to a spoken one. The bodily mimesis hypothesis fares better with this objection, since it (a) emphasizes the non-conventionality and non-systematicity of bodily mimetic signaling, (b) posits a long biocultural spiral of conventionalization and adaptation for speech, and (c) insists that the transition to speech should be seen as only partial. Following Brown (2012), a cognitive–semiotic explanation can further be given as to why speech has eventually taken on increasingly higher communicative load: Vocalization is intrinsically less capable of iconic representation, and given a multimodal gestural–vocal communicative signal, the vocal element is bound to eventually take on the role of symbolic representation, involving higher levels of conventionality and systematicity.

Keywords Cognitive semiotics • Conventionalization • Gesture • Iconicity • Intersubjectivity • Mimesis • Multimodality • Speech • Symbols

1 Introduction

It is now generally accepted that the human capacity to imitate bodily actions far outstrips that of other animals, including apes (Custance et al. 1995; Call 2001). Another capacity, closely related to imitation, in which human beings excel, is

J. Zlatev (✉)
Centre for Languages and Literature, Department of Cognitive Semiotics,
Lund University, Lund, Sweden
e-mail: jordan.zlatev@ling.lu.se

M. Pina and N. Gontier (eds.), *The Evolution of Social Communication in Primates*, Interdisciplinary Evolution Research 1, DOI: 10.1007/978-3-319-02669-5_9,

intersubjectivity or empathy (Hurely and Chater 2005; Zlatev et al. 2008). Jointly, imitation and empathy function as springboards for the development of uniquely human capacities for intentional communication in childhood (Piaget 1962; Tomasello 1999; Zlatev 2013). Considerations such as these have given rise to the *bodily mimesis hypothesis*, stating that an adaptation for improved volitional control of the body gave our ancestors advantages in the domains of imitation, empathy, and (gestural) intentional communication. It is assumed that this paved the way for the evolution of language, with no other biological adaptations being required apart from improved vocal control (Donald 1991, 2001; Zlatev 2008a, b).

The first aim of this chapter is to spell out this hypothesis in some more detail and to sum up the empirical evidence in its favor. To some degree, both the hypothesis and the evidence for it overlap with so-called gesture-first theories of language origins (Hewes 1973; Corbalis 2002, 2003; Arbib 2003, 2005), but there are some important differences, making bodily mimesis less vulnerable to the most common counterargument to gesture-first theories: Why are all current languages of hearing people predominantly spoken rather than gestural, like the signed languages of deaf communities?

The second aim of the chapter is therefore to elaborate on the possible transition from a predominantly mimetic form of communication to a predominantly symbolic one, using the vocal channel. The hurdle has appeared as so great for conceptual as well as empirical reasons, i.e., treating human language as a purely symbolic ("arbitrary") code. It will be argued that the explanatory task appears differently, and as more manageable, if we rather acknowledge the inherently multimodal nature of linguistic communication, with differential roles for speech and gesture, and furthermore see speech itself not as completely arbitrary, but with a considerable degree of sound symbolism (Ahlner and Zlatev 2010).

2 Bodily Mimesis

Donald (1991) initially proposed that bodily mimesis served a crucial role in evolution in his general theory of human cognitive–semiotic origins, defining mimesis as "the ability to produce conscious, self-initiated, representational acts that are intentional but not linguistic" (ibid: 168). In another characterization, he explicates that "it manifests in pantomime, imitation, gesturing, shared attention, ritualized behaviors, and many games. It is also the basis of skill rehearsal, in which a previous act is mimed, over and over, to improve it" (Donald 2001: 240). Crucially, it allowed a qualitatively new form of culture to emerge: "Mimesis served as a mode of cultural expression and solidified a group mentality, creating a cultural style that can still be recognized as typically human" (ibid: 261). Thus, mimesis is manifested in the evolution of the following cognitive–semiotic capacities or functions, in ways that are uniquely human.

(1) Functions of bodily mimesis are as follows:

- *Learning*: through imitation and teaching
- *Skill*: through conscious rehearsal
- *Imagination and planning*: through re-enactment
- *Communication*: through pantomime and other kinds of gesture
- *Culture*: through shared practices, concepts, and beliefs.

What has made the bodily mimesis hypothesis attractive is that evidence from a number of different sources can be said to converge toward it. Donald (1991) appealed to the paleoanthropology, neuroscience, and gesture studies of his day. In addition, evidence from human ontogeny (Zlatev 2007), comparative psychology, "mirror neuron" neuroscience (Zlatev 2008b), and experimental semiotics (Brown 2012) has been argued to support the hypothesis as well. What follows is an updated summary of this supportive evidence.

2.1 Paleoanthropology

The hominin species with which bodily mimesis is most strongly associated is *Homo ergaster*, appearing about 1.8 mya in Africa, and the Asian version of this species, *Homo erectus*, attested between 1.5 and 0.1 mya: "the first universally accepted member of our own genus" (Fitch 2010: 265). The body size of *H. erectus* had increased at least twice compared to the earlier *australopithecines* and the brain size even more, to almost modern proportions. The shape of the body had changed as well, giving rise to complete bipedalism, with the capacity for efficient long-distance running—highly adaptive for hunting and/or scouting (Cela-Conde and Ayala 2007). In terms of technology, there was a qualitative shift in style and complexity from older Oldowan to the larger symmetrical hand axes of Achulean technology, requiring considerable skill, practice, and pedagogy. These biological and cultural adaptations, including the domestication of fire, from at least 400,000 mya (Weiner et al. 1998), made migration to most parts of Eurasia possible.

Yet, it is not clear whether all these achievements coincided with the evolution of the vocal control necessary for speech. One possible marker of such control in the fossil record is an extended thoracic canal, needed for controlling breathing during speech (or singing). Based on earlier evidence, it was concluded that *H. erectus* still had a thoracic canal in the range of *australopithecines* (MacLarnon and Hewitt 1999). This has been contested on the basis of more recent and extensive evidence, suggesting that the species may have had a thoracic canal in the range of modern humans (Gómez-Olivencia et al. 2007). The debate continues, but it remains that while it is clear that *H. erectus* must have had improved volitional control of the body and unprecedented level of culture, there is no firm evidence for the simultaneous evolution of speech. Bodily mimesis thus stands as the likely basis for achievements that are both remarkable, compared to those of earlier hominins, and yet limited compared to those of *Homo sapiens*.

2.2 Mirror Neuron Systems

Gestural/bodily theories of language origins received a major boost with the discovery of so-called mirror neurons, responding both to one's own and to others' hand movements, in the 1990s. One argument for their relevance for language was that they were initially found in area F5 in the premotor cortex of the macaque brain, which appears to be homologous to the left inferior frontal gyrus of the human brain, corresponding to the well-known "Broca's area" (Arbib 2003, 2005). Extensive studies, using various imaging methods, confirmed that BA 44 and 45 (≈Broca's area) and BA 22, 39, 40 (≈Wernicke's area) overlap extensively with the (extended) human "mirror neuron system" (MNS) and are activated in tasks involving action recognition, imitation, pantomime, and iconic gestures (Iacoboni 2008).

Early enthusiasm that this would be sufficient to explain both the neural mechanisms of language and its evolution (Rizzolatti and Arbib 1998) was, however, rather premature. Admittedly, there is a major gap between the "parity" of action recognition and that of shared symbolic meanings (Hurford 2004). In response to such criticism, Arbib (2003, 2005) proposed a more elaborated scenario for how the MNS was gradually extended over evolution from serving the function of action recognition (in monkeys), to "simple imitation" (in apes) and to "complex imitation" and pantomime in early *Homo*, to "protosign" and eventually to speech. Apart from the stage of "protosign," consisting of "elements for the formation of compounds which can be paired with meanings in a more or less arbitrary fashion" (Arbib 2003: 195), the model is consistent with the bodily mimesis hypothesis (Zlatev 2008b). For example, BA 4 and BA 6 are not credited with being part of the human MNS, but they have been shown to activate during the perception and production of meaningless syllables (Wilson et al. 2004), and BA 44 and 45 likewise are differentially associated with speech. All this is consistent with the hypothesis that speech was only gradually recruited for intentional communication, "atop" older systems serving action, imitation, and gesture.

2.3 Comparative Psychology

One of the primary types of evidence used by Hewes (1973) in arguing for a gestural origin of language was the recent for the time findings of relative success in "ape language" studies using a simple form of American Sign Language (ASL). The large controversies that surrounded these studies have made it clear that apes indeed have highly limited abilities to use manual signs compositionally and "declaratively" (i.e., to provide information rather than to request an action), but also that they are capable of learning manual and other forms of non-vocal signs and to use these flexibly, with close attention to the addressee's state of attention (cf. Zlatev 2008a). These conclusions have also been confirmed by a number of naturalistic studies of spontaneous bodily communication in great apes, living both in the wild and in captivity (cf. Call and Tomasello 2007). Tomasello (2008: 54) summarizes the contrast between the vocal and gestural modalities in fairly categorical terms: "... primate gestures are individually learned and flexibly produced communicative acts. [...] vocal displays are mostly unlearned, genetically fixed, emotionally urgent, involuntary, and inflexible. [...] They are broadcast mostly indiscriminately." Since extant great apes are our best approximate model for the last common ancestor (LCA) of hominins and apes, it is reasonable that the LCA had similar skills and that gesture/bodily mimesis was therefore within its "zone of proximal evolution" (Donald 2001), unlike speech. While several researchers have argued that such an appraisal underestimates chimpanzee vocal capacities and their communicative functions (Slocombe and Zuberbuehler 2005), it seems clear that there is at least a quantitative if not qualitative difference between the flexibility, volitional control, and referentiality of ape gestures as opposed to vocalizations (Pika 2008). Thus again, producing signs with the body was more "at hand" than with the voice.

Looked from the other direction, what are the main differences between ape and human cognition, leaving language aside? It has been popular for some time to downplay such differences (cf. Tallis 2011), but in a recent extensive review article, Vaesen (2012) examines the evidence from nine cognitive domains (including language) related to tool production and use and concludes that "striking differences between humans and great apes stand firm in eight out of nine of these domains" (ibid: 203). The seven non-linguistic domains in which human capacities clearly exceed those of apes according to this review are as follows: (a) hand–eye coordination, (b) causal reasoning, (c) functional representations (e.g., for tools), (d) executive control (e.g., inhibition and planning), (e) social learning (e.g., imitation), (f) teaching, and (g) social intelligence (e.g., passing false-belief tasks). Rather than considering one of these as the crucial difference, Vaesen concludes that "no individual cognitive trait can be singled out as the key trait differentiating humans from other animals" (ibid: 203). This claim is quite in line with the bodily mimesis hypothesis, since mimesis is polyfunctional. Indeed, there is a close correspondence between the functions associated with bodily mimesis under (1) and the features in Vaesen's list given above, especially when the latter are grouped as (a) motoric, (b–d) cognitive, and (e–g) social–cognitive.

In such a manner, the bodily mimesis hypothesis of the origins of human uniqueness can help generalize over a number of findings from comparative psychology.

2.4 Gestures and Ontogeny

Several decades of extensive research on the spontaneous gestures of adults and their development in children have shown that gestures are ubiquitous in all human cultures and that they align temporally and semantically with speech, at least in adult language use (Kendon 2004; McNeill 2005). The explanations of these findings, however, differ. While McNeill (1985, 2005, 2012) considers speech and gesture (production) to be two parts of a single system, others point out that there are good reasons to regard them as two closely interacting, but distinct systems. The resolution of this controversy has direct implications for evolutionary hypotheses.

It is now generally accepted that gestures share semantic properties with what is being said and that speakers of different languages gesture somewhat differently, in ways that can be related to the semantics of the respective languages (Kita and Özyürek 2003). However, speakers also use gestures to represent objects and events iconically in ways that go beyond what is said and in ways that are similar across languages (Zlatev and Andrén 2009). This is consistent with a model of "the two qualitatively different representations [which] are adjusted with respect to each other and co-evolve" (Kita and Özyürek 2003: 30). Careful analyses have also shown that co-speech gestures synchronize with features of the interaction as a whole, including the responses of the addressee (Sikveland and Ogden 2012) and are thus not automatically tied to speech production itself.

The developmental evidence also appears to support an analysis in terms of two interacting systems rather than a completely inseparable speech–gesture bond of the kind that McNeill envisages. On the one hand, there is general agreement that there is close interaction between gesture and speech in language development (Volterra et al. 2005; Goldin-Meadow 1998; Andrén 2010). Still, it appears that both pointing and iconic gestures emerge prior to speech, at around 9–12 months, and play an essential role for the development of language (Bates et al. 1979; Liszkowski et al. 2012; Lock and Zukow-Goldring 2012). Speech and gesture become gradually integrated in ontogeny, with at least some analyses showing "a gradual specialization from unimodal forms of communication, less demanding in cognitive, social and semiotic terms, to multimodal patterns involving the coordination of specific gestures and vocalizations" (Murillo and Belinchón 2012: 31).

Of course, such apparent gestural primacy in ontogeny is not a strong argument for a corresponding primacy in evolution, since the old principle of "recapitulation" cannot be accepted without prior justification. Still, if gesture plays a scaffolding role for language in development, it is reasonable to suppose that it played an analogous role in evolution as well, since in both ontogeny and phylogeny, (a) bodily movement comes under volitional control earlier than vocalization, as argued in Sect. 2.3, and (b) gesture affords a greater degree of *iconicity* than speech.

The last point, i.e., the iconic (resemblance-based) relation between at least some gestures and their meanings, has been a rather controversial topic. Intuitively, communicating with the whole body should be easier than only with the voice when lacking a common language, since this is indeed what people do when they need to communicate in such cases. On the other hand, many gestures are conventionalized, and some researchers have even argued that iconicity plays hardly any role at all in gestural communication (Streek 2009). This controversy can be in part resolved by turning to semiotics, where the topic of iconicity has been thoroughly investigated.

2.5 Semiotic Analysis and Experiments

Semiotics is the interdisciplinary field investigating commonalities and differences between different communicative systems, such as visual representations, speech, and gestures (in both spontaneous and artistic forms), and their dependence on and interaction with cognitive capacities including perception, movement, and consciousness (cf. Sonesson 1989). While traditional semiotics was based almost entirely on a form of conceptual analysis and was often quite speculative, modern approaches of experimental (Galantucci and Garrod 2010) and cognitive semiotics (Zlatev 2012) are considerably more empirical. It is the combination of conceptual (intuition-based) analysis and experimental validation that makes semiotics so useful in addressing controversial topics such as the iconicity of gestures.

First of all, it is important to recognize that iconicity and conventionality (as well as the third type of expression–meaning relation known as indexicality, which is contiguity-based) do not stand in a mutually exclusive relation, as pointed out by several of the classics of the field:

> One of the most important features of Peirce's semiotic classification is … that the difference between the three basic classes of signs is merely a difference in relative hierarchy. It is not the presence or absence of similarity or contiguity between the signans and signatum, nor the … habitual connection between both constituents which underlies the division of signs into icons, indices and symbols, but merely *the predominance of one of these factors over the others*. (Jakobson 1965: 26, my emphasis)

Furthermore, in his defense of the iconicity of pictures, Sonesson established a useful conceptual distinction between *primary iconicity*, where "the perception of an iconic ground obtaining between two things is one of the reasons for positing the existence of a sign function joining two things together as expression and content," and *secondary iconicity*: "the knowledge about the existence of a sign function between two things […] is one of the reasons for the perception of an iconic ground between these same things" (Sonesson 1997: 741). The iconicity of a typical picture (Fig. 1a) is primary, whereas that of a more abstract representation such as that shown in Fig. 1b is secondary: First, when we are told that this represents, e.g., a man in a telephone booth playing a trombone, we can see the resemblance.

The question concerning gestures can now be reformulated along the lines of Jakobson (1965): Does iconicity "predominate" over conventionality at least in some cases and in the style of Sonesson (1997): is it of the primary kind? A recent experimental study by Fay et al. (2013) suggests positive answers to both questions. The researchers asked pairs of participants to play a game in which a "director" had to communicate 24 different concepts, divided in the categories emotion, action, and object, to a "matcher," without using language, by one of three means: vocalization, gesture, or a combination of both. The results showed that in all cases, matching was above chance and that for the emotion class, the vocalization-only group managed fairly well (ca. 70 %). However, (pantomimic) gestures with or without vocalization had a clear advantage, with success rates approaching ceiling level. The authors conclude that "gesture outperforms non-linguistic vocalization because it

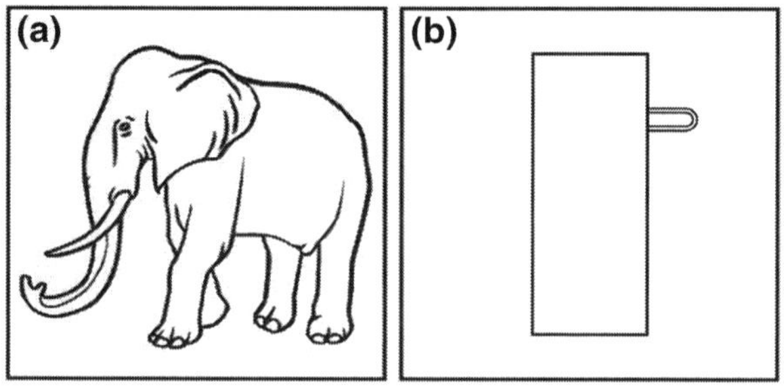

Fig. 1 An example of (**a**) primary versus (**b**) secondary iconicity (borrowed from Ahlner and Zlatev 2010)

lends itself more naturally to the production of motivated signs" (ibid: 1). Since the game was played a number of times by each pair, a degree of simplification and conventionalization of the gestures occurred, but in no point did they become "arbitrary," or their iconicity purely secondary. On the other hand, the success rates for vocalization-only increased considerably with use, suggesting that conventionalization played a stronger role for successful communication in that medium. This leads to an important conclusion: While both the bodily/gestural and vocal modalities can be used for signs that are fully conventionalized, to the extent of losing all traces of iconicity and indexicality and thus becoming "arbitrary," the bodily/gestural modality is *intrinsically* more suited for motivated signs, while the vocal modality is less so. This difference is crucial to explain both why bodily mimesis and gesture are advantageous for establishing a sign system initially and why with time there will be a shift toward the vocal modality, i.e., speech, as argued below.

3 But Why Speech?

The different kinds of evidence discussed in the previous section are supportive not only of the bodily mimesis hypothesis, but also of gesture-first theories of language evolution in general. The proposal of a "gestural stage" in language evolution has always been found appealing to some, but objectionable to others who have theorized about language origins. The major objection can be formulated tersely: Why speech? Even authors who are very well aware of the importance of gesture in human communication find this objection (nearly) "fatal" or "insuperable":

> The gestural theory has one nearly fatal flaw. Its sticking point has always been the switch that would have been needed to move from a visual language to an audible language. (Burling 2005: 123)
>
> Several different lines of evidence, then, can be added up to support the hypothesis that the first step in the evolution towards linguistic expression was taken with the employment of visible action, or gesture, for referential expression. Yet, as has often been pointed out, this seemingly attractive hypothesis faces [...] an insuperable problem: Languages are overwhelmingly *spoken*. (Kendon 2008: 12)

In his critical review of "gestural protolanguage theories," Fitch (2010, Chap. 13) argues convincingly that appealing to ecological factors is not sufficient to explain the transition to speech, since "each posited advantage can be paired with a similar selective force that would oppose them" (ibid: 443). Communicating in the dark may be beneficial, but silent gesturing is clearly safer in an environment of extensive predation. Speech may be "freeing the hands" for other purposes while communicating, but then it "burdens the mouth," making communication somewhat difficult and even dangerous during a common communal activity: eating. Analogously, vocal communication may free visual attention, but it burdens auditory attention, and furthermore, in all cultures, linguistic communication is predominantly conducted "face to face," involving multimodal perception (Kendon 2004).

As Fitch points out, Hewes (1973) did not appeal to such factors but rather to what he then thought to be certain linguistic disadvantages of signed languages compared to speech: having a limited vocabulary, lacking duality of patterning, i.e., the equivalent of phonemes, and being slower. However, such claims have been disproved since then. As even the currently popular praxis of parallel translation between spoken and signed languages shows, signed languages have the full linguistic functionality of spoken languages. This has made them a potent argument against an initial "gestural protolanguage": If everything that can be said can be just as easily signed, then why turn to speech? Furthermore, as recent studies of emerging signed languages show, modern human beings are capable of spontaneously constructing a signed language from the pantomimic kind of gestures typical of bodily mimesis over the span of a few generations (Senghas et al. 2005; Sandler 2012).

The why-speech argument is indeed damaging to some proposals of gestural primacy, but not to all. On the one hand, proposals differ with respect to how exactly the "gestural protolanguage" is conceived of. Corballis sees it as "a form of signed language similar in principle, if not in detail, to the signed languages that are used today by the deaf" (Corballis 2003: 125). Arbib, it will be remembered, breaks up the evolutionary process in several stages, and preceding speech, there is "proto-sign: a manual-based communication system, breaking the fixed repertoire of primate vocalizations to yield a combinatorially open repertoire [...] elements for the formation of compounds which can be paired with meanings in a more or less arbitrary fashion" (Arbib 2003: 195). Bodily mimesis, on the other hand, corresponds to neither: Its virtue (as well as its ultimate disadvantage) is that the type of signs (in the semiotic sense) that it gives rise to is precisely *not* conventionalized, arbitrary, and combinatorial (Zlatev 2008a).

Furthermore, very few if any of the proponents of gestural primacy in evolution view the transition to speech as a discrete "switch," but rather as a process that was both gradual and, given the ubiquity of co-speech gesture, still remains only partial:

> While human primates must have been at first better at transmitting information through gesture than through voice, at some point voice became the preferred vehicle. But what if this "point" was a transitional period of over half a million years, say, from the appearance of *Homo erectus* to that of archaic *Homo sapiens*? And what if, during all this time, humans regularly communicated bi-modally, only gradually shifting from a code that foregrounded gesture to one that foregrounded voice? (Collins 2013: 136)

In general, the less prelinguistic gestural communication is thought of as a "language," and the less modern the spoken languages are conceived of as purely vocal, the less problematic the why-speech argument appears. While it is indeed damaging for scenarios that frame the transition as one "from hand to mouth" (Corballis 2002), they are not if stated in the much less idiomatic "from body to mouth and body" (Zlatev et al. 2010), that is, from whole-body communication supported by the human-specific capacity for bodily mimesis to the multimodal system of linguistic communication which we use today, involving both speech and gesture.

Thus, the typical counterargument against gesture-first theories is not in principle "fatal" or "insuperable" for the bodily mimesis hypothesis of human cognitive, and linguistic, origins. Still, a more explicit account of how and why the transition has taken place is due. In a recent doctoral dissertation, Brown sets herself this task precisely:

> A major step in the evolutionary process by which human communication could have emerged has been proposed in the bodily mimesis hypothesis. … This ability provides a foundation from which symbolic communication can arise, but how such a transition would have taken place has not been fully examined. This thesis examines the gap between bodily mimesis and symbolic communication. (Brown 2012: 1)

Brown reviews different gesture-first theories of language origins and concludes, similarly to Fitch (2010), that those that posit some form of "switch" between an already conventionalized (proto) language and speech (e.g., Corballis 2002; Arbib 2005) fail to provide an adequate explanation for this switch. In addition to the issues discussed in Sect. 3, Brown argues that an intermediary stage of arbitrary gestures, e.g., corresponding to Arbib's notion of "protosign," would have minimized support for the stabilization of a conventional code: "the conventionalization process requires a rich and supportive communicative infrastructure in which novel arbitrary signs can be used … so that the intended form-meaning relationships could be correctly interpreted" (Brown 2012: 81). This conclusion is supported by computational models of language evolution, showing that the stabilization of a conventional language across a greater number of speakers requires factors such as extensive corrective feedback or restricted context—neither of which is characteristic of actual communication—or support from parallel non-arbitrary signals.

While theories that posit that "multimodal referential communication was a combination of arbitrary and non-arbitrary representation from inception" (ibid: 116), such as that of McNeill (2012), avoid the need to explain any switch, they face complementary problems since they both predict a stronger degree of speech–gesture unity that appears to be the case (cf. Sect. 2.4) and underestimate the degree of non-arbitrariness in speech.

By method of exclusion, Brown concludes that theories that propose a gradual and only partial transition from mimesis/gesture to speech (e.g., Zlatev 2008b; Collins 2013) are most plausible, but objects that they "do not provide a reason why one modality is now predominantly symbolic and not the other" (Brown 2012: 120), i.e., why speech has undergone a greater degree of conventionalization, showing less iconicity, than gesture.

The answer proposed by Brown is both simple and ingenious: "the vocal modality would have become predominantly symbolic because its lower non-arbitrary capacity increases the likelihood that vocalizations are perceived as arbitrary" (ibid: 134).

This conclusion is supported by the methods of experimental semiotics (cf. Sect. 2.5), showing that the gestural modality carries more "communicative load" than the vocal modality when communication is restricted to non-conventional signaling and furthermore that iconic gestures help the audience to interpret novel vocalizations as meaningful words, even when the latter are perceived as "arbitrary." Supported by a combination of semiotic experimentation and computational modeling, Brown concludes that in multimodal gesture–vocalization communication, there will be an automatic pull toward increased arbitrariness with the need to communicate a larger and more diverse set of concepts and that this would take place in the vocal modality.

Taken along with the scenario suggested by Collins (2013) of a gradual shift of communicative load from gesture to speech over the duration of "over half a million years" thus gives a plausible answer to the why-speech question: Due to the diversification of hominin cultures, a less iconic (=more symbolic) code would have been beneficial, and since the vocal modality affords less iconicity than the manual/bodily one, it became naturally "recruited" to the task. The supposition that this took place from the emergence of *H. erectus* at 1.5 mya to *H. sapiens* at 0.2 mya gives more than sufficient time for necessary biological adaptations necessary for increased vocal control to take place. The answer is consistent with evidence for bodily mimesis summarized earlier and with the increasing evidence for the partial non-arbitrariness of speech (Ahlner and Zlatev 2010).

4 Conclusions

This chapter reviewed some of the confirming evidence for the bodily mimesis hypothesis, much of which can be also brought in favor of gesture-first theories of language origins. Unlike some recent and well-known proposals of a "gestural protolanguage," however, bodily mimesis is both a more general adaptation, since it concerns the volitional use of the body for other means than gestural communication as well, and less language-like. Hence, it was argued that it fares much better against the argument typically bought against gesture-first theories: How to explain the switch from a gestural (proto) language to a spoken one. It does so since (a) it emphasizes the non-conventionality and non-systematicity of bodily mimetic signaling, (b) it rejects the notion of a switch and instead posits a long biocultural spiral of conventionalization and adaptation for speech, and (c) it insists that the "transition," which is possibly the wrong word, should be seen as only partial, given all the evidence for the adaptive role of gesture in language development and face-to-face communication.

What Brown's theorizing and evidence add to this is a cognitive–semiotic explanation for why speech has during this process taken an increasingly higher communicative load: Bodily movement and vocalization do not differ in their capacity to represent meaning purely conventionally, but vocalization is intrinsically less capable of doing so iconically. Given a multimodal gestural–vocal communicative signal, the vocal element is bound to be less iconic than the gestural and thus to differentiate more clearly between an extensive set of concepts, even when their referents are visually similar.

In sum, the transition from communication based on bodily mimesis to relatively "arbitrary" speech was made possible by the multimodal character of human communication, through a prolonged process of increased articulation and conventionalization, but without language cutting off its bodily roots.

References

Alhner F, Zlatev J (2010) Cross-modal iconicity: a cognitive semiotic approach to sound symbolism. Sign Syst Stud 38(1/4):298–348

Andrén M (2010) Children's gestures between 18 and 30 months. Media Tryck, Lund

Arbib M (2003) The evolving mirror system: a neural basis for language readiness. In: Christiansen M, Kirby S (eds) Language evolution. Oxford University Press, Oxford, pp 182–200

Arbib M (2005) From monkey-like action recognition to human language: an evolutionary framework for neurolinguistics. Behav Brain Sci 28:105–168

Bates E, Benigni L, Bretherton I, Camioni L, Volterra V (1979) The emergence of symbols: cognition and communication in infancy. Academic Press, New York

Brown JE (2012) The evolution of symbolic communication: an embodied perspective. PhD thesis. University of Edinburgh, Edinburgh

Burling R (2005) The talking ape. Oxford University Press, Oxford

Call J (2001) Body imitation in an enculturated orangutan (*Pongo pygmaeus*). Cybern Syst 32:97–119

Call J, Tomasello M (2007) The gestural communication of apes and monkeys. Lawrence Erlbaum, London

Cela-Conde CJ, Ayala FJ (2007) Human evolution, trains from the past. Oxford University Press, Oxford

Collins C (2013) Paleopoetics. The evolution of the literary imagination. Columbia University Press, New York

Corballis MC (2002) From hand to mouth: the origins of language. Princeton University Press, Princeton

Corballis MC (2003) From hand to mouth: the gestural origins of language. In: Christiansen M, Kirby S (eds) Language evolution. Oxford University Press, Oxford, pp 201–218

Custance D, Whiten A, Bard K (1995) Can young chimpanzees (Pan troglodytes) imitate arbitrary actions? Behav 132:837–859

Donald M (1991) Origins of the modern mind: three stages in the evolution of human culture. Harvard University Press, Cambridge

Donald M (2001) A mind so rare: the evolution of human consciousness. Norton, New York

Fay N, Arbib M, Garrod D (2013) How to bootstrapp a human communication system. Cogn Sci 37(7):1356–1367

Fitch WT (2010) The evolution of language. Cambridge University Press, Cambridge

Galantucci B, Garrod S (2010) Experimental semiotics: a new approach for studying the emergence and the evolution of human communication. Interact Stud 11:1–13

Goldin-Meadow S (1998) The development of gesture and speech as an integrated system. Jossey-Bass, San Francisco
Gómez-Olivencia A, Carretero MJ, Arsuaga L, Rodríguez-García JL, García-González R, Martínez I (2007) Metric and morphological study of the upper cervical spine from the Sima de los Huesos site (Sierra de Atapuerca, Burgos, Spain). J Hum Evol 5:6–25
Hewes G (1973) Primate communication and the gestural origins of language. Curr Anthropol 14:5–24
Hurely S, Chater N (2005) Perspectives on imitation. From neuroscience to social science, vol I & II. MIT Press, Cambridge
Hurford JR (2004) Language beyond our grasp. In: Oller K, Griebel U, Plunkett K (eds) Evolution of communication systems: a comparative approach. Cambridge University Press, Cambridge, pp 297–313
Iacoboni M (2008) Mirroring people: the new science of how we connect with others. Farrar, Straus & Giroux, New York
Jakobson R (1965) Quest for the essence of language. Diogenes 13:21–38
Kendon A (2004) Gesture: visible action as utterance. Cambridge University Press, Cambridge
Kendon A (2008) Signs for language origins? Pub J Semiot 2(1):2–27
Kita S, Özyurek A (2003) What does cross-linguistic variation in semantic coordination of speech and gesture reveal?: Evidence for an interface representation of spatial thinking and speaking. J Mem Lang 48:16–32
Liszkowski U, Brown P, Callaghan T, Takida A, de Voc C (2012) A prelinguistic gestural universal of human communication. Cogn Sci 1–16
Lock A, Zukow-Goldring P (2012) Preverbal communication. In: Bremner J, Wachs T (eds) The Wiley-Blackwell handbook of infant development. Willey-Blackwell, Oxford, pp 394–425
MacLarnon AM, Hewitt GP (1999) The evolution of human speech: the role of enhanced breathing control. Am J Phys Anthropol 109:341–363
McNeill D (1985) So you think gestures are nonverbal? Psychol Rev 92:350–371
McNeill D (2005) Gesture and thought. University of Chicago Press, Chicago
McNeill D (2012) How language began: gesture and speech in human evolution. Cambridge University Press, Cambridge
Murillo E, Belinchón M (2012) Gestural-vocal coordination. Gesture 12(1):16–39
Piaget J (1962) Play, dreams, and imitation in childhood. Norton, New York
Pika S (2008) What is the nature of gestural communication in great apes? In: Zlatev J, Racine T, Sinha C, Itkonen E (eds) The shared mind: perspectives on intersubjectivity. Benjamins, Amsterdam, pp 165–186
Rizzolatti G, Arbib M (1998) Language within our grasp. Trends Neurosci 362:188–194
Sandler W (2012) Dedicated gestures, and the emergence of sign language. Gesture 12(3):265–307
Senghas R, Senghas A, Pyers J (2005) The emergence of Nicaraguan sign language: questions of development, acquisition and evolution. In: Langer J, Parker S, Milbrath C (eds) Biology and knowledge revisited: from neurogenesis to psychogenesis. Lawrence Erlbaum, Mahwah, pp 287–306
Sikveland RO, Ogden R (2012) Holding gestures across turns. Gesture 12(2):166–199
Slocombe K, Zuberbuehler K (2005) Functionally referential communication in a chimpanzee. Curr Biol 15:1779–1784
Sonesson G (1989) Pictorial concepts. Inquiries into the semiotic heritage and its relevance for the analysis of the visual world. Aris/Lund University Press, Lund
Sonesson G (1997) The ecological foundations of iconicity. In: Rauch I, Carr GF (eds) Semiotics around the world: synthesis in diversity. Mouton de Gruyter, Berlin, pp 739–742
Streek J (2009) Gesturecraft: the manufacture of meaning. Benjamins, Amsterdam
Tallis R (2011) Aping mankind. Neuromania, darwinitis and the misrepresentation of humanity. Acumen, Durham
Tomasello M (1999) The cultural origins of human cognition. Harvard University Press, Cambridge

Tomasello M (2008) The origins of human communication. MIT Press, Cambridge
Vaesen K (2012) The cognitive bases of human tool use. Behav Brain Sci 35:203–262
Volterra V, Caselli M, Caprici O, Pizzuto E (2005) Gesture and the emergence and development of language. In: Tomasello M, Slobin D (eds) Beyond nature-nurture: essays in honor of Elisabeth Bates. Lawrence Erlbaum, Mahwah, pp 3–40
Weiner S, Xu Q, Goldberg P, Liu J, Bar-Yosef O (1998) Evidence for the use of fire at Zhoukoudian, China. Science 281:251–253
Wilson SM, Saygin AP, Sereno MI, Iacoboni M (2004) Listening to speech activates motor areas involved in speech production. Nat Neurosci 7(7):701–712
Zlatev J (2007) Intersubjectivity, mimetic schemas and the emergence of language. Intellectica 46(2–3):123–152
Zlatev J (2008a) The coevolution of intersubjectivity and bodily mimesis. In: Zlatev J, Racine T, Sinha C, Itkonen E (eds) The shared mind: Perspectives on intersubjectivity. Benjamins, Amsterdam, pp 215–244
Zlatev J (2008b) From proto-mimesis to language: evidence from primatology and social neuroscience. J Physiol—Paris 102:137–152
Zlatev J (2012) Cognitive semiotics: an emerging field for the transdisciplinary study of meaning. Pub J Semiot 4(1):2–24
Zlatev J (2013) The mimesis hierarchy of semiotic development: five stages of intersubjectivity in children. Pub J Semiot 4(2):47–70
Zlatev J, Andrén M (2009) Stages and transitions in children's semiotic development. In: Zlatev J, Andrén M, Johansson-Falck M, Lundmark C (eds) Studies in language and cognition. Cambridge Scholars, Newcastle, pp 380–401
Zlatev J, Racine T, Sinha C, Itkonen E (2008) The shared mind: perspectives on intersubjectivity. Benjamins, Amsterdam
Zlatev J, Donald M, Sonesson G (2010) From body to mouth (and body). In: Smith A, Schouwstra M, deBoer B, Smith K (eds) The evolution of language. World Scientific, London, pp 527–528

From Grasping to Grooming to Gossip: Innovative Use of Chimpanzee Signals in Novel Environments Supports Both Vocal and Gestural Theories of Language Origins

David A. Leavens, Jared P. Taglialatela and William D. Hopkins

Abstract The unique challenges posed by ecologically novel situations can illuminate the limits of flexibility in animal signalling systems. Here we describe the innovative application of species-typical calls by chimpanzees exposed to the novel circumstances in which the animals are dependent upon others to act on the world for them, what we have previously termed "the Referential Problem Space". When chimpanzees are put into the Referential Problem Space, they display attention-getting calls and other auditory signals, and they tailor these signals to the state of attention of the receiver. Here we report that the kinds of calls that chimpanzees use in these evolutionarily novel circumstances are, generally, amplified versions of the same calls that they display in grooming contexts. Thus, this class of auditory signals, used in affiliative, grooming contexts, is chosen overwhelmingly by chimpanzees for application towards novel ends. This is consistent with Dunbar's (Grooming, gossip and the evolution of language. Faber and Faber, London, 1996) hypothesis that early humans substituted auditory contact for manual grooming as group sizes exceeded ca. 150 people. Moreover, these calls are primarily produced by supralaryngeal modulation of the airstream. This is consistent with Corballis's (From hand to mouth: The origins of language. Princeton University Press, Princeton, 2002) hypothesis that intentional

D. A. Leavens (✉)
School of Psychology, University of Sussex, Falmer, East Sussex,
Brighton BN1 9QH, UK
e-mail: davidl@sussex.ac.uk

J. P. Taglialatela
Department of Biology and Physics, Kennesaw
State University, Kennesaw, GA, USA
e-mail: jtaglial@kennesaw.edu

W. D. Hopkins
Neuroscience Institute and Language Research Center,
Georgia State University, Atlanta, GA, USA
e-mail: whopkins4@gsu.edu

W. D. Hopkins
Yerkes National Primate Research Center, Emory University, Atlanta, GA, USA

M. Pina and N. Gontier (eds.), *The Evolution of Social Communication in Primates*,
Interdisciplinary Evolution Research 1, DOI: 10.1007/978-3-319-02669-5_10,

communication in humans moved from the hands to the mouth and then into the larynx. In contemporary chimpanzees, we find intentional modulation of calls focused at the fronts of their oral cavities, for most grooming calls.

Keywords Cerebral asymmetries • Chimpanzee calls • Grooming • Intentionality • Language origins • Multimodal signals

1 Introduction

Relatively little scientific attention has been directed towards the innovative use of animal signals in novel environmental contexts (Hopkins et al. 2007a, b). In particular, primate calls have been long characterized as inflexible, reflexive, biologically determined systems over which animals exert little to no voluntary control (e.g., Arbib et al. 2008; Hauser 1996; Smith 1977). Indeed, it is this alleged inflexibility of calls that is taken as evidence for various versions of the gestural theory of language origins (e.g., Arbib 2005; Arbib et al. 2008; Corballis 1991, 2002; Donald 1991; Hewes 1973). According to these theorists, it is only in the manual gestures of great apes, communicating in the visual modality, that we find evidence for the levels of voluntariness in signalling that approach the voluntary control manifest in modern human speech.

Contrasting with this view are a variety of vocal origins theories, which either argue or assume that language evolved within the auditory–vocal modality (e.g., Deacon 1997; Dunbar 1996; Fitch 2000; Knight 2008; Zuberbühler 2005). Of particular relevance to our argument, Dunbar (1996) has postulated that, as social group sizes increased in hominid evolution, our ancestors developed "vocal grooming" to service affiliative relationships beyond the relatively limited numbers of individuals who could be physically groomed, given the chronic constraints on time. Among primates, species with larger group sizes tend to have larger vocal repertoires, with many researchers noting the strong relationship between measures of social complexity and complexity in call systems (e.g., Freeberg et al. 2012; McComb and Semple 2005). With respect to the origins of language, the unprecedented social complexity of our hominid ancestors is considered to have been an ecological determinant of increased vocal complexity in our lineage (Dunbar 1996).

Dunbar's (1996) hypothesis was developed to address one of the many questions pertaining to the earliest origins of human language: what was its adaptive function? As Bateson (1972a, b) pointed out so many years ago, the function of language cannot be, in any simple sense, "to communicate". As Bateson (1972a, p. 411) put it, "There is a general popular belief that in the evolution of [humans], language replaced the cruder [communicative] systems of the other animals. I believe this to be totally wrong". Based on cybernetic and evolutionary principles, Bateson argued that if language had evolved to supplant the functions of communication, then humans' non-verbal expressive repertoires would necessarily decay under the repeated scything of natural selection. Yet, in reality, humans have exquisitely subtle capacities for non-verbal expression, through facial expressions, dance, gestures, and systems of touch (e.g., massage techniques). Moreover, it is not the case that animal communication systems are, inherently, maladaptively primitive. Hence, language poses a deep paradox: it is used to communicate, but it has not functionally replaced non-symbolic modes of communication. The adaptive context in which language arose, thus, remains mysterious from a functional point of view: on the one hand, animals communicate perfectly well without language or speech; on the other hand, humans can also communicate effectively, with great impact, without speech, so what is language "for?" Dunbar's (1996) insight echoes Bateson's frequent refrain that communication, writ large, is about relationships (e.g., Bateson 1972b).

Dunbar's theory arose from his study of gelada baboons; he noticed that these social animals, which live in very large, multilevel communities of several hundred individuals, spend up to 20 % of their waking time grooming each other (e.g., Dunbar 2010, for review). Grooming is, fundamentally, about relationships. Social grooming (allogrooming) is a universal feature of the social lives of primates (e.g., Smuts et al. 1987). Grooming is a significant component in the maintenance of friendly social relationships and in the resolution of conflict between erstwhile combatants (Aureli and de Waal 2000). Grooming is much more than merely a mechanism for maintaining skin and coat: there are profound endocrinological (hormonal) effects of receiving grooming from others (Keverne et al. 1989). Dunbar noticed that as group size increases among social primates, so does the amount of time devoted to grooming. Above a certain community group size (~150 individuals, in Dunbar's thesis), the demands of maintaining relationships begin to conflict with other survival needs, such as time spent foraging. Dunbar postulated that affiliative intentions could, in this circumstance, be communicated via vocal–auditory means, leaving the hands free for foraging. Thus, Dunbar's theory describes a functional replacement of the grasping and stroking manual actions deployed in grooming bouts to a call-mediated system of relationship maintenance.

Corballis has long been concerned with the functional neuroscience of manual activity and the cerebral, asymmetrical specializations for speech (e.g. Corballis 1991, 2002). For Corballis, there is no coincidence that (a) the vast majority of humans are right-handed (and therefore left-hemisphere dominant for manual action) and (b) the vast majority of humans are left-hemisphere dominant for speech. Corballis (2002) posits that, in the human lineage, the intentional control of manual gestures (and other manual actions) that is manifest among great apes—and therefore presumably the last common ancestor of humans and the other great apes—was acquired by components of the oral cavity, from the lips to the tongue and, eventually, in our larynxes (voice boxes). Indeed, for Corballis, phonemes (the constituent sound units of speech) are occult gestures. Thus, while Dunbar emphasizes the transition from manual grooming to speech, Corballis emphasizes a transition from manual gesturing to speech. Both theories ground the origins of human speech in intentional manual action.

In this chapter, we elaborate the multimodal theory of speech origins through exploration of an intriguing intersection between Corballis's (2002) theory of the gestural origins of language and Dunbar's (1996) theory of the origins of speech as the vocal maintenance of grooming/affiliative relationships in complex social environments. Rather than focusing exclusively on either a vocal or gestural origins view of language evolution, some researchers, including ourselves, posit various versions of multimodal (vocal–auditory; visual–gestural) origins hypotheses of language origins (e.g., Hopkins et al. 2007a, b; Hurford 2007; Leavens 2003; Leavens et al. 2004; McNeill 1992; Taglialatela et al. 2011). In contrast to the strictly gestural origin or vocal origin hypotheses, multimodal origin hypotheses of language origins posit that signalling in the vocal and gestural domain coevolved as a single signalling mechanism for intraspecific communication. After a brief review of call production, we will turn to several of the lines of evidence that support a multimodal origin of language.

2 Primate Calls

Calls are produced by air inhaled or expelled through the pharyngeal (oral or nasal) cavities. The primary mechanical engine for inhalations and exhalations is the diaphragm. The air stream produced can be modulated at numerous places in the laryngopharyngeal column, including vibrations at the vocal folds, and a variety of compressions of the airstream in the supralaryngeal (above the larynx) cavities. For example, the lips might be compressed during exhalation, creating a sputtering sound, or during inhalation, creating a kissing or squeaking kind of sound. In human speech, many different consonants are created by different parts of the tongue impacting against different parts of the hard and soft palates (Fitch 2000; Owren and Rendall 2001).

The preponderance of current opinion is that the primary human/non-human animal difference in the control of this articulatory apparatus is that humans display a unique and very high degree of voluntary control over both (a) the emission of sounds with vocal cord vibrations (a.k.a. "voicing") and (b) the degree of modulation of the airstream in the supralaryngeal cavities (e.g., Fitch 2000; Owren and Rendall 2001; but see, e.g., Lemasson 2011; Owren et al. 2011; Snowdon 2009, for recent reviews of evidence for vocal plasticity in non-human primates). Both claims have been challenged by recent findings. With respect to the assumption that primates lack control over voiced calls, we found, for example, that some chimpanzees display an apparently voluntary extended grunt—a voiced call—to attract attention to themselves (Leavens et al. 2004; Russell et al. 2013; Taglialatela et al. 2012). In addition, a recent study of a gibbon demonstrated apparent voluntary control over the physical properties of the animal's larynx (Koda et al. 2012). Thus, emerging evidence suggests that some apes do display some apparently voluntary control over voiced calls, in some circumstances (also see Owren et al. 2011).

However, here we are concerned with the second assumption, the idea that humans have a unique ability to voluntarily modulate calls in the supralaryngeal cavities. In a recent review, Owren and his colleagues (Owren et al. 2011) have suggested that other primates do display apparent voluntary control of mostly non-voiced calls. This conclusion is consistent with our own findings that chimpanzees display a spontaneous and manifest choice over the sensory modality of their signalling behaviour, in some experimental circumstances (Hopkins et al. 2007a, b; Leavens et al. 2004, 2010). Moreover, several other lines of evidence converge on the conclusion that great apes have voluntary control over some of their calls [see Hopkins et al. (2011), for a recent review].

3 Evidence from Attention-Getting Behaviour

Evidence supporting the idea of voluntary control over calling behaviour in great apes includes the tactical deployment of both calls and manual gestures by chimpanzees who are exposed to humans in experimentally manipulated states of visual attentiveness. Thus, chimpanzees will display attention-getting calls or other sounds, if an experimenter is facing away from them, but then switch to manual gestures or other visual signals when the experimenter turns to look directly at them (Bodamer and Gardner 2002; Hostetter et al. 2001; Krause and Fouts 1997; Leavens et al. 2004, 2010; McCarthy et al. 2013). Moreover, chimpanzees choose from qualitatively different categories of calls depending on the specific circumstances; if presented with a banana placed outside their cage, but no human, they display species-typical food calls, but if an inattentive human is also present with a banana, the apes display a variety of attention-getting behaviours, including a number of calls that have not been described in these kinds of contexts in wild great apes (Hopkins et al. 2007b). Captive apes frequently face a situation in which they can see desirable items (often, but not always food), but are literally barred from directly reaching out and acquiring these items. Apes in these situations develop tactics for capturing the attention of any humans present and redirecting their attention to the desired entities, for example through pointing. These communicative tactics permit the apes to exert influence beyond the boundaries of their enclosures. Indeed, we have argued that these kinds of contexts, which we have termed the Referential Problem Space, are almost completely absent from the environments of wild apes (e.g., Leavens et al. 1996, 2005, 2008)—wild chimpanzees are only rarely subject to situations in which their instrumental goals on distal objects, such as object retrieval, can only be met through the communicative manipulation of other chimpanzees [see Hobaiter et al. (2013), for rare examples of such contexts among wild chimpanzees]. In contrast, both captive apes and human infants face long daily epochs in which they are physically restrained, and in this context, in the Referential Problem Space, both apes and human children develop communicative tactics for the manipulation of social agents to meet their instrumental goals (Leavens et al. 1996, 2005, 2008). Thus, chimpanzees choose the modality of their signals in accordance with context-specific communicative demands, using auditory signals to capture the attention of visually inattentive humans.

4 Evidence from Attention-Getting Calls

There is considerable inter-individual variability in the attention-getting calls that chimpanzees use when they are soliciting the attention of humans (reviewed by Hopkins et al. 2010, 2011). Recently, Taglialatela et al. (2012) have demonstrated that offspring of captive chimpanzees tend to acquire and use the attention-getting calls of their mothers—significantly more so than their siblings, who were equally related to their mothers, but raised apart from them. Taglialatela (2012) identified six attention-getting calls in their sample (see their Table 1, p. 499):

extended grunts (voiced, atonal sounds produced by the chimpanzees with an open mouth);
kisses (produced by inhaling air through pursed lip);
lip smacks (produced by placing upper and lower lips tightly together then pulling them apart quickly, making an audible "pop" sound);
pants (audible, rapid, rhythmic sequence of inhaling and exhaling);
raspberries (produced by blowing air out through pursed lips); and
teeth chomps (produced by clacking teeth together so that the hitting together of upper and lower jaws is audible).

For purposes of the present argument, note that only extended grunts appear to be voiced, whereas the other 5 call types are all produced by supralaryngeal modification of the airstream. The most significant aspect of these calls, from the standpoint of this chapter, is that, with the exception of the extended grunts, *they are used both in the wild and captivity in association with grooming* (e.g., Ghiglieri 1988; Goodall 1986; de Waal 1982). Although we currently lack the data to address this question directly, our impression is that these calls, when used in attention-getting contexts, are amplified versions of the softer calls used during grooming sessions by chimpanzees. In more recent work, Russell et al. (2013) demonstrated that some chimpanzees can be trained to display novel attention-getting calls; thus, not only is there a growing body of a posteriori evidence consistent with the view that attention-getting calls are socially learned, this latest study is a direct, prospective experimental demonstration of this capacity in chimpanzees.

5 Evidence from Patchy Distribution of Calls

Another category of evidence for flexibility in calls is the emerging evidence for geographical differences in call repertoires (van Schaik et al. 2003; Wich et al. 2012). The inclusion of calls in some locations, and its absence in others, in the same species, suggests that there is a social, learned component to some calls. This is a different phenomenon from group-based geographical differences in the acoustic structures of calls that are, themselves, displayed across groups, which is well established among some birds (e.g., Barrington 1773; Darwin 1871) and has more recently been widely reported among primate species (e.g., Crockford et al. 2004; Green 1975; Marshall et al. 1999; Wich et al. 2008). There is increasing evidence that call repertoires are geographically distinct in two distinct ways: categorically different call repertoires, in which specific calls are present or absent in different populations, and contextually different uses of calls in different populations.

Among orangutans, for example, van Schaik, Wich and their colleagues have demonstrated that three calls, raspberries, kiss squeak with hands, and kiss squeak with leaves, are patchily distributed among disparate study populations (van Schaik et al. 2003; Wich et al. 2012). For example, raspberries, which are bilabial fricatives associated in this species with nest-building, are reportedly absent from four of six sites studied, but present in two sites, one on the island of Sumatra the other on Borneo (van Schaik et al. 2003). Hence, these calls, made by expelling or inhaling air through slightly compressed lips, are modulated supralaryngeally; they are not automatic emissions tied to particular contexts in this species. More recent work has demonstrated that these calls are distributed independently of the genetic relatedness of individuals who display them (Wich et al. 2012). Wich et al. (2012, p. e36180) concluded that "[o]rang-utans occasionally invent calls with an arbitrary acoustic structure".

6 Evidence From Language-Trained Apes

Early scientific attempts to teach apes to speak were largely ineffective. In the late nineteenth century, Garner (1896) reported that a chimpanzee could articulate the French word, "fue". Witmer (1909) reported that a chimpanzee named Peter could, with difficulty, articulate the word "Mama", on demand. Similarly, Furness (1916) described an orangutan, also named Peter, that could articulate "Papa" and "cup". Hayes and Hayes (1954) reported that a chimpanzee named Viki could articulate four words, "Papa", "Mama", "cup", and "up". These very modest findings underscore the apparent difficulty apes have in displaying speech, but they do also highlight that apes can produce novel articulations on demand.

Hopkins and Savage-Rumbaugh (1991) demonstrated that Kanzi, a language-trained bonobo, displayed a vocal repertoire that differed acoustically from those of other captive, but not language-trained bonobos. More recently, Taglialatela et al. (2003) identified semantic categories in Kanzi's idiosyncratic vocalizations. Thus, Kanzi displays substantial innovation in his use of vocal signals.

There are numerous and long-standing reports of apes smoking (e.g., Kearton 1925), and, more recently, Perlman et al. (2012) have documented the ability of another language-trained ape, the gorilla, Koko, to make sounds with such musical instruments as harmonicas and recorders. Recently, Kanzi has demonstrated the ability to inflate balloons by mouth (Daily Mail 2010). This body of evidence demonstrates that apes have voluntary control over their breathing apparatus, the engine for making sounds, and the fronts of their oral cavities.

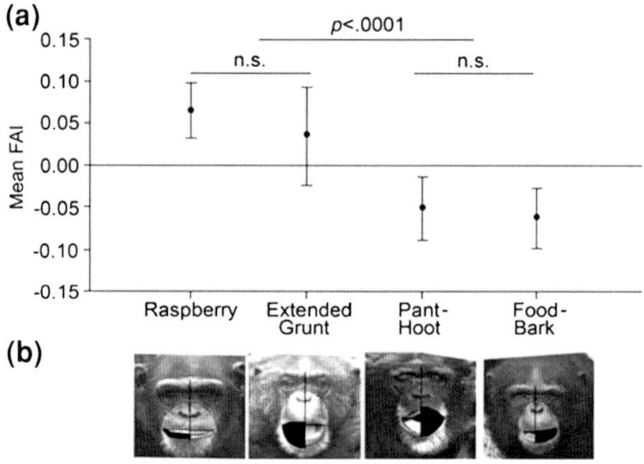

Fig. 1 The least squares mean facial asymmetry index (FAI) and 95 % confidence intervals for four calls, including two calls used in captive circumstances to capture the attention of humans (raspberry, extended grunt) and two calls not used in this specific, ecologically novel context (pant-hoot, food bark). See Losin et al. (2008) for complete method, but, in short, this technique involves measuring the areal asymmetries in the left and right sides of the oral cavity at the point of its maximum opening; thus, negative numbers reflect greater oral exposure on the left side of the face, implicating right-hemisphere dominance, and positive numbers, conversely, imply left-hemisphere dominance. Reprinted with permission from Losin et al. (2008, p. e2529; doi:10.1371/journal.pone.0002529.g002)

7 Evidence from Oro-Facial Asymmetries

Many calls are associated with expressive facial expressions that typically accompany those calls. One tactic to assess asymmetries in cerebral function is to evaluate asymmetries in the facial expressions that accompany particular calls. For example, Hauser (1993) reported more rapid retraction of the lips on the left side of the faces of rhesus monkeys, compared to the right side, during emotionally aggressive facial expressions, implying right-hemisphere dominance in these facial expressions (see also Hook-Costigan and Rogers 1998). Among great apes, Hopkins and his colleagues have reported similar asymmetries, demonstrating apparent right-hemisphere dominance during emotional displays (e.g., Fernández-Carriba et al. 2002).

Some calls, however, are associated with oro-facial asymmetries in the opposite direction, implicating left-hemisphere dominance (Losin et al. 2008). In particular, as noted above, this class of calls is distinguished by use as attention-getting signals in ecologically novel, captive environments. Thus, the facial expressions associated with the calls that captive chimpanzees use to attract the attention of human experimenters who are looking away from them tend to display a strikingly different pattern of oro-facial asymmetry, compared to most other calling contexts.

Two of these calls are the raspberry and the extended grunt. While these calls have been reported in ape repertoires in the wild (chimpanzees: Goodall 1986; orangtuans: van Schaik et al. 2003), they have not been reported to have an attention-getting function. Figure 1 depicts this pattern of left-hemisphere cerebral lateralization reported for the faces of chimpanzees displaying these attention-getting calls.

8 Evidence from the Neuro-Functional Foundations of Attention-Getting Calls

Positron emission tomography (PET) studies of chimpanzee brains during communication have revealed activation of the left inferior frontal gyrus (IFG), including regions identified in human brains as Broca's area, among other areas (Taglialatela et al. 2008, 2011). Broca's area has long been identified as a crucial component of humans' ability to produce articulate speech. In the first of these studies, (Taglialatela et al. 2008) reported that chimpanzees displayed activation of these anatomical homologues of human speech production during vocal and gestural communication, although the independent contributions of vocal and gestural signalling to this activation could not be identified. Subsequently, these same authors compared two chimpanzees who displayed gestures, but not any of the attention-getting calls identified in the previous section with two chimpanzees who did display these attention-getting sounds (Taglialatela et al. 2011). They found that the chimpanzees who displayed attention-getting calls also showed more activation in the left IFG, relative to the two chimpanzees who did not display attention-getting calls, suggesting a unique association of attention-getting calls with a region of the brain that, in humans, is devoted to intentional communication.

9 Evidence from Cerebral Asymmetries

It has been known for a long time that in human populations, Broca's area and Wernicke's area, critical for production and comprehension of speech, respectively, are usually larger in the left cerebral hemisphere than in the right cerebral hemisphere (e.g., Foundas et al. 1998; Geschwind and Levitsky 1968). Hopkins and his colleagues (Cantalupo and Hopkins 2001; Hopkins et al. 1998) and others (e.g., Gannon et al. 1998) have demonstrated that these "language" areas can be identified in the brains of great apes, and they are also asymmetrically larger, on average, in the left cerebral hemispheres of these close human relatives, although not every study of Broca's area homologues in great ape brains finds this asymmetry, suggesting that the degree of asymmetry, here, is less robust in great apes than in humans (e.g., Meguerditchian et al. 2012; Schenker et al. 2010). In related work, there is some evidence that chimpanzees display a weak but significant right-hand bias for bimanual grooming, implicating a left-hemisphere dominance for this activity (e.g., Hopkins et al. 2007a).

10 Evidence from Comparative Neurobiological Studies

There are important cortical regions, nuclei and cranial nerves involved in orofacial motor control and control of vocal folds. Notably, the trigeminal, facial and hypoglossal nuclei directly innervate the muscles, and recent comparative studies in primates have shown that there are qualitative changes in their volume and architecture between humans and apes compared to monkeys. For example, Sherwood et al. (2005) compared the volume and grey level index (GLI) of these three nuclei in a sample of 47 species of primates and found that for facial nuclei, great apes and humans (after scaling for overall medulla size differences) were significantly larger than predicted for all primates. These authors suggested that these differences may be related to potential differences in oro-facial motor control associated with communication or emotional expressions. In a related study, Sherwood et al. (2004) examined the laminar distribution and density of Brodmann's area 4 (BA4) in several catarrhine primate species including macaques, baboons, apes, and humans. BA4s located within the ventral portion of the precentral gyrus has been implicated in oro-facial motor control. Humans and great apes showed relatively greater thickness within layer III and lower cell volume densities compared to the Old World monkeys. The lower cell densities were interpreted to suggest that there was greater spacing between neurons within the region providing for greater cortical–cortical connectivity between BA4 and other brain regions. The collective findings suggest that there is enhanced neural representation of cortical control of the oro-facial musculature of chimpanzees, relative to other primates. We suggest that this increased cortical representation may allow for chimpanzees and other great apes to learn new sounds such as the attention-getting sounds discussed in this chapter.

11 Summary of Evidence and Relation to Corballis and Dunbar

Thus, there is a class of calls displayed by chimpanzees that consist of apparently voluntary control over the respiratory apparatus, apparently voluntary control over a variety of post-laryngeal modifications of the airstream, are apparently amplified versions of sounds made during grooming, display a reverse patterning of cerebral dominance, compared with most vocalizations, and are ontogenetically adapted for use in ecologically novel experimental contexts in which chimpanzees are dependent upon humans to act on the world for them, the Referential Problem Space. There are at least two theoretically significant aspects of this pattern of empirical results.

The first significant implication of this pattern of findings is that these sounds are amplified versions of sounds that chimpanzees make during grooming episodes. When chimpanzees groom each other, they might repetitively chomp their teeth, display low-level sputtering, smack their lips together or pant repeatedly. Dunbar (1996) has suggested that when social networks become too large for one-to-one grooming to support those networks, then calling behaviour fulfils that role and sees in this postulate a possible socioecological mechanism that might have fostered oral communication in our hominid ancestors. The patterns we reviewed are consistent with Dunbar's hypothesis: we find that even in the absence of natural selection, a relatively simple set of changes to chimpanzees' ecological circumstances elicits remarkable innovation in call use, when these apes are dependent upon others to act on the world outside their cages. We propose that there is substantial, yet heretofore underappreciated flexibility in the call systems of great apes, and we think that it is possibly no coincidence that the calls associated with grooming—crucial for developing and maintaining affiliative social relationships—are the calls that display the most flexibility in use. Grooming is used strategically—and therefore apparently intentionally—in great apes (e.g., Aureli and de Waal 2000; de Waal 1982). The flexibility in these call systems is manifested in the Referential Problem Space. We have previously argued that this socioecological circumstance, in which an organism is dependent upon another to act on the world for them, characterized the early developmental environments of our hominid ancestors, when babies began to be born too weak and helpless to cling to their mothers throughout the infancy period (Leavens et al. 2008, 2009). In contemporary chimpanzees, newborns are similarly weak and helpless, but rapidly develop the capacity to cling to their mothers during locomotion, and this occurs early in infancy—indeed, chimpanzees are capable of independent locomotion by approximately 5 months of age. In contrast, human babies lack this clinging capacity throughout their infancy period, and locomotor development is an extremely protracted process with a duration of several years (Adolph and Berger 2005).

The second aspect of theoretical interest is that these calls are, except for pants and extended grunts, modulated at the very top of the supralaryngeal cavity, specifically at the lips. This is consistent with the evolutionary scenario for language origins proposed by Corballis (2002); in his view, the evolution of language proceeded in our own lineage according to the following order: gestures from the hands to gestures of the mouth and then, finally, to occult gestures of the larynx in contemporary speech. It is, therefore, really quite remarkable that the flexibility in calls that we find in these close relatives of humans is largely manifested at the front of the mouth. We interpret this to be consistent with Corballis's suggestion, and, moreover, we think that this supports the view that our hominid ancestors were preadapted for supralaryngeal modulation of calling, in the sense that Hauser et al. (2002) proposed that humans share a mosaic of communication characteristics with other mammals. Corballis's long-standing concern with the left hemisphere as being preadapted for linguistic communication is, we think, supported by the evidence suggestive (a) of right-hand dominance for manual gestures in chimpanzees, particularly when the animals are simultaneously calling (Hopkins and Leavens 1998; Hopkins and Cantero 2003), (b) of right-hand dominance for bimanual grooming in chimpanzees (Hopkins et al. 2007a, b), and (c) of left-hemisphere dominance for speech in humans. This pattern supports the idea that the last common ancestor of great apes and humans were already left-hemisphere dominant for manual grooming, and when the later Pleistocene growth in mean group size in the human lineage exerted the adaptive effects on relationship maintenance postulated by Dunbar—capping, in effect, the amount of time available for relationship maintenance through grooming—the left hemisphere was already preadapted for this affiliative function.

Thus, in modern apes, we find an unanticipated intersection between Dunbar's (1996) gossip-as-grooming hypothesis and Corballis's (2002) hand-to-mouth hypothesis. The former implies that grooming calls are those most readily adapted to new ecological circumstances, while the latter implies that the mouth is the next most flexible site for intentionally communicative signalling, after the hands. The evidence suggests that:

(a) apes in ecologically novel circumstances tend to adapt grooming calls to novel ends, particularly when attempting to gain the attention of an otherwise inattentive social partner;
(b) the intentionality of these attention-getting calls is well established, suggesting that the last common ancestor of apes and humans was preadapted for intentional signalling;
(c) the left-hemisphere dominance associated with the production of these attention-getting calls presages the later left-hemisphere dominance for speech found for most humans;
(d) the last common ancestor of humans and apes had substantial voluntary control over both their manual and oral gestures.

Hence, on the basis of these premises, we suggest that in the evolution of speech, voluntary control over significant aspects of both visual and auditory communication was already possessed by the last common ancestor of extant non-human and human apes. This ancestor was an ape that lived in the Late Miocene. If this is true, then the epoch of time required to develop the apparently uniquely human, rapid-fire, dynamic control over the larynx and the tongue is greatly increased in duration over most contemporary scenarios for the evolution of speech (e.g., Arbib 2005, Arbib et al. 2008; Corballis 1991, 2002). If the last common ancestor of humans and the other apes already had intentional control over the most rostral portion of the oral cavity, then there are approximately 6.5 million years in which to evolve the further specialized control over lingual and laryngeal structures evinced by our species. Others have noted the relative paucity of appropriate studies of wild apes to address the questions of intentional calling in great apes (e.g., Burling 1993; Owren et al. 2011; Zuberbühler 2005), but recent fieldwork on chimpanzees is beginning to demonstrate substantial apparent volitional control over their calls (see, e.g., Schel et al. 2013a, b). To the extent that this scenario is correct is the extent to which the evolution of speech becomes more of an evolutionarily adaptive solution and less of a *deus ex machina*.

References

Adolph KE, Berger SE (2005) Physical and motor development. In: Bornstein MH, Lamb ME (eds) Developmental science: an advanced textbook, 5th edn. Lawrence Erlbaum Associates, Mahwah

Arbib MA (2005) From monkey-like action recognition to human language: an evolutionary framework for neurolinguistics. Behav Brain Sci 28:105–167

Arbib MA, Liebal K, Pika S (2008) Primate vocalization, gesture, and the evolution of human language. Curr Anthro 49:1053–1076

Aureli F, de Waal FBM (eds) (2000) Natural conflict resolution. University of California Press, Berkeley

Barrington D (1773) Experiments and observations on the singing of birds. Phil Trans Roy Soc 63:249–291

Bateson G (1972a) Redundancy and coding. In: Bateson G (ed) Steps to an ecology of mind. Ballantine Books, New York. (Original chapter published 1968 in Sebeok TA (ed) Animal communication: techniques of study and results of research. Indiana University Press, Bloomington, IN)

Bateson G (1972b) Form, substance, and difference. In: Bateson G (ed) Steps to an ecology of mind. Ballantine Books, New York. (Originally delivered January 9, 1970 as the Nineteenth Annual Korzybski Memorial Lecture)

Bodamer MD, Gardner RA (2002) How cross-fostered chimpanzees Pan troglodytes initiate and maintain conversations. J Comp Psych 116:12–26

Burling R (1993) Primate calls, human language, and nonverbal communication. Curr Anthro 34:25–53

Cantalupo C, Hopkins WD (2001) Asymmetric Broca's area in great apes. Nature 414:505

Corballis MC (1991) The lopsided ape: evolution of the generative mind. Oxford University Press, New York

Corballis MC (2002) From hand to mouth: the origins of language. Princeton University Press, Princeton

Crockford C, Herbinger I, Marchant L, Boesch C (2004) Wild chimpanzees have group specific calls: a case for vocal learning? Ethology 110:221–243

Daily Mail (Anonymous reporter) (2010, October 6) Kanzi the amazing ape who can 'speak' with humans and understands 450 words makes Oprah appearance. The Daily Mail. Retrieved from: http://www.dailymail.co.uk/sciencetech/article-1317838/Kanzi-amazing-ape-speak-humans-understands-450-words.html

Darwin C (1871) The descent of man and selection in relation to sex. John Murray, London

de Waal FBM (1982) Chimpanzee politics: power and sex among apes. Harper and Row, New York

Deacon TW (1997) The symbolic species: the co-evolution of language and the brain. Norton, New York

Donald M (1991) Origins of the modern mind: three stages in the evolution of culture and cognition. Harvard University Press, Cambridge

Dunbar RIM (1996) Grooming, gossip and the evolution of language. Faber and Faber, London

Dunbar RIM (2010) The social role of touch in humans and primates: behavioural function and neurobiological mechanisms. Neurosci Biobehav Rev 34:260–268

Fernández-Carriba S, Louches A, Morcillo A, Hopkins WD (2002) Asymmetry in facial expression of emotions by chimpanzees. Neuropsycholog 40:1523–1533

Fitch WT (2000) The evolution of speech: a comparative review. Trends Cog Sci 4:258–267

Foundas AL, Eure KF, Luevano LF, Weinberger DR (1998) MRI asymmetries of Broca's area: the pars triangularis and pars opercularis. Brain and Lang 64:282–296

Freeberg TM, Dunbar RIM, Ord TJ (2012) Social complexity as a proximate and ultimate factor in communicative complexity. Phil Trans Roy Soc, B 367:1785–1801

Furness WH (1916) Observations on the mentality of chimpanzees and orangutans. Proc Amer Phil Soc 55:281–290

Gannon PJ, Holloway RL, Broadfield DC, Braun AR (1998) Asymmetry of chimpanzee planum temporale: humanlike pattern of Wernicke's brain language area homolog. Science 279: 220–222 January 9

Garner RL (1896) Gorillas and chimpanzees. Osgood, McIlvane, and Co, London

Geschwind N, Levitsky W (1968) Human brain: left-right asymmetries in temporal speech region. Science 161:186–187 July 12

Ghiglieri MP (1988) East of the mountains of the moon. The Free Press, New York

Goodall J (1986) The chimpanzees of Gombe: patterns of behavior. Belknap Press, Cambridge

Green S (1975) Dialects in Japanese monkeys: vocal learning and cultural transmission of local-specificity behavior? Zeit Tierpsych 38:304–314

Hauser MD (1993) Right hemisphere dominance in the production of facial expression in monkeys. Science 261:475–477

Hauser MD (1996) The evolution of communication. The MIT Press, Cambridge
Hauser MD, Chomsky N, Fitch WT (2002) The faculty of language: what is it, who has it, and how did it evolve? Science 28:1569–1579 November 22
Hayes KJ, Hayes C (1954) The cultural capacity of chimpanzee. Human Biol 26:288–303
Hewes GW (1973) Primate communication and the gestural origin of language. Curr Anthro 14:5–24
Hobaiter C, Leavens DA, Byrne RW (2013) Deictic gesturing in wild chimpanzees? Some possible cases. J Comp Psych. (Published online in advance of print)
Hook-Costigan MA, Rogers LJ (1998) Lateralized use of the mouth in production of vocalizations by Marmosets. Neuropsychol 36:1265–1273
Hopkins WD, Cantero M (2003) From hand to mouth in the evolution of language: the influence of vocal behaviour on lateralized hand use in manual gestures by Chimpanzees Pan troglodytes. Dev Sci 6:55–61
Hopkins WD, Leavens DA (1998) Hand use and gestural communication in Chimpanzees Pan troglodytes. J Comp Psych 112:95–99
Hopkins WD, Savage-Rumbaugh ES (1991) Vocal communication as a function of differential rearing experiences in Pan paniscus: a preliminary report. Int J Primat 12:559–583
Hopkins WD, Marino L, Rilling JK, MacGregor LA (1998) Planum temporale asymmetries in great apes as revealed by magnetic resonance imaging (MRI). NeuroRep 9:2913–2918
Hopkins WD, Russell JL, Remkus M, Freeman H, Schapiro SJ (2007a) Handedness and grooming in Pan troglodytes: comparative analysis between findings in captive and wild individuals. Int J of Primat 28:1315–1326
Hopkins WD, Taglialatela JP, Leavens DA (2007b) Chimpanzees differentially produce novel vocalizations to capture the attention of a human. Anim Behav 73:281–286
Hopkins WD, Taglialatela JP, Leavens DA, Russell JL, Schapiro S (2010) Behavioral and brain asymmetries in chimpanzees Pan troglodytes. In: Lonsdorf E, Ross S, Matsuzawa T (eds) The mind of the chimpanzee. The University of Chicago Press, Chicago
Hopkins WD, Taglialatela JP, Leavens DA (2011) Do chimpanzees have voluntary control of their facial expressions and vocalizations? In: Vilain A, Schwartz J-L, Abry C, Vauclair J (eds) Primate vocalizations and human language: vocalisation, gestures, imitation and deixis in humans and non-humans. Benjamins, Amsterdam
Hostetter AB, Cantero M, Hopkins WD (2001) Differential use of vocal and gestural communication in chimpanzees in response to the attentional status of a human audience. J Comp Psych 115:337–343
Hurford JR (2007) The origins of meaning: language in the light of evolution. Oxford University Press, Oxford
Kearton C (1925) My friend Toto: the adventures of a chimpanzee, and the story of his journey from the Congo to London. Arrowsmith, London
Keverne EB, Martensz ND, Tuite B (1989) Beta-endorphin concentrations in cerebrospinal fluid of monkeys are influenced by grooming relationships. Psychoneuroend 14:155–161
Knight C (2008) Honest fakes and language origins. J Consc Stud 15:236–248
Koda H, Nishimura T, Tokuda IT, Oyakawa C, Nihonmatsu T, Masataka N (2012) Soprano singing in gibbons. Amer J Phys Anthro. doi:10.1002/ajpa.22124
Krause MA, Fouts RS (1997) Chimpanzee Pan troglodytes pointing: hand shapes, accuracy, and the role of eye gaze. J Comp Psych 111:330–336
Leavens DA (2003) Integration of visual and vocal communication: evidence for Miocene origins. Behav Brain Sci 26:232–233
Leavens DA, Hopkins WD, Bard KA (1996) Indexical and referential pointing in chimpanzees Pan troglodytes. J Comp Psych 110:346–353
Leavens DA, Hostetter AB, Wesley MJ, Hopkins WD (2004) Tactical use of unimodal and bimodal communication by chimpanzees, Pan troglodytes. Anim Behav 67:467–476
Leavens DA, Hopkins WD, Bard KA (2005) Understanding the point of chimpanzee pointing: epigenesis and ecological validity. Curr Dir Psych Sci 14:185–189

Leavens DA, Hopkins WD, Bard KA (2008) The heterochronic origins of explicit reference. In: Zlatev J, Racine T, Sinha C, Itkonen E (eds) The shared mind: perspectives on intersubjectivity. Benjamins, Amsterdam

Leavens DA, Racine TP, Hopkins WD (2009) The ontogeny and phylogeny of non-verbal deixis. In: Botha R, Knight C (eds) The prehistory of language. Oxford University Press, Oxford

Leavens DA, Russell JL, Hopkins WD (2010) Multimodal communication by captive chimpanzees Pan troglodytes. Anim Cogn 13:33–40

Lemasson A (2011) What can forest guenons "tell" us about the origin of language? In: Vilain A, Schwartz J-L, Abry C, Vauclair J (eds) Primate communication and human language: vocalisation, gestures, imitation and deixis in humans and non-humans. Benjamins, Amsterdam

Losin EAR, Russell JL, Freeman H, Meguerditchian A, Hopkins WD (2008) Left hemisphere specialization for oro-facial movements of learned vocal signals by captive chimpanzees. PLoS One 3:e2529

Marshall AJ, Wrangham RW, Arcadi AC (1999) Does learning affect the structure of vocalizations in chimpanzees? Anim Behav 58:825–830

McCarthy MS, Jensvold MLA, Fouts DH (2013) Use of gesture sequences in captive chimpanzee Pan troglodytes play. Anim Cogn 16:471–481

McComb K, Semple S (2005) Coevolution of vocal communication and sociality in primates. Biol Lett 1:381–385

McNeill D (1992) Hand and mind: what gestures reveal about thought. University of Chicago Press, Chicago

Meguerditchian A, Gardner MJ, Schapiro SJ, Hopkins WD (2012) The sound of one-hand clapping: handedness and perisylvian neural correlates of a communicative gesture in chimpanzees. Proc Roy Soc B 279:199–1966

Owren MJ, Rendall D (2001) Sound on the rebound: bringing form and function back to the forefront in understanding nonhuman primate vocal signaling. Evol Anthro 10:58–71

Owren MJ, Amoss RT, Rendall D (2011) Two organizing principles of vocal production: implications for nonhuman and human primates. Amer J Primatol 73:530–544

Perlman M, Patterson P, Cohn R (2012) The human-fostered gorilla Koko shows breath control in play with wind instruments. Bioling 6:433–444

Russell JL, McIntyre JM, Hopkins WD, Taglialatela JP (2013) Vocal learning of a communicative signal in captive chimpanzees, Pan troglodytes. Brain and Lang 127:520–525

Schel AM, Machanda Z, Townsend SW, Zuberbühler K, Slocombe KE (2013a) Chimpanzee food calls are directed at specific individuals. Anim Behav. (Published online in advance of print)

Schel AM, Townsend SW, Machanda Z, Zuberbühler K, Slocombe KE (2013b) Chimpanzee alarm call production meets key criteria for intentionality. PLoS One 8:e76674

Schenker NM, Hopkins WD, Spocter MA, Garrison AR, Stimpson CD, Erwin JM, Hof PR, Sherwood CC (2010) Broca's area homologue in chimpanzees Pan troglodytes: probabilistic mapping, asymmetry, and comparison to humans. Cereb Cort 20:730–742

Sherwood CC, Holloway RL, Erwin JM, Schleicher A, Zilles K, Hof PR (2004) Cortical orofacial motor representation in old world monkeys, great apes, and humans. I. Quantitative analysis of cytoarchitecture. Brain Behav Evol 63:61–81

Sherwood CC, Hof PR, Holloway RL, Semendeferi K, Gannon PJ, Frahm HD, Zilles K (2005) Evolution of the brainstem orofacial motor system in primates: a comparative study of trigeminal, facial, and hypoglossal nuclei. J Human Evol 48:45–84

Smith WJ (1977) The behavior of communicating: an ethological approach. Harvard University Press, Cambridge

Smuts B, Cheney D, Seyfarth R, Wrangham R, Struhsaker T (eds) (1987) Primate societies. University of Chicago Press, Chicago

Snowdon CT (2009) Plasticity of communication in nonhuman primates. In: Naguib M, Janik VM (eds) Advances in the study of behavior, vol 40. Academic Press, Burlington

Taglialatela JP, Savage-Rumbaugh ES, Baker LA (2003) Vocal production by a language-competent Pan paniscus. Int J Primat 24:1–17

Taglialatela JP, Russell JL, Schaeffer JA, Hopkins WD (2008) Communicative signalling activates 'Broca's' homolog in chimpanzees. Curr Biol 18:343–348

Taglialatela JP, Russell JL, Schaeffer JA, Hopkins WD (2011) Chimpanzee vocal signaling points to a multimodal origin of human language. PLoS One 6:e18852

Taglialatela JP, Reamer L, Schapiro SJ, Hopkins WD (2012) Social learning of a communicative signal in captive chimpanzees. Biol Lett 8:498–501

van Schaik C, Ancrenaz M, Borgen G, Galdikas B, Knott CD, Singleton I, Suzuki A, Utami SS, Merrill M (2003) Orangutan cultures and the evolution of material culture. Science 299:102–105 January 3

Wich SA, Schel AM, de Vries H (2008) Geographic variation in Thomas Langur (Presbytis thomasi) loud calls. Amer J Primat 70:566–574

Wich SA, Krützen M, Lameira AR, Nater A, Arora N, Bastian ML, Meulman E, Morrogh-Bernard HC, Utami Atmoko SS, Pamungkas J, Perwitasari-Farajallah D, Hardus ME, van Noordwijk M, van Schaik CP (2012) Call cultures in orang-utans? PLoS One 7:e36180

Witmer L (1909) A monkey with a mind. The Psych Clin III 3:179–205

Zuberbühler K (2005) The phylogenetic roots of language: evidence from primate communication and cognition. Curr Dir Psych Sci 14:126–130

Reevaluating Chimpanzee Vocal Signals: Toward a Multimodal Account of the Origins of Human Communication

Adam See

Abstract The vocalizations of chimpanzees have long been thought to be largely genetically predetermined and therefore unlearnable, involuntarily produced, and broadcast indiscriminately. Tomasello (2008) has recently written that, while chimpanzee vocalizations share these constraints and limitations with the vocal displays of all other non-human animals, the attention-getting gestures of chimpanzees are an "evolutionary novelty" because they are, in his estimation, capable of being produced intentionally. As such, chimpanzee gestures are highly significant to discussions of animal cognition and the evolution of human communication. This chapter challenges Tomasello's grounds for restricting this evolutionary novelty to the gestural modality. I argue that, in fact, recent evidence suggests that there is a significant functional difference between certain chimpanzee vocalizations and the vocal displays of other animals and that, based on Tomasello's own criteria for intentionality, gestures do not appear to have a monopoly on intentional communication in chimpanzees. Ultimately, this chapter aims to provide grounds for a multimodal account of the evolution of human communication. I conclude by suggesting that although there is reason to doubt that chimpanzees can communicate intentionally, there is no *more* reason to doubt this ability in the vocal modality than there is in the gestural modality.

Keywords Tomasello • Vocalizations • Intentionality • Chimpanzee • Animal communication • Evolution of language

---

A. See (✉)
Department of Philosophy, Graduate Center of the City University of New York, 365 Fifth Avenue, New York, NY 10016, USA
e-mail: asee@gc.cuny.edu

M. Pina and N. Gontier (eds.), *The Evolution of Social Communication in Primates*, Interdisciplinary Evolution Research 1, DOI: 10.1007/978-3-319-02669-5_11,

1 Questioning an "Evolutionary Novelty"

It has long been recognized that chimpanzees, along with other great ape species, possess a remarkable capacity for gestural communication. While the vast majority of non-human communicative acts are inextricably bound up with specific emotions, contexts, and environmental cues (Janik and Slater 1997; Cheney and Seyfarth 2010), the past 30 years of work in primatology have provided suggestive evidence that certain gestures of both wild and captive chimpanzees are produced voluntarily and with great circumstantial flexibility [see Pollick and de Waal (2007), Arbib et al. (2008), and Tomasello (2008) for reviews]. Further, much work has been done to substantiate the view that novel gestures are capable of being learned [see Tomasello (1996, 2008) for reviews], socially inherited (Pollick and de Waal 2007), and combined to construct a "simple syntax" (reviewed in Tomasello 2008). Perhaps most impressively, certain chimpanzee gestures appear to be produced "dyadically," i.e., with sensitivity to the attentional state of the recipient. Liebal et al. (2004) found that, when gesturing to both humans and conspecifics, chimpanzees will reliably exercise the following process: Attempt one gesture, monitor the receiver's response, and if necessary, walk around the receiver and repeat the gesture or try a different one. As Tomasello (2008: 30) notes, "This shows persistence to a goal with adjusted means as necessary—the prototype of intentional action." The fact that chimpanzees appear to employ "practical reasoning" in gestural communication strongly suggests that they possess a theory of mind, i.e., that they attribute mental states such as attention to others (Tomasello 2008; Premack and Woodruff 1978).[1] Chimpanzee gestures are, therefore, highly significant to discussions about animal minds and the evolution of human communication. According to Tomasello, "attention to the attention of the other during communication is unprecedented in non-primate, and maybe even non-ape, communication" (33).

It is because of this fact that Tomasello (2008) draws a "sharp contrast" between the attention-getting gestures of chimpanzees (which he calls "intentional signals"), and the mere "communicative displays" that encompass all other acts of animal communication. Setting aside any contentious terminological issues for the moment, Tomasello's view is that there is no *functional* difference between deer horns, peacock tails, bee dances, teeth-bearing, and, salient to this chapter, all animal vocalizations including those of non-human primates, song birds, and cetacea. Tomasello's claim that the gestures of chimpanzees are an "evolutionary novelty" is therefore a very strong one. It is also, perhaps surprisingly, not very

[1] Since Premack and Woodruff's original discussion, "theory of mind" has been used to describe and/or explain an extensive range of phenomena. In the interest of clarity, I purport here to use the term exactly as Tomasello (2008) does in its relationship to what he calls "intentional signals." For a discussion of Tomasello's understanding of theory of mind and its significance to intentional signals see Sect. 2 below.

controversial in the modern literature. The view that chimpanzee vocalizations are, for the most part, unlearnable, inflexibly tied to emotions, involuntarily produced, genetically predetermined, and broadcast indiscriminately, is in fact widely accepted [see Arbib et al. (2008), Hammerschmidt and Fischer (2008), and Cheney and Seyfarth (2010) for reviews].

There are, however, studies as recent as this year that support a contrary position on the communicative potential of chimpanzee vocalizations. In response to Tomasello's view, I argue here that there *is* a significant functional difference between certain vocalizations of chimpanzees and the vocal displays of other animals. Gestures do not have a monopoly on intentional communication in non-human primates (henceforth primates). This chapter suggests both that certain vocalizations of chimpanzees satisfy all of Tomasello's criteria for intentional signals and that any skepticism pertaining to the intentionality of these vocalizations applies equally to their gestural counterparts. If this claim can be substantiated, Tomasello's (2008) argument that the origins of human communication emerged primarily from primate gestures may demand revision.

The most common argument against the evolutionary relevance of the great ape vocal modality is a physiological one [see Hammerschmidt and Fischer (2008) for a review]. Like all non-human animals, great apes have very limited flexibility in vocal production (Cheney and Seyfarth 2010). So although chimpanzees have, for instance, been shown to socially inherit novel whistling behavior (Crockford et al. 2004) and flexibly employ goal-oriented "raspberry" lip-purses to achieve communicative ends (Leavens et al. 2004; Russell et al. 2005; Hostetter et al. 2007), in this chapter I follow Hopkins et al. (2007) in distinguishing "vocalizations" from mere "sounds" such as lip-smacks and whistles and restrict the evidence considered to vocal signals generated with use of the vocal cords.

The general trajectory of this chapter proceeds as follows. Since Tomasello (2008) is not explicit about the criteria he uses to distinguish "communicative displays" from "intentional signals," I begin by deriving three general criteria from his argument. These are the presence of (1) social and asocial learning in the ontogenetic development of signaling behavior, and (2) contextual flexibility and (3) attention to the attention of the receiver during signal production. I then provide what I hope to be compelling evidence that great apes may be capable of vocalizations that meet all of these criteria and thus should, by Tomasello's own account, be classified as intentional signals that differ significantly from common vocal displays.

It is imperative to note that my argument here is *not* that the vocal signals of chimpanzees are as flexible or even as evolutionarily significant as chimpanzee gestures, but rather that Tomasello's arguments against their communicative potential are unjustified. To this end, I conclude by suggesting that although there is reason to doubt that chimpanzees can communicate intentionally, the fact that they utilize attention-getting (henceforth AG-) vocalizations in ways that are functionally equivalent to AG-gestures demonstrates that there is *no more* reason to doubt intentionally in the vocal modality than in the gestural modality.

2 "Communicative Displays" and "Intentional Signals"

Tomasello (2008) unfortunately does not list any specific criteria that distinguish "communicative displays" from "intentional signals." Rather, he tends toward defining gestural signals negatively in relation to what he calls "vocal displays." This section aims to identify and make explicit the criteria that appear to be tacitly operating in his argument in order to apply them in subsequent sections.

Defined as broadly as possible, a signal is any sign or indication of a given state of affairs. In the animal kingdom, pelt coloration, nest construction style, beak size, vocalizations, and gestures are all signals because they all convey information. There is therefore a very important sense in which all animal displays simply *are* signals. That said, discussions of the evolution of human communication will inevitably demand more precise terminology. This is because intentionality, voluntarism, and developmental and contextual flexibility now become highly salient traits in the psychology of both senders and receivers. Tomasello largely avoids the task of psychologically carving up the world of animal communication by drawing a very thick line in the sand. Any signal that is not produced *intentionally* is a display:

> Communicative displays are prototypically physical characteristics that in some way affect the behavior of others, such as large horns which deter competitors or bright colors which attract mates. Functionally, we may group with displays reflexive behaviors that are invariably evoked by particular stimuli or emotional states and over which the individual has no voluntary control. Such inflexible physical and behavioral displays, created and controlled by evolutionary processes, characterize the vast majority of communication in the biological world (14).

In "sharp contrast," Tomasello writes that "intentional signals"…

> …are chosen and produced by individual organisms flexibly and strategically for particular social goals, adjusted in various ways for particular circumstances. These signals are *intentional* in the sense that the individual controls their use flexibly toward the goal of influencing others (14).

It is worth highlighting here that Tomasello—presumably for clarity—is using the word "signal" to demarcate volitional or intentional modes of communication, while "display" is used exclusively in reference to communicative signs that the signaler has no control over. I will stick to Tomasello's terminology in what follows. Though Tomasello does not explicitly classify primate vocalizations as displays, textual evidence that he classifies them as such is evident in the following "(many gestures) are as genetically fixed and inflexible as primate vocalizations—and thus should be called displays—an important subset are individually learned and flexibly used, especially in the great apes, and so may be properly called intentional signals" (20). It is clear then that while displays and intentional signals both share the function of influencing the behavior of others, the latter possess two inextricable qualities: *volitional flexibility,* i.e., "the individual controls their use," and *agent-directed behavior* with "the goal of influencing others." Note that for Tomasello a theory of mind is implicit in the latter. In order to intentionally alter "the attentional state of the recipient, (…) the communicator needs some kind of cognitive model of how the recipient perceives the signal and acts as a result" (45). It is therefore evident that for Tomasello, the production of intentional signals involves second-order intentionality (Dennett 1987). Namely, in order for their signals to modify not just the behavior, but also the mental states of others, the signaler must possess an understanding of both their own mental states and those of others, e.g., alarm calls are intended to alter an ignorant receiver's *knowledge* of the situation, and cause them to flee because of this new information. Second-order intentionality is contrasted with first-order intentionality, where signals are produced with a desire to influence the behavior of others but, crucially, the signaler need not understand the mind of the receiver in order to predict and recognize the behavioral effects of their signal, e.g., the receiver flees. Tomasello (2008) appears to suggest that chimpanzee vocalizations, like all "communicative displays," are produced with *zero-order* intentionality; namely, the signaling behavior is purely reflexive and is involuntarily produced with zero mentality attributed to the receiver(s). Since chimpanzee gestures are often produced with persistence and attention to the attention of the receiver—behavior that best exemplifies second-order intentionality—Tomasello uses behavior of this nature as his primary criterion for intentional signals.

On a more foundational level, Tomasello (2008: 21) identifies a strong connection between volitional flexibility and advanced capacities for signal learning. Whether or not a given behavioral trait requires learning can be a strong indication as to whether that trait is genetically determined, and as a result, the extent to which its production may be voluntary or intentional. Tomasello thus uses the presence of *social and/or asocial learning* in gestural communication, and the apparent lack of this characteristic in the chimpanzee vocal modality, as further evidence for the lack of intentionality in the latter.

It is clear then that if a signal is capable of being learned and utilized flexibly with persistence and attention to the attention of the other, that signal satisfies Tomasello's criteria for intentional production. As mentioned above, this chapter confronts these criteria beginning with social learning, continuing on to communicative flexibility, and finally engaging intentional production. It is worth stressing that Tomasello denies that the vocalizations of great apes satisfy any of these criteria. According to Tomasello, the modality of ape gestures "contrasts totally with their unlearned, inflexible, and emotional vocalizations indiscriminately broadcast to the world" (320).

3 Social and Asocial Learning

In *Origins of Human Communication*, Tomasello spends a considerable amount of time providing evidence that chimpanzees are capable of learning novel gestures—a feat unavailable to the vocal modality. This is presumably because, according to Hammerschmidt and Fischer (2008: 93), a "prerequisite for a high degree of (communicative) flexibility is learning, in terms of both production and comprehension." The capacity for social and/or asocial learning is therefore a logical precondition for both flexibility and intentionality in primate vocalizations. The overarching goals of this section are to demonstrate that the existing evidence in support of vocal learning in chimpanzees is (1) comparable to the evidence in support of learning in chimpanzee gestures, and (2) currently insufficient to justify Tomasello's conclusions on the matter.

The parameters of which acts of behavioral transmission should be classified as instances of "social learning" have been the subject of considerable debate. As Galef (1976) observed, interactions among conspecifics in several species are known to affect the acquisition and expression of complex behaviors. There is clearly a difference, however, between a social influence on the *use* or *application* of "innate" behaviors and behavior that either is *itself* modified or that would not exist if the subject were not socialized to exhibit it. In their oft-cited paper on social learning in animals, Janik and Slater (1997) usefully distinguish between "contextual learning" and "production learning." The former refers to learned modifications in the contextual usage of the signal, and the latter "refers to instances where the signals themselves are modified in form as a result of experience with those of other individuals." *Vocal learning*, they argue, "is defined by production learning in the vocal domain." In order for a vocalization to be classified as "socially learned," then, I submit that it must either be (1) a novel vocalization that is only used in particular groups of conspecifics, (2) a vocalization that individuals do not develop when raised in isolation, or (3) an idiosyncratic vocalization shared primarily by the mother and her own offspring. While species that meet any or all of these criteria are extremely rare [see Cheney and Seyfarth (2010) for comments], the well-documented case of the zebra finch has, for instance, demonstrated that production learning of vocalizations does occur outside of the human domain. The juvenile zebra finch will learn the complex, idiosyncratic calls of its mother during its first few weeks and then will begin to lose them with age as its hearing deteriorates (Fehér et al. 2009). The zebra finch is thus a paradigm case of vocal production learning in the animal kingdom.

Though he is not explicit about this, Tomasello appears to use "production learning" as his criterion for social learning. Support for this claim is may be identified in Tomasello's two major arguments about the relative uniqueness of chimpanzee gestures. First, "individuals with significant human contact invent or learn different kinds of novel gestures quite easily," and second, "there are many and very large individual differences in the gestural repertoires of different individuals of the same species, even within the same group, including some idiosyncratic gestures produced by single individuals" (21). Tomasello then contrasts these points with claims that "within any monkey and ape species all individuals have the same basic vocal repertoire, with essentially no individual differences in repertoire," and that "monkeys raised in social isolation and monkeys cross-fostered by another species (…) still produce their same basic species-typical vocalizations" (16).

I submit that Tomasello's first claim is false and that his second claim is deceptively used and largely irrelevant to his argument. I will confront these two claims in reverse-order beginning with the latter. In their comprehensive overview of research in primate vocalizations, Hammerschmidt and Fischer (2008: 94) observe that “Most of the evidence accumulated about vocal development comes from studies on monkeys, while little is known about the vocal development of apes.” Tomasello himself actually recognizes this strong asymmetry in the available data (Tomasello and Zuberbuhler 2002). While it is true that monkey vocalizations almost certainly do not demonstrate production learning—e.g., squirrel monkeys deafened at birth acquire structurally equivalent calls to normal monkeys (Winter et al. 1973)—notably, the *same is true of monkey gestures,* which are largely identical across species (Arbib et al. 2008). Indeed, the general consensus among primatologists is that monkey and ape species differ significantly in cognitive capacity and communicative modality [see Arbib et al. (2008) for a review]. Tomasello’s use of the limitations of monkey vocal development is therefore not sufficient to support his claim that *ape* vocalizations are “unlearned” or “not individually learned” (Tomasello 2008: 320, 33). This claim brings me to my second point, which is that though they are few in number, there do exist studies that suggest a marked capacity for social and asocial vocal production learning in chimpanzees.

Tomasello’s claim that within ape species “all individuals have the same basic vocal repertoire” is challenged on a number of levels. First, Leavens et al. (2004) demonstrated that captive chimpanzees produce “novel” grunts that vary in tone, timbre, and length previously unheard in the wild.[2] A number of studies since have corroborated Leavens et al.’s interpretation that these vocalizations appear to be used exclusively among captive chimpanzees toward humans for “attention-getting” purposes (Russell et al. 2005; Hostetter et al. 2007; Hopkins et al. 2007; Taglialatela et al. 2012). Until recently, primatologists have been in the dark as to potential causal explanations for the emergence of these signals unique to apes raised in captivity. One plausible hypothesis is offered by Leavens et al. (2010), who argue that some captive apes learn, asocially, how to solve a problem unique to their environment. In the wild, chimpanzees can attain food whenever they want; only in captivity do they often need to *ask* for it. It is therefore possible that captive chimps are spontaneously learning to use AG-vocalizations (and humans) as tools to achieve desired ends outside their natural environment.

In a recent paper entitled “Social learning of a communicative signal in captive chimpanzees,” Taglialatela et al. (2012) offer data in support of social transmission of AG-vocalizations. This study found that juveniles raised by their biological mothers in captivity are far more likely to exhibit these novel vocalizations than those juveniles raised by humans in a nursery environment. Further, a strong correlation exists

[2] Note that I am referring here to what are commonly referred to as “extended food grunts” and not “raspberry” sounds since only the former employ the vocal chords. Though raspberry sounds have never been observed in the wild and serve the same AG-function as novel grunts, they involve only a pursing of the lips. The appropriate place of whistles and lip-smacks in discussions of the evolution of human communication is certainly interesting yet remains beyond the scope of this chapter.

between *which* of the two basic AG-vocalizations the mother typically uses and the one picked up and utilized by the infant. From these discoveries the authors conclude that, "These data support the hypothesis that social learning plays a role in the acquisition and use of communicative vocal signals in chimpanzees" (3). Captive chimpanzees therefore exhibit novel vocalizations previously unheard among conspecifics in the wild and there is a strong correlation between idiosyncratic vocal production in mothers and their own offspring. As evidence for vocal production learning in captive chimpanzees, these studies challenge Tomasello's claim that while ape gestures are "individually learned," "this is not true of ape vocalizations" (33).

There have also been studies that suggest production learning in the vocalizations of wild chimpanzees. Using spectrographic analyses in two separate studies, Mitani et al. (1999) and Mitani and Nishida (1993) found significant differences in the vocal calls of adjacent and distal chimpanzee groups in East Africa. Crockford et al. (2004), moreover, found that the pant hoots of male chimpanzees living in three adjacent communities along the Ivory Coast differ more strongly from each other than either of them do from those of a community over 70 km away. They conclude that since neither habitat nor genetics appear to account for this difference, these chimpanzees seem to have "actively modified" their pant hoot structure to better facilitate in-group identification. While "actively modified" is surely a contentious interpretation of the data, note that even the more modest interpretation, i.e., unconscious "call convergence," still demonstrates the learning and dissemination of idiosyncratic call *structures* (Marshall et al. 1999). Marler (1991), for instance, has suggested that "action-based learning" can selectively reinforce structural call variants in chimpanzees. Regardless of how one interprets this data, these findings are clearly more suggestive of Janik and Slater's notion of *production learning* than they are of *contextual learning.*

Since the literature on this subject is rare and still in its nascent stages, Tomasello is unjustified in his conclusion that while ape gestures are "individually learned," "this is not true of ape vocalizations" (33). Contrary to Tomasello's view, chimpanzees appear to be capable of socially acquiring vocalizations as per the stringent criteria of Janik and Slater (1997) *and* in comparison with Tomasello's data on ape gestures. For instance, Tomasello's claim that apes with significant human contact learn "different kinds of novel gestures" applies equally to vocalizations. Also, his claim that various chimpanzee groups exhibit different gestural repertoires is also evident in chimpanzee vocalizations. As I mentioned above, this is not to suggest that novel vocalizations are as easily and flexibly acquired as gestures, but rather that the production of gestures and vocalizations do not differ in their general capacity to be learned by chimpanzees.

4 Contextual Flexibility and Smart Receivers

As mentioned above, Tomasello (2008) never offers a positive definition of volitional flexibility in animal minds. Instead, he tends to define volition negatively in relation to inflexible "vocal displays" (16). This tactic is slightly misleading since a negative definition is hardly a definition at all. It is, however, understandable. As Povinelli and collaborators (Povinelli and Eddy (1996), Povinelli and Vonk (2006)) have repeatedly pointed out, studies of the workings of animal minds are almost exclusively limited to interpretations of animal behavior. While the literature is slowly incorporating more work in neuroscience, Tomasello does not present any evidence suggesting that chimpanzees lack neurological capacities for "volitional" communication. Tomasello's means of defining volitional flexibility is therefore to contrast various communicative *behaviors*: A "signaler has intentional control over the signal" (13) rather than being "controlled by evolutionary processes" (14) if the signal is not "ritualized" or accompanied by an external stimulus or an emotion (16–17). As such, he claims, AG-gestures are ideal candidates for volitional flexibility because they occur in a wide variety of contexts, most of which are non-urgent and divorced from impulsive emotions. Tomasello thus argues that while vocalizations have an evolutionary history of association with urgent functions such as alarm calls and mediating conflict, chimpanzees occasionally use AG-gestures to call attention to displays that initiate common, everyday activities such as sex, play, nursing, begging, and grooming (20). Tomasello's principle criterion for volitional flexibility is therefore the capacity to use a signal in contexts that are not ritualized or emotionally urgent.

Tomasello understands the apparent contextual limitations of chimpanzee vocalizations and the non-emotional expression of these vocalizations to be intimately related: "the connection between a vocal call and its eliciting emotion or situation is mostly very tightly fixed; non-human primates do not vocalize flexibly by adjusting to the communicative situation" (16–17). For concrete evidence of this claim, Tomasello again relies almost exclusively on monkey data (18–20). For reasons discussed in the previous section, this move is ultimately inadequate to make his point. The only additional evidence that Tomasello provides with respect to chimpanzees is a field observation made by Jane Goodall in the 1980s that "The production of a sound in the *absence* of the appropriate emotional state seems to be an almost impossible task for a chimpanzee" (17) and the fact that chimpanzees make pant hoot calls in the presence of food even when everyone else is already present. However, as the authors of this latter study (Clark and Wrangham 1994) suggest these pant hoots are more likely signals of *status* than alarm calls. Tomasello is therefore unjustified in using this example alongside those of alarm calls to (surreptitiously) suggest the emotional inflexibility of chimpanzee vocalizations.

These two sources are the extent of Tomasello's evidence with respect to chimpanzees. The rest of his evidence is taken from studies on monkey alarm calls wherein, according to Seyfarth and Cheney (2003: 168), "Listeners acquire information from signalers who do not, in the human sense, intend to provide it." In these cases, "alarms" appear to be involuntary, emotional vocalizations produced reflexively in response to stimuli that nearby animals merely "eavesdrop" on (Cheney and Seyfarth 2005, 2007). Monkey alarm calls are therefore unlikely to be flexibly or intentionally produced. Macaque mothers, for instance, do not vocalize when predators approach their young so long as they are themselves a safe distance away (Cheney and Seyfarth 2005).

Both Cheney and Seyfarth (2005) and Tomasello (2008) explain this surprising behavior by suggesting the absence of a theory of mind in macaques. Presumably, if the mother understood her juvenile's *perspective,* i.e., as being *unaware* of the predator, she would have informed her of its presence (Cheney and Seyfarth 2005). Regardless of the strength of this reasoning—or, for that matter, whether or not certain monkey species possess a theory of mind[3]—Tomasello's argument here draws attention to another serious flaw in his reliance on monkey data to make claims about apes: If one does not believe that monkeys have a theory of mind, then it is unfair to use them as evidence for the lack of cognitive capacities in a species that one believes *does* possess a theory of mind. This criticism is particularly damning for Tomasello's argument that chimpanzee vocalizations are incapable of intentional production—a capacity, recall, that Tomasello claims requires a theory of mind. Notably, there do not appear to be any available studies that demonstrate analogous behavior with respect to chimpanzee alarm calls. If there is, such studies do not appear in any of the major review papers on the subject, i.e., Cheney and Seyfarth (2010), Hammerschimdt and Fischer (2008), or Arbib et al. (2008).

However, even if such evidence did exist its utility would ultimately be circular. This is because alarm calls are *exclusively* sounded in urgent, emotionally charged contexts. Another significant flaw in Tomasello's argument that chimpanzee vocalizations are inextricable from emotions is therefore that the only examples he provides involve contexts where this must be the case. For evidence of vocalizations in non-urgent contexts, I refer back to the attention-getting vocalizations described in the previous section. Like ape gestures, the AG-vocalizations used exclusively with humans have been utilized in non-predatory and non-competitive contexts without direct emotional stimulation. In fact, Russell et al. (2005) and Hopkins et al. (2007) conducted experiments to control for precisely this factor. Whereas Leavens et al.'s original (2004) study used visible food to elicit AG-vocalizations, Russell's team found that chimpanzees will utilize AG-vocalizations to solicit a necessary tool from a human in order to attain food that is out of sight. This finding suggests not only that these vocalizations are not automatic, emotional reactions to stimuli (food), but also that they are used selectively to communicate with others in instances of practical reasoning. According to Tomasello, "practical reasoning" about others' perspectives, e.g., a theory of mind, "underlies flexible communication" and is fundamental to intentional communication (48-9).

Further support for the claim that chimpanzees are capable of selectively using AG-vocalizations may be found in the results of Hopkins' group. Hopkins et al. (2007) recorded the frequencies of both traditional "food" vocalizations and AG-vocalizations when the chimpanzees were presented with the food alone, the human alone, or the food visible with a human. They found that chimpanzees produce significantly more "food" vocalizations with food alone than in the other

[3] See Flombaum and Santos (2005) for evidence suggesting that rhesus macaques can be sensitive to the attention of others and engage in practical reasoning about others' perspectives. Further, Wich and de Vries (2006) offer evidence that Thomas langur monkeys possess the capacity for goal-directed vocal production.

two conditions and significantly more AG-vocalizations when food was presented along with a human. Hopkins et al. conclude that "chimpanzees recognize functional differences" between various calls and can "selectively produce" AG-vocalizations in particular contexts.

Recent studies of chimpanzees in their natural habitat corroborate these findings, further challenging Tomasello's claim that chimpanzee vocalizations are "broadcast indiscriminately." In line with the studies on captive chimpanzees, wild chimpanzees have been documented producing agent-directed vocalizations to conspecifics [Goodall (1986), Townsend et al. (2008), Slocombe et al. (2010), Laporte and Zuberbühler (2010), Schel et al. (2013a, 2013b); also see Halloran (2012) for spectrographic analyses of context-specific, agent-directed AG vocalizations among conspecifics in captivity]. Contrary to Tomasello's questionable interpretation of Clark and Wrangham's (1994) findings on chimpanzee food calls, a recent study by Slocombe et al. (2010) suggests that wild male chimpanzees "produce food-associated calls selectively in the presence of important social partners." The authors claim that the stable, long-term relationships formed by male chimpanzees play an important role in whether or not food vocalizations are produced. Further, Schel et al. (2013b) have more recently provided evidence of food calls by wild male chimpanzees being directed at specific individuals based on rank and friendship. They conclude by suggesting that "chimpanzee food calls are not simply reflexive responses to food, but can be selectively directed at socially important individuals." They continue: "Our findings are thus inconsistent with traditional views of primate vocalizations as inflexibly and indiscriminately produced. Instead, our results indicate that great apes can produce semantically meaningful calls in a highly selective, recipient-directed manner." Audience composition therefore appears to have a marked influence on not only *when* male chimpanzees produce food vocalizations, but also to *whom* these calls are directed.

Laporte and Zuberbühler (2010) likewise provide field data suggesting that female chimpanzees "selectively deploy vocal signals depending upon the social context." The authors reported that female chimpanzees are more likely to produce agent-directed vocalizations toward lower-ranked males in the absence of the group's alpha male. Complementary findings can be found in an earlier paper entitled "Female Chimpanzees Use Copulation Calls Flexibly to Prevent Social Competition," wherein Townsend et al. (2008) report that female chimpanzees called significantly more around high-ranking males and suppressed vocalizations if high-ranking females were within earshot. According to the authors, since competition among females is dangerously high in the wild, "Copulation calling may therefore be one potential strategy employed by female chimpanzees to advertise receptivity to high-ranked males, confuse paternity and secure future support from these socially important individuals." Townsend et al. claim that their findings contradict an established hypothesis that copulation calls are merely a sexually selected trait to instigate male–male competition. Females in fact appear to have considerable selective control over when they vocalize and whom they vocalize to—an hypothesis reinforced by a hormone analysis that demonstrated that their subjects vocalized at times unrelated to their fertile period.

These results are further corroborated by studies measuring the effects of audience composition on both the selection and structure of chimpanzee vocalizations during inter-group conflict (Wilson et al. 2001; Slocombe and Zuberbühler 2007). Slocombe and Zuberbühler (2007), for instance, found that victims of agonistic interactions "alter the acoustic structure of their screams depending on the severity of aggression experienced, providing nearby listeners with important cues about the nature of the attack," and that if one audience member matched or surpassed the aggressor in rank, these victims reliably "exaggerated the true level of aggression experienced." The authors therefore conclude that "chimpanzees possess sophisticated understanding of third-party relationships, so-called triadic awareness, and that this knowledge influences their vocal production." Significantly, this data suggests that Tomasello is incorrect that the fact that chimpanzees vocalize most often in emotional contexts necessarily reflects negatively on the agent-directedness and contextual flexibility of their calls.

Notably, all of the studies discussed in this section suggest that vocalizations are used to facilitate intentional, goal-directed behavior. As will become relevant in the next and final section, this behavior can also be explained by employing learned behavioral rules rather than, as it might initially appear, a theory of mind. As I aim to show, however, the same criticism applies to analogous studies of chimpanzee gestures.

Before turning to this issue, it is important to note that the studies discussed thus far also suggest that a limited call repertoire can nonetheless demonstrate flexibly in function when semantic context is taken into account. In his analysis of primate vocalizations, Tomasello restricts his discussion to the vocal capacities of the sender, but once context is taken into consideration, it becomes clear that one does a great disservice to the communicative potential of not only primates, but all vocalizing species, to discount the role of the receiver in information transmission. The context in which vocalizations are produced can provide "smart receivers" with a wealth of salient information. In their survey of animal vocal communication, Cheney and Seyfarth (2010: 93) claim that even though vocal production is largely restricted across the animal kingdom, one should not assume that these physiological limitations correspond to limitations in signal *content:*

> Many species of birds and mammals have only a small repertoire of acoustically fixed vocalizations. However, because calls are individually distinctive and each call type is predictably linked to a particular social context, this limited call repertoire can nonetheless provide listeners with an open-ended, highly modifiable, and cognitively rich set of meanings, because individuals can continue to learn new sound-meaning pairs throughout their adult lives. As a result, listeners can potentially acquire a huge number of messages from a finite number of call types.

Similarly, Hammerschmidt and Fischer (2008: 113) conclude their (primarily negative) summary of vocal limitations in primates with the following: "It might in fact be the case that smart listeners lift some of the pressure off senders because they not only are able to perceive fine-grained differences among calls but also are able to make rich interpretations of calls—in other words, to attribute meanings to them." Because primate species typically live in large communities where the social hierarchy is strictly observed and—as in the case of chimpanzees—often shifting, the *context* of a given vocalization might be thought to carry just as much or even more information than the *type* of vocalization produced. Playback experiments by Cheney and Seyfarth (1990, 2005, 2007) show that baboons recognize specific breaks in social hierarchy and therefore deduce complex "social narratives" from calls. This is a significant feat when one considers that baboon troops live in communities as large as 150 members. The fact that monkeys have the cognitive capacity to identify the specific vocalizations of individual group members has also been demonstrated in ape species [see Berlin (2006) for a review]. Kojima et al. (2003), for instance, found that captive chimpanzees can not only successfully match the pant hoots, grunts, and screams of conspecifics to their photographs, but can also identify both individuals during "duets." Further, Levréro and Mathevon (2013) identified distinctive vocal "signatures" in adult and infant chimpanzees that were consistently present in a variety of different types of call from the same individual. The authors suggest that chimpanzee vocal signatures have tremendous evolutionary value as they, among other functions, facilitate in-group identification and aid mothers in keeping track of their offspring.

In response to their findings on baboons, Cheney and Seyfarth (2005: 149–150) express their surprise that "an animal that can learn to associate hundreds of sounds and symbols with objects and events find it so difficult to produce novel calls or create novel call combinations." While this may be true with monkey species, there do exist studies demonstrating the use of "creative call combinations" by chimpanzees (Crockford and Boesch 2003, 2005). For instance, the same year that Cheney and Seyfarth made this claim, Crockford and Boesch (2005: 397) conducted a spectrographic analysis of calls made by chimpanzees in the Tai Forest finding that over half of their vocalizations "occurred in combination with other vocalizations or with drumming," and that overall, these chimpanzees utilize a total of "88 different types of combinations" each produced in "specific contexts." They therefore conclude that vocal call combinations are an "important" part of chimpanzee communication that "increase message complexity" by increasing "the range of information that can be decoded by listeners."

Both the controlled and field studies discussed in this section strongly suggest that chimpanzees not only have voluntary control over their ability to vocalize, but also *which* vocalizations they use and *who* they vocalize to. We may conclude, then, that Goodall's field observation that vocalizing in the absence of emotions appears to be an "impossible" task for a chimpanzee is false. Further, when combined with the notion of smart receivers and the extensive use of call combinations, chimpanzee vocalizations have impressive potential for contextual flexibility *even when* produced in emotional contexts. It is therefore reasonable to conclude that these vocalizations meet Tomasello's second criterion for contextual flexibility in both signal production and utility. While this is certainly significant, the criterion that truly separates "intentional signals" from common displays is the third and final criterion: that apes vocalize referentially and discriminately, i.e., with attention to the attentional state of the receiver.

5 Intentionality, Theory of Mind, and the Logical Problem

Though certainly significant in its own right, the evidence and argumentation produced thus far have been intended as foundational to a considerably larger project toward which much future work needs to be done. By attempting to substantiate the hypothesis that chimpanzee vocalizations have the potential to be produced intentionally with volitional flexibility, I have taken a bottom-up approach to suggesting that certain types of their vocalizations are socially learned (and thus are not genetically predetermined) and can be flexibly and selectively produced in non-emotional and non-ritualized contexts. Although Tomasello denies chimpanzee vocalizations even these more basic qualities, they are not terribly rare among vertebrates in the animal kingdom (Cheney and Seyfarth 2010). What remains to be demonstrated is that apes possess the extraordinary ability of using their vocalizations *intentionally*—a capacity that, Tomasello claims, outside human communication only ape gestures appear to facilitate.

In the previous section, I cited studies suggesting that chimpanzees are capable of "selectively choosing" their vocalizations in interactions with humans and conspecifics and that they use novel "attention-getting" vocalizations specifically in these contexts. What is unclear is whether they have learned new behavioral rules, i.e., use vocalization *A* in situation *B* to receive a given reward, or whether they are choosing particular vocal signals intentionally with the goal of altering not just the *behavior* (first-order intentionality) but the mental state, i.e., the "attention" (second-order intentionality), of the human and therefore possess a theory of mind.

It is important to stress here that these two explanations, i.e., theory of mind and behavioral rules, are "functionally equivalent" (Cheney and Seyfarth 2005: 138). In other words, they both achieve the same result, which, in the controlled studies above, is receiving food. This fact has led to perhaps the most highly debated issue in the animal minds literature, which is often referred to as the logical problem. The logical problem states that since all we can observe is an animal's behavior, there is no conceivable way to distinguish whether that animal is employing a theory of mind or a behavioral rule when interacting with others [see Povinelli and Vonk (2006) and Lurz (2011) for comments]. Because there is no functional difference between theory of mind and behavioral rules, and because the latter are far less cognitively taxing, it is scientifically irresponsible to assume that non-human animals possess a theory of mind (Povinelli and Vonk 2006). In this section, I remain neutral on the debate itself and argue that the force of the logical problem applies equally to claims of second-order intentionality in both vocal and gestural signals.

I will begin with a brief overview of Tomasello's argument which again relies almost exclusively on data taken from monkey species in the context of alarm calls. According to Tomasello (2008), since primate vocalizations are "broadcast indiscriminately to everyone nearby," this means that "psychologically" the signaler "need not pay any attention to the recipients, and indeed cannot easily direct vocal calls to selected individuals to the exclusion of others" (18). Elsewhere he claims that, besides vocalizing in slightly modified forms to distinguish different amounts of food, "great apes do not produce any referentially specific calls" (16). Tomasello's argument for intentional communication in ape gestures ultimately boils down to the fact that, when gesturing, apes occasionally pay attention to the attention of the receiver. Chimpanzees are known to alternate their gaze between food and human subjects while gesturing (Leavens and Hopkins 1998) and, monitoring the receiver's response, "repair" communication when it has failed by moving closer and/or trying different gestures (Liebal et al. 2004; Leavens et al. 2005). These studies, Tomasello notes, "show persistence to a goal with adjusted means as necessary—the prototype of intentional action."

What Tomasello (2008) conspicuously does not mention is that the cited Leavens and Hopkins study notes that both gestures *and* vocalizations were utilized during gaze alteration. Indeed, they conclude their paper as follows:

> This gesturing or vocalizing while alternating the gaze between the communicative interactant and the object of interest cardinally represents what has been called in both the human development literature and the literature pertaining to communication in apes, intentional communication (819).

Tomasello's selective emphasis is also demonstrated in his reference to the Leavens et al. (2004) study, which found that chimpanzees frequently use their AG-vocalizations as *alternatives* to physical readjustment in situations when an inattentive human is facing the wrong direction. In fact, several studies found that chimpanzees modulate their vocal production depending on the orientation (Hostetter et al. 2001) and visual attention of the human receiver (Krause and Fouts 1997; Theall and Povinelli 1999; Bodamer and Gardner 2002; Hostetter et al. 2007; Hopkins et al. 2007). Two of these studies (Bodamer and Gardner 2002; Leavens et al. 2004) found that when one vocalization failed to catch the attention of a human, chimpanzees were *more likely* to try a different vocalization than when the original AG-vocalization was successful. In their analysis of data from nine studies on the use of AG-vocalizations, Hopkins et al. (2007) conclude that the general findings suggest that chimpanzees differentially produce vocalizations according to different attentional cues. The fact that all of these studies were released prior to *Origins of Human Communication* makes it surprising that Tomasello does not at least acknowledge them.

In the human development literature, tactics of persistence and elaboration have frequently been noted as indicators of intentional communication (Bates et al. 1975; Golinkoff 1993). We have seen that Tomasello also clearly classifies signals that demonstrate these communicative tactics as being indicative of intentional production. In line with this literature, Leavens et al. (2005, 2010) found that chimpanzees used tactics of persistence and elaboration to vary "their signals within a modality that was appropriate to the attentional status of a human" to "rapidly accommodate changes" in the human's attention. In a study of 110 chimpanzees, Leavens et al. (2010) found that both their AG-gestures and AG-vocalizations "followed a logical and efficient pattern of modality-specific permutations" as the chimpanzees continued "to elaborate in a tactically efficient manner throughout a minute-long episode in which their communication was having no apparent effect." The results of these studies certainly suggest, in Tomasello's words, "persistence to a goal with adjusted means as necessary."

Not only do all of these studies suggest that chimpanzee vocalizations can be produced, suppressed, and elaborated upon depending on the attentiveness of a human receiver but, more significantly, that ape vocalizations can, like gestures, be produced intentionally. This is, at least, the conclusion that one is committed to if, like Tomasello, one is convinced that when gestures are used in the same context they exhibit "the prototype of intentional action."

This is precisely the conclusion that Schel et al. (2013a) arrive at in their recent study on wild chimpanzees entitled "Chimpanzee Alarm Call Production Meets Key Criteria for Intentionality" which assessed whether chimpanzees recognize mental states of knowledge and ignorance in conspecifics and then utilize that information selectively in their alarm calls. Schell et al. claim that "In contrast to gestural research, intentionality has rarely been the focus of primate vocal research, thus it is vital that directly comparable evidence is gained to empirically test whether great ape vocal production engages first-order intentionality." Their experiment elaborated upon findings published the previous year by Crockford et al. (2012). Crockford's group ran a field study that demonstrated that chimpanzees were more likely to produce alarm calls in response to a predator when in the presence of unaware or ignorant group members then they were when in the presence of group members with clear knowledge of the threat. In this experiment, a model viper was placed on the projected travel path of a group of 33 chimpanzees, whose vocalizations were then recorded. The authors found that "alarm calls were significantly more common if the caller was with group members who had either not seen the snake or had not been present when alarm calls were emitted," thus concluding that "chimpanzees monitor the information available to other chimpanzees and control vocal production to selectively inform them." Though the results of this study are strongly suggestive of intentional communication, the experiment suffers from one potential oversight: in each iteration, the signaler was also ignorant of the predator, so although *more* calls were given in the presence of unaware group members, the signaler's initial calls may have been produced automatically upon discovery of the snake. This complementary explanation would suggest that the signalers original intention was not to alert group members after all but was rather a reflexive expression of their own fear, i.e., zero-order intentionality.

In their own version of this experiment, Schel et al. (2013a) corrected for this oversight and in doing so "tested the production of chimpanzee vocal signals across multiple markers of intentionality, in a comparable manner to chimpanzee gestures," a task never attempted in previous studies. They begin by distinguishing three distinct types of alarm calls: "soft huus" (SH), "alarm huus" (AH), and "waa barks" (WB). Upon encountering the snake alone in the absence of other group members, subjects only elicited the comparably low, short-ranged SH vocalizations. The other two, much louder and abrasive alarm calls, were discovered to be reserved exclusively for the presence of group members, and their production "exhibited characteristics previously used to argue for intentionality in gestural communication." Firstly, production of AH and WB vocalizations were more likely to accompany the arrival of a friend or a dominant than non-friends or lower-ranking individuals. In line with studies discussed above, these findings suggest that certain alarm calls are produced tactically and discriminately toward significant individuals. Secondly, signalers who produced AH and WB calls visually monitored significant individuals in the audience and continually alternated their gaze between these individuals and the snake. Thirdly, signalers persisted in their calls until all group members were a safe distance from the threat, suggesting goal- and agent-directedness in their calls. Schell et al. conclude their paper as follows: "We interpret these patterns as evidence that chimpanzee alarm calling meets the key diagnostic features of intentional signal production. Although each of these behaviors can be explained separately as the product of less complex cognitive processes, the combined overall pattern is more consistent with the hypothesis that call production is both socially directed and goal-directed." When taken alongside the plethora of complementary studies discussed thus far, this conclusion is decidedly substantive and serves to effectively refute Tomasello's hypothesis that the AG-gestures of chimpanzees are, at a foundational level, functionally different from AG-vocalizations.

Over the past three sections, I have attempted to show that the only truly salient difference between the vocalizations and gestures of chimpanzees is that the latter possess a more open-ended plasticity in production. However, because of the logical problem, the question as to whether this production is *intentional* applies equally to both modes of communication. To see this, consider the following complementary explanation of the Liebal et al. (2004) study on "attention-monitoring" during gestural communication. From a purely behavioral perspective, two correlations have been observed. First, chimpanzees observed a correlation between the production of specific gestures and an action on the part of the experimenter that leads to a reward. Second, in order to explain the physical-reorientation behavior, it is reasonable to assume chimpanzees recognize a correlation between the facial or bodily orientation of the experimenter and specific actions that, combined with gestures, tend to bring about the desired effect. Similar behavioral rules can be applied to the analogous studies on ape vocalizations. In short, these chimpanzees did not need to attribute mental states of attention and inattention to humans in order to achieve their goal of receiving food. Judging from their behavior alone, they could have merely identified and took advantage of correlations between the actions of an experimenter, their own actions, and the appearance of desired rewards.

The fact that primatologists and philosophers refer to these gestures and vocalizations as "attention-getting" is therefore in a sense functionally accurate, but as the logical problem highlights, the mere *function* of a behavior does not allow one to distinguish its underlying and/or accompanying cognitive processes. I am suggesting that if apes *do* possess a theory of mind—and there is strong evidence in favor of this—then, when combined with the fact they can use vocalizations in ways functionally equivalent to gestures, there is no salient reason to doubt that a theory of mind is operating in these circumstances as well. By Tomasello's own criteria, great apes have been observed to use vocalizations that may justifiably be referred to as "intentional signals."

6 Conclusions

Contrary to the judgment passed not only by Tomasello but many primatologists [see Arbib et al. (2008), Hammerschmidt and Fischer (2008), and Cheney and Seyfarth (2010) for reviews] chimpanzees are capable of using vocalizations in ways considerably more advanced than the inflexible behavioral displays commonly observed in the biological world. In this chapter, I have used current research to provide what I hope to be a comprehensive, bottom-up account of the communicative potential of chimpanzee vocalizations. Beginning with evidence that chimpanzee vocal production can be both socially and asocially learned, I have offered reason to believe that, like their gestures, not all chimpanzee vocalizations are genetically pre-determined and evoked by particular stimuli and emotional states. To the contrary, chimpanzees appear to have some voluntary control over *which* vocalizations they employ, *when* they employ them, and to *whom* they are directed. Chimpanzees are therefore capable of producing vocalizations with a considerable degree of contextual flexibility. This potential is amplified tremendously when smart receivers and call combinations are considered and, further, if some of their vocalizations are utilized dyadically, i.e., with sensitivity to the attentional state of the recipient. Although there is some reason to doubt this capacity, I have shown that there is no *more* reason to doubt it for gestures than for vocalizations.

I noted above that Tomasello (2008) does not present any neurobiological evidence suggestive of his view that manual gestures are produced volitionally and vocalizations are not. Tomasello's hypothesis regarding the "gestural origins" of human communication is grounded exclusively on observations of chimpanzee behavior. In this chapter, I have gone to lengths to argue that chimpanzees use gestures and vocalizations in ways that are functionally equivalent from a behavioral point of view. I have suggested that this evidence provides a foothold for a multimodal account of the evolution of human communication. There is, however, recent neurobiological evidence that both adds support this hypothesis and, as such, further contradicts Tomasello's "gestural origins" thesis. In a recent paper entitled "Chimpanzee Vocal Signaling Points to a Multimodal Origin of Human Language," Taglialatela et al. (2011) present evidence that AG-vocalizations selectively activate the Broca's area homolog in chimpanzees. The Broca's area has long been recognized as an area of the human brain critical for vocal planning and production. Taglialatela et al. note that, prior to their study, it was known that the Broca's area homolog in chimpanzees was activated by manual gestures as well as the combination of gestures and sounds. Their new study revealed that while isolated *non*-AG-vocalizations fail to get a response from the Broca's area homolog, AG-vocalizations in the absence of gestures *do* activate area of the brain. According to Taglialatela et al. (2011), "the activity observed in the Broca's area homolog reflects the production of vocal signals by the chimpanzees, (suggesting) that this critical human language region was involved in vocal signaling in the common ancestor of both modern humans and chimpanzees."

All of this suggests that the vocalizations of great apes, both in addition to and as distinct from ape gestures, can be hypothesized to have played a significant role in the evolution of human communication. Though such a discussion would clearly far exceed the space available here, I will conclude with some brief thoughts on the subject. First, the capacity to recognize individual voices on their own and in a crowd can be a powerful communicative ability, especially when combined with theory of mind. This capacity would not only aid the evolution of speech and language, but may facilitate in-group identification and therefore the evolution of specific dialects. Second, unlike gestural communication, vocalizations have the potential to facilitate communication at long distances and, as I suggested above, aid in developing idiosyncratic group identities by distinguishing the calls of neighboring communities. This ability would also presumably be important in the facilitation of "between-group" communication as it avoids the tensions that can emerge in direct physical confrontation. Thirdly, like gestures, vocal signals have been shown to be combinable to increase message complexity. While the combination of vocal signals may not be as flexible as the "simple syntax" in the gestures of "linguistic apes" proposed by Tomasello (2008), it is notable that since we presently lack the ability to effectively "decode" primate call combinations, something approximating a "simple syntax" might in fact be the case among groups of conspecifics.

As was mentioned above, there is currently a scarcity of research on great ape vocal communication (Hammerschmidt and Fischer 2008; Schel et al. 2013a). This is especially true in comparison with studies on ape gestures. Consequently, we should not overlook the fact that the current asymmetry in the research on ape gestures and vocalizations corresponds to the general perception of their respective communicative potential and roles in the origins of human communication.

References

Arbib MA, Liebal K, Pika S (2008) Primate vocalization, gesture, and the evolution of human language. Curr Anthropol 49:1053–1076

Bates E, Camaioni L, Volterra V (1975) The acquisition of performatives prior to speech. Merrill Palmer Q 21:205–226

Berlin P (2006) Voice processing in human and non-human primates. Philos Trans R Soc B 361:2091–2107

Bodamer MD, Gardner RA (2002) How cross-fostered chimpanzees (Pan troglodytes) initiate and maintain conversations. J Comp Psychol 116:12–26

Cheney DL, Seyfarth RM (1990) How monkeys see the world. University of Chicago Press, Chicago

Cheney DL, Seyfarth RM (2005) Constraints and preadaptations in the earliest stages of language evolution. Linguist Rev 22:135–159

Cheney DL, Seyfarth RM (2007) Baboon metaphysics: the evolution of a social mind. University of Chicago Press, Chicago

Cheney DL, Seyfarth RM (2010) Production, usage, and comprehension in animal vocalizations. Brain Lang 115:92–100

Clark AP, Wrangham RW (1994) Chimpanzee arrival pant-hoots: do they signify food or status? Int J Primatol 15:185–205

Crockford C, Boesch C (2003) Context-specific calls in wild chimpanzees, Pan troglodytes verus: analysis of barks. Anim Behav 66:115–125

Crockford C, Boesch C (2005) Call combinations in wild chimpanzees. Anim Behav 142:397–421
Crockford C, Herbinger I, Vigilant L, Boesch C (2004) Wild chimpanzees produce group-specific calls: a case for vocal learning? Ethology 110:221–243
Crockford C, Wittig RM, Mundry R, Zuberbühler K (2012) Wild chimpanzees inform ignorant group members of danger. Curr Biol 22:142–146
Dennett DC (1987) The intentional stance. MIT Press, Cambridge
Fehér O, Wang H, Saar S, Mitra PP, Tchernichovshi O (2009) De novo establishment of wild-type song culture in the zebra finch. Nature 459:564–568
Flombaum JI, Santos LR (2005) Rhesus monkeys attribute perceptions to others. Curr Biol 15:447–452
Galef BG (1976) Social transmission of acquired behavior: a discussion of tradition and social learning in vertebrates. In: Rosenblatt JS, Hinde RA, Shaw E, Beer C (eds) Advances in the study of behavior, vol 6. Academic Press, New York, pp 77–100
Golinkoff RM (1993) When is communication a 'meeting of minds'? J Child Lang 20:199–207
Goodall J (1986) The chimpanzees of Gombe: patterns of behavior. Harvard University Press, Cambridge
Halloran AR (2012) The song of the ape: understanding the language of chimpanzees. St. Martin's Press, New York
Hammerschmidt K, Fischer J (2008) Constraints in primate vocal production. In: Griebel U, Oller K (eds) The evolution of communicative creativity: from fixed signals to contextual flexibility. MIT Press, Cambridge, pp 93–119
Hopkins WD, Taglialatela JP, Leavens DA (2007) Chimpanzees differentially produce novel vocalisations to capture the attention of a human. Anim Behav 73:281–286
Hostetter AB, Cantero M, Hopkins WD (2001) Differential use of vocal and gestural communication by chimpanzees (Pan troglodytes) in response to the attentional status of a human (Homo sapiens). J Comp Psychol 115:337–343
Hostetter AB, Russell JL, Freeman H, Hopkins WD (2007) Now you see me, now you don't: evidence that chimpanzees understand the role of the eyes in attention. Anim Cogn 10:55–62
Janik VW, Slater PJB (1997) Vocal learning in mammals. Adv Study Behav 26:59–99
Kojima S, Izumi A, Ceugniet M (2003) Identification of vocalizers by pant hoots, pant grunts and screams in a chimpanzee. Primates 44:225–230
Krause MA, Fouts RS (1997) Chimpanzee (Pan troglodytes) pointing: hand shapes, accuracy, and the role of eye gaze. J Comp Psychol 111:330–336
Laporte MNC, Zuberbuhler K (2010) Vocal greeting behaviour in wild chimpanzee females. Anim Behav 80:467–473
Leavens D, Hopkins WD (1998) Intentional communication by chimpanzees: a cross-sectional study of the use of referential gestures. Dev Psychol 34(5):813–822
Leavens DA, Hostetter AB, Wesley MJ, Hopkins WD (2004) Tactical use of unimodal and bimodal communication by chimpanzees, Pan troglodytes. Anim Behav 67:467–476
Leavens DA, Russell JL, Hopkins WD (2005) Intentionality as measured in the persistence and elaboration of communication by chimpanzees (Pan troglodytes). Child Dev 76:291–306
Leavens DA, Russell JL, Hopkins WD (2010) Multimodal communication by captive chimpanzees (Pan troglodytes). Anim Cogn 13:33–40
Levréro F, Mathevon N (2013) Vocal signature in wild infant chimpanzees. Am J Primatol 75:324–332
Liebal K, Pika S, Call J, Tomasello M (2004) To move or not to move: how apes adjust to the attentional state of others. Interact Stud 5:199–219
Lurz R (2011) Mindreading animals: the debate over what animals know about other minds. MIT Press, Cambridge
Marler P (1991) Song learning behavior: the interface with neuroethology. Trends Neurosci 5:199–206
Marshall AJ, Wrangham RW, Arcadi AC (1999) Does learning affect the structure of vocalisations in chimpanzees? Anim Behav 58:825–830

Mitani JC, Hunley KL, Murdoch ME (1999) Geographic variation in the calls of wild chimpanzees: a reassessment. Am J Primatol 47:133–151

Mitani JC, Nishida T (1993) Contexts and social correlates of long-distance calling by male chimpanzees. Anim Behav 45:735–746

Pollick AS, de Waal FBM (2007) Ape gestures and language evolution. PNAS. http://www.pnas.org/content/104/19/8184.long. Accessed 24 Oct 2013

Povinelli DJ, Eddy TJ (1996) What young chimpanzees know about seeing. Monogr Soc Res Child Dev 61:56–97

Povinelli DJ, Vonk J (2006) We don't need a microscope to explore the chimpanzee's mind. In: Hurley S (ed) Rational animals. Oxford University Press, Oxford

Premack D, Woodruff G (1978) Does the chimpanzee have a theory of mind? Behav Brain Sci 1:515–526

Russell JL, Braccini S, Buehler N, Kachin MJ, Schapiro SJ, Hopkins WD (2005) Chimpanzees (Pan troglodytes) intentional communication is not contingent upon food. Anim Cogn 8:263–272

Schel AM, Townsend SW, Machanda Z, Zuberbuhler K, Slocombe KE (2013a) Chimpanzee alarm call production meets key criteria for intentionality. PLoS ONE 8(10):e76674

Schel AM, Machanda Z, Townsend SW, Zuberbühler K, Slocombe KE (2013b) Chimpanzee food calls are directed at specific individuals. Anim Behav 86(5):955–965

Seyfarth RM, Cheney DL (2003) Signalers and receivers in animal communication. Annu Rev Psychol 54:145–173

Slocombe KE, Kaller T, Turman L, Townsend SW, Papworth S et al (2010) Production of food-associated calls in wild male chimpanzees is dependent on the composition of the audience. Behav Ecol Sociobiol 64:1959–1966

Slocombe KE, Zuberbühler K (2007) Chimpanzees modify recruitment screams as a function of audience composition. Proc Natl Acad Sci 104:17228–17233

Taglialatela JP, Reamer L, Schapiro SJ, Hopkins WD (2012) Social learning of a communicative signal in captive chimpanzees. Biol Lett. http://www.ncbi.nlm.nih.gov/pubmed/22438489. Accessed 24 Oct 2013

Taglialatela JP, Russell JL, Schaeffer JA, Hopkins WD (2011) Chimpanzee vocal signaling points to a multimodal origin of human language. PLoS ONE 6(4):e18852

Theall LA, Povinelli DJ (1999) Do chimpanzees tailor their gestural signals to fit the attentional state of others? Anim Cogn 2:207–214

Tomasello M (1996) Do apes ape? In: Heyes CM, Galef BG (eds) Social learning in animals: the roots of culture. Academic Press, San Diego, pp 319–346

Tomasello M (2008) Origins of human communication. MIT Press, Cambridge

Tomasello M, Zuberbuhler K (2002) Primate vocal and gestural communication. In: Bekoff M, Allen C, Burghardt G (eds) The cognitive animal: empirical and theoretical perspectives on animal cognition. MIT Press, Cambridge

Townsend SW, Deschner T, Zuberbühler K (2008) Female chimpanzees use copulation calls flexibly to prevent social competition. PLoS ONE 3(6):e2431

Wich SA, de Vries H (2006) Male monkeys remember which group members have given alarm calls. Proc R Soc B 273:735–740

Wilson ML, Hauser MD, Wrangham RW (2001) Does participation in intergroup conflict depend on numerical assessment, range location, or rank for wild chimpanzees? Anim Behav 61:1203–1216

Winter P, Handley P, Ploog D, Schott D (1973) Ontogeny of squirrel monkey calls under normal conditions and under acoustic isolation. Anim Behav 47:230–239

Part IV
Evolutionary Origins of Human Language

Communication and Human Uniqueness

Ian Tattersall

Abstract Modern human communication is dominated by language, an extremely unusual mode that appears to be intimately tied to our equally unusual symbolic form of thought as well as to our unique speech apparatus. Some view language as gradually acquired under natural selection, others as a sudden and recent acquisition. The disagreement arises because language leaves no direct traces in the material record, and anatomical proxies for speech such as cranial base or hyoid architecture have proven equivocal. Similarly, even sophisticated Paleolithic stone tools cannot be taken as proxies for symbolic thought, as cognitively complex as their makers may have been. Unequivocal evidence for symbolic thought—and by extension, for language—is only found in overtly symbolic objects, which first occur significantly *after* the appearance of *Homo sapiens*. This suggests that the biological substrate for symbolic thought resulted from the major developmental reorganization that gave rise to our anatomically distinctive species, but that the new potential was not exploited until it was exaptively released by a cultural stimulus, plausibly the invention of language. By this time, the vocal apparatus necessary for speech was already in place.

Keywords Communication • Human evolution • Language • Symbolism

1 Introduction

All primates are social, even if they are not gregarious. And therefore they communicate. Primates have many means of communication: vocal, postural, olfactory, gestural, and (among catarrhines) through facial expression. We human

---

I. Tattersall (✉)
Division of Anthropology, American Museum of Natural History,
Central Park West at 79th Street, New York, NY 10024, USA
e-mail: iant@amnh.org

M. Pina and N. Gontier (eds.), *The Evolution of Social Communication in Primates*, Interdisciplinary Evolution Research 1, DOI: 10.1007/978-3-319-02669-5_12,

beings use all of these modes of communication, although we place a premium on our vocal abilities. Or at least, we think we do. In reality, much of what we communicate is conveyed through body language, manual gesture, and facial expression. Most of what is expressed in this way is, however, emotional content: feelings and reactions of the kind shared with our living primate relatives, and presumably also with our now-extinct predecessors. And the reason we think of our communication as being overwhelmingly vocal is that our vocalizations, or at least, the symbolic meanings they represent, have uniquely been co-opted into the way we process information about ourselves, about each other, and about the world beyond: literally, into the way we think.

As a result of our long evolutionary history we share almost all of our attributes with a widening circle of organisms, and this is as true of our basic communication and interaction systems as it is of the nucleic acids that bond us to all other living organisms, or of the body structure that tells us we are tetrapods, or of the clever hands that make us higher primates. But, as is also true of every other species, we nonetheless possess features that make us different. To the eye, the most obvious of our unusual features are those directly related to our odd form of upright bipedal locomotion: the large, short-faced head, balanced precariously atop a vertical spine; the long legs and short, broad pelvis; the strange, stiff feet. Still, although our bizarre form of carriage has freed our grasping hands to allow us to develop and practice the dexterousness that is so essential to our adopted lifeway—without which we could never have become the kind of creature we are—it is not our weird body structure that gives us our acute feeling of alienation from the rest of the living world, our sense of being so *different* from everything else. It is the way we process information in our minds.

This is not to disparage the intelligence of other animals. It is perfectly possible to be very complex cognitively without dealing with knowledge in the particular way we do. But as far as we can tell, all other living organisms react more or less directly to the stimuli that impinge on them from the outside world, even if those reactions may be mediated in very sophisticated ways that involve multiple stored memories. In contrast, modern human beings literally re-create that world in their minds: So much so, that much of the time we live in the world not as it literally is, but as we have reconstituted it in our heads.

We are able to do this because of our symbolic faculty, our ability to deconstruct our inner and outer worlds into a vocabulary of symbols. Once this is done, we can shuffle those intangible symbols around, according to rules, to create new possibilities in our minds and to imagine things that we have never seen or otherwise experienced. Of course, our close relatives the bonobos and chimpanzees are highly complex creatures that can also recognize verbal and visual symbols (see Jensvold, this volume); they can combine those symbols in an additive way to make and to respond to simple statements such as "take … red … ball … outside." But there is a limit to the complexity or the creativity of any statement made by simply adding symbols in this manner; and, no matter how extensive the undoubted behavioral similarities between us and them, it is obvious when all is said and done that a chimpanzee's apprehension of the world is very different from our own.

2 Biology of the Human Capacity

The neural underpinnings of our capacity to manipulate symbols in an intricate fashion—what Marshack (1985) neatly termed "the human capacity"—are poorly understood. The notably large size of our brains has something to do with it, of course. But it is not the whole story, since as we shall see you can have a large brain and not be symbolic in the human sense, or at least not leave any evidence of being so. What is more, among individuals of *Homo sapiens* brain sizes vary hugely without any correlation to intelligence, however measured (Holloway et al. 2004). The volumes of our vast, globular cranial vaults are thus poor proxies for the functioning of the brains inside them. What is more, the same may be said for all other aspects of our bony structure—including the inner contours of the skull vault, from which endocasts representing the external morphology of the brain can be made. Paleoneurologists have debated the significance of variations in fossil endocast morphologies for years (Falk 1992; Holloway et al. 2004), without reaching any consensus. Such speech-associated and externally visible brain areas as Broca's cap have been identified in some very ancient members of the genus *Homo* (e.g., Walker and Leakey 1993); but it remains highly arguable whether the presence of such structures is in itself indicative of language, or even of speech. Most likely, they simply form part of a much larger complex of structures and interior brain connections that are all involved in the production of normal speech: They need to be there if you are going to speak, but do not by themselves indicate that you possess speech.

3 History of the Human Clade

Members of the quite speciose genus *Homo* have had the same tall, upright-striding basic body anatomy for at least the last 1.6 million years (Walker and Leakey 1993)—although there was a significant shift to lighter and slenderer build at the origin of our anatomically distinctive species *H. sapiens* (Tattersall 2012). The most striking osteological changes among Pleistocene hominids belonging to our clade (roughly, those living during the past two million years) occurred in the skull, as average sizes of the cranial vault grew and faces became reduced and less prognathous, eventually to become retracted under the front of the braincase as ours is today. The archeological record makes it clear that the expanding brain is telling us, in some way, about how hominids became, in a very general sense, more intelligent—perhaps "cognitively complex" is a better term—over the course of the Pleistocene. But unfortunately, it does not tell us anything very specific about how this quality developed, or how it expressed itself at any particular juncture in human evolution. What's more, it tells us nothing about precisely how our own particular lineage of hominids became more cognitively complex. This is because brain enlargement over time was actually a property of the entire genus *Homo*, having taken place independently within the genus in at least three separate lineages (the ones leading to *H. sapiens* in Africa, to *Homo neanderthalensis* in Europe, and to late *Homo erectus* in eastern Asia).

Putatively more informative than the cranial vault is the base of the skull, which also happens to be the roof of the upper vocal tract. As such, the skull base can at least potentially tell us something about the range of sounds early hominids used in their vocal communication (Laitman et al. 1979). It was even thought at one time that cranial base morphology might reflect the moment at which our forebears became capable of modern speech, something that in turn might relate to critical aspects of their cognitive potential. Here is how the reasoning went. The larynx (voice box) is a critical structure in modulating the column of air that generates the sounds we use in speech today. During the developmental period in which modern human infants begin to learn how to speak, the larynx moves from a position high up below the cranial base, behind the oral cavity, to a lower position in the neck. At the same time, the bony cranial base, which starts off as a flattish plane, becomes progressively more concave as maturity is achieved. In apes the cranial base stays flat throughout life, and the larynx remains high. So, in theory at least, the amount of cranial flexion in a fossil hominid skull will tell you the degree to which the individual involved was able to produce the sounds that are essential to producing language (Laitman 1984). In the event, though, it turned out that basicranial flexion among fossil hominids is wildly variable, although full flexion is typically only achieved in anatomically modern *H. sapiens*. Many Neanderthals, for example, turned out to have flattish cranial bases, arguing against their ability to produce the formant frequencies used in speech. Awkwardly, though, others turned out to have a noticeable degree of flexion, leaving the issue in question.

As it became obvious that the cranial base would remain an area of contention, many paleoanthropologists began to hope that fossil evidence of the larynx itself would help decide the matter. But when a Neanderthal hyoid bone, part of the largely cartilaginous laryngeal apparatus, was finally found (Arensburg et al. 1989), it turned out to be very similar to that of a modern human. Similarly, it was later discovered that archaic Neanderthal relatives from Spain possessed a middle ear apparatus that was able to process the range of sounds that we use today in speech (Martinez et al. 2008), a finding that was taken as *prima facie* evidence that these hominids could potentially speak. But there is a continuing problem with evidence of this kind, analogous to the one encountered with the discovery that Neanderthals possessed the modern variant of the FOXP2 gene, malfunctions in which impede the production of speech: (Krause et al. 2007). Because while, in all these cases, the modern conformation may be *necessary* for speech production or comprehension, in none of them can its possession be considered a *sufficient* condition for inferring speech—or, by extension, language. What is more, it has recently been persuasively argued (Lieberman 2007) that approximately equal proportions of the vertical and horizontal parts of the upper vocal tract are required to produce the sounds of speech; and this is a requirement that is basically fulfilled among fossil hominids only by early members of our species *H. sapiens*.

4 Archeological Evidence

Anatomical proxies for cognition and language have thus so far proven something of a disappointment as a means of pinpointing when speech and language appeared in our lineage. For more reliable putative indicators of language use by extinct hominid species, we thus have to turn to the archeological record. This furnishes us with the tangible evidence for early hominid behaviors, at least following the first deliberate manufacture of stone tools at about 2.5 million years ago (Semaw et al. 1997). Comparative evidence of primate and particularly of ape vocalizations (discussed by Tattersall, this volume) suggests that ancestral hominids already possessed a rich vocal repertoire, which was undoubtedly supplemented in the context of communication by the gestural and body language components mentioned at the beginning of this essay. The resulting complex ancestral substrate gives us a starting point as we begin to seek archeological indicators for the evolution of communication systems, and eventually for the appearance of language, in the human lineage.

Before looking for proxies for language in the archeological record, though, it is important to emphasize that language is a very special form of communication and that it is not simply a more complex extrapolation of whatever it was that preceded it in this role. This is because language is intimately associated with the symbolic faculty to which I have already referred. Words are symbols; and indeed, language maps very closely on to thought as we experience it today. For, while thought may additionally have a strong intuitive component, its expression is entirely dependent on moving around those intangible symbols. It is virtually impossible to conceive of symbolic thought in the absence of language, and vice versa.

It is a truism that language per se does not preserve in the Paleolithic archeological record. Prior to the advent of writing systems, nothing in archeology records anything about phonology, or about syntactic capacities. But the mutual interdependence between symbolic thought and language does allow us to seek significant Paleolithic proxies for language in the form of overtly symbolic objects. And it turns out that such items show up in the material record at a remarkably late date. What is more, their appearance also announces a very significant change in the tempo of technological innovation: A change so radical that it strongly implies a fundamental shift in the way in which the hominids concerned were processing information in their minds.

For the first 2.4 million years of the archeological record, significant technological innovations were both highly sporadic and rare (Tattersall 2008, 2012). There was a million-year wait before the production of the initial Mode 1 stone tools began to be supplemented by that of Mode 2 implements; and it was as long again before Mode 3 stone-working techniques appeared (Klein 2009). Throughout this long period, nothing was produced that can convincingly be interpreted as the product of a modern symbolic human sentience. Beginning about 400,000 years ago, during the tenure of *Homo heidelbergensis*, more elaborate technologies appear. These include such sophisticated activities as the hafting of stone tools, the construction of free-standing shelters, and the routine use of fire. But, sophisticated as those behaviors were, none of them can be taken alone as convincing proxies for the workings of symbolic minds.

The same is even true for the productions of the highly encephalized *H. neanderthalensis*. As members of an incredibly egotistical and egocentric species, many paleoanthropologists have over the years had difficulty in believing that it is possible for a big-brained hominid to be sophisticated and cognitively very complex, yet not like us. As a result, it has regularly been proposed that one or another putatively Neanderthal artifact reflected an essentially modern sensibility. Recent studies have, however, cast huge doubt upon the actual association of the most convincing such items with *H. neanderthalensis* (Bar-Yosef and Bordes 2010; Higham et al. 2010). And, once such things are removed from contention, there is little remaining to suggest any symbolic component in Neanderthal behavior. Even burial of the dead, which Neanderthals occasionally and with great simplicity carried out, probably implies nothing more than that these hominids possessed (in common with chimpanzees, as well as with us) a sense of empathy in addition to complex intuitive cognition.

Perhaps more remarkably yet, the same may be said of the rather sketchy archeological record that accompanies the earliest *H. sapiens* fossils known. These come from sites in Ethiopia between 200 and 160,000 years old (White et al. 2010; McDougall et al. 2005), and the associated artifacts are remarkably archaic (Clark et al. 2003; Klein 2009). So much so, indeed, that we can conclude with some confidence that the first members of our species functioned cognitively on a level broadly comparable to Neanderthals. It is not until anatomically recognizable *H. sapiens* had been around for close to 100 millennia that we begin to pick up any overt archeological evidence for symbolic activities.

5 Origin of the Human Capacity

At about 100,000 years ago, the piercing and ocher-staining of marine tick-shell "beads" at sites in Africa and nearby (Bouzouggar et al. 2007; d'Errico et al. 2010; Henshilwood et al. 2004; Vanhaeren et al. 2006) appears to announce the advent of bodily ornamentation, a practice universally associated in historically documented societies with status, social role, and other symbolic issues. And for more overtly symbolic artifacts the wait is not long. By a little under 80,000 years ago, plaques engraved with deliberate geometric designs (affirmed as symbolic, rather than as mere doodlings, by their repetition at different Middle Stone Age sites) had begun to appear in South Africa (Henshilwood et al. 2002; Texier et al. 2010). At around the same time, complex technologies requiring elaborate forward planning appeared (Brown et al. 2009), and the tempo of technological innovation changed dramatically. From this point on change itself, previously very rare, became the norm. Something cognitively radical was stirring among those Middle Stone Age humans, and it clearly involved the mental manipulation of symbols. Once the new mind-set had become entrenched, *H. sapiens* emerged definitively from Africa, rapidly replacing resident hominid species throughout the Old World. By

40,000 years ago, cave artists in Europe were producing some of the most powerful art ever made: the most eloquent testimony possible to the arrival of the modern human sensibility (White 1986).

What does this sequence of events imply for the emergence of language? Language is at base a biological property; and almost certainly, its enabling biology was acquired at the point when *H. sapiens* emerged as an anatomically distinctive entity some 200,000 years ago. The new skeletal structure then acquired speaks to a major developmental reorganization, presumably due to a change in gene expression that had major cascading consequences throughout the structures of the body. Plausibly, these would have extended to the brain, introducing a new cognitive potential for symbolic thought that, among other things, involved enhanced association capabilities in the neocortex. However, this new potential evidently lay unused for a short but significant period of time, until it was realized through the action of what was necessarily a cultural stimulus. And, given the intimate relationship between thought and language, that stimulus was very plausibly the invention of the latter, something we already know can happen spontaneously among members of a biologically enabled species (Kegl et al. 1999). What is more, if this event took place in a population of anatomically modern *H. sapiens*, the individuals concerned would have already possessed the peripheral vocal apparatus necessary to express articulate language, having acquired it in some other context entirely.

If this scenario is correct, we can eliminate the intrinsically implausible possibility that, as many have liked to believe, language and symbolic thought were slowly driven into existence, over the eons, by the action of natural selection on a central hominid lineage (Tobias 1991; Deacon 1997; Holloway et al. 2004). Instead, we can look for the origin of our altogether unprecedented cognitive capacities (as of our ability to speak) in a routine evolutionary event of exaptation. In other words, we can look for the origin of symbolic thought and language jointly, in the co-option of already existing anatomical systems to a radically new use. There is nothing particularly special about this. Ancestral birds, for example, had feathers for many millions of years before using them to fly, while tetrapod ancestors acquired their limbs in the oceans, long before using them to drag their bodies up on land. In other words, remarkable as we may justifiably pride our species *H. sapiens* on being, in evolutionary terms the process that produced us was an entirely ordinary one.

Acknowledgment I thank my colleagues Natalie Gontier and Marco Pina for kindly inviting me to contribute these thoughts to this fascinating collection.

References

Arensburg B, Tillier A-M, Vandermeersch B, Duday H, Schepartz LA, et al (1989) A Middle Palaeolithic human hyoid bone. Nature 338:758–760

Bar-Yosef O, Bordes J-G (2010) Who were the makers of the Chatelperronian culture? J Hum Evol 59:586–593

Bouzzougar A, Barton N, Vanhaeren M, d'Errico F, Collcutt S, Hodge E, Highham T, Parfitt S, Rhodes E, Schwenninger JL, Stringer C, Turner E, Ward S, Moutmir A, Stambouli A (2007) 82, 000-year-old shell beads from North Africa and implications for the origins of modern human behavior. Proc Nat Acad Sci USA 104:9964–9969

Brown KS, Marean CW, Herries AJR, Jacobs Z, Tribolo C, Braun D, Roberts DL, Meyer MC, Bernatchez J (2009) Fire as an engineering tool of early modern humans. Science 325:859–862

Clark JD, Beyene Y, WoldeGabriel G, Hart WK, Renne PR, Gilbert H, Defleur A, Suwa G, Katoh S, Ludwig KR, Boisseri JR, Asfaw B, White TD (2003) Stratigraphic, chronological and behavioural contexts of Pleistocene *Homo sapiens* from Middle Awash, Ethiopia. Nature 423:747–752

Deacon TW (1997) The symbolic species: the co-evolution of language and the brain. W. W Norton, New York

d'Errico F, Salomon H, Vignaud C, Stringer C (2010) Pigments from Middle Paleolithic levels of es-Skhūl (Mount Carmel, Israel). J Archaeol Sci 37:3099–3110

Falk D (1992) Braindance: new discoveries about human origins and brain evolution. Henry Holt, New York

Henshilwood C, d'Errico F, Yates R, Jacobs Z, Tribolo C, Duller GAT, Mercier N, Sealy JC, Valladas H, Watts I, Wintle AG (2002) Emergence of modern human behavior: Middle Stone Age engravings from South Africa. Science 295:1278–1280

Henshilwood C, d'Errico F, Vanhaeren M, van Niekerk K, Jacobs Z (2004) Middle Stone Age shell beads from South Africa. Science 304:404

Higham T, Jacobi R, Julien M, David F, Basell L, Wood R, Davies W, Ramsey CB (2010) Chronology of the Grotte du Renne (France) and implications for the context of ornaments and human remains within the Chatelperronian. Proc Natl Acad Sci USA 107:20234–20239

Holloway RL, Broadfield DL, Yuan, MS (2004) The Human Fossil Record. Brain endocasts: the paleoneurological evidence, Vol 3.

Hoboken, NJ: Wiley-Liss. Kegl J, Senghas A, Coppola M (1999) Creation through contact: sign language emergence and sign language change in Nicaragua. MIT Press, Cambridge

Klein R (2009) The human career, 3rd edn. University of Chicago Press, Chicago

Krause J, Lalueza-Fox C, Orlando L, Enard W, Green RE, Burbano HA, Hublin JJ, Hänni C, Fortea J, de la Rasilla M, Bertranpetit J, Rosas A, Pääbo S (2007) The derived FOXP2 variant of modern humans was shared with Neanderthals. Curr Biol 17:1908–1912

Laitman JT (1984) The anatomy of human speech. Nat Hist 84(8):20–27

Laitman JT, Heimbuch RC, Crelin ES (1979) The basicranium of Plio-Pleistocene hominids as an indicator of their upper respiratory systems. Am J Phys Anthropol 59:323–343

Lieberman P (2007) The evolution of human speech: its anatomical and neural bases. Curr Anthropol 48:39–66

Marshack A (1985) Hierarchical evolution of the human capacity. American Museum of Natural History James Arthur Lectures, New York

Martinez I, Ros M, Arsuaga JL, Quam R, Lorenzo C, Gracia A, Carretero JM, de Bermudez Castro JM, Carbonell E (2008) Auditory capacities in Middle Pleistocene humans from the Sierra de Atapuerca in Spain. Proc Natl Acad Sci USA 101:9976–9981

McDougall I, Brown FH, Fleagle JG (2005) Stratigraphic placement and age of modern humans from Kibish, Ethiopia. Nature 433:733–736

Semaw S, Renne P, Harris JWK, Feibel CS, Bernor RL, Fesseha N, Mowbray K (1997) 2.5 million-year-old stone tools from Gona, Ethiopia. Nature 385:333–336

Tattersall I (2008) An evolutionary framework for the acquisition of symbolic cognition by *Homo sapiens*. Comp Cogn Behav Revs 3:99–114

Tattersall I (2012) Masters of the planet: the search for our human origins. Palgrave Macmillan, New York

Texier PJ, Porraz G, Parkington J, Rigaud JP, Poggenpoel C, Miller C, Tribolo C, Cartwright C, Coudenneau A, Klein R, Steele T, Verna C (2010) A Howiesons Poort tradition of engraving ostrich eggshell containers dated to 60,000 years ago at Diepkloof Rock Shelter, South Africa. Proc Natl Acad Sci USA 107:6180–6185

Tobias PV (1991) Olduvai Gorge. The skulls, endocasts and teeth of *Homo habilis*, Vol 4. Cambridge: Cambridge University Press

Vanhaeren M, d'Errico F, Stringer C, James SL, Todd JA, Mienis HK (2006) Middle Paleolithic shell beads in Israel and Algeria. Science 312:1785–1788

Walker AC, Leakey REF (1993) The Nariokotome *Homo erectus* skeleton. Harvard University Press, Cambridge

White R (1986) Dark caves, bright visions: life in Ice age Europe. WW Norton, New York

White TD, Asfaw B, DeGusta D, Gilbert H, Richards GD, Suwa G, Howell FC (2003) Pleistocene *Homo sapiens* from Middle Awash, Ethiopia. Nature 423:742–747

How did Humans become Behaviorally Modern? Revisiting the "Art First" Hypothesis

Rita Nolan

Abstract Numerous impressive proposals addressing the title question have been made that appear to be rich, coherent, and stimulating but are either theoretically incomplete or empirically weak. It is proposed here that recent scientific results provide the elements for an empirically robust hypothesis that fills some theoretical gaps. Prompted by Henshilwood's archeological discovery at Blombos SA and drawing from Tomasello's results in comparative psychology on shared intentionality and from Rizzolotti and Arbib's results in neuropsychology on mirror neurons, and adapting Sterelny's notion of *decoupling*, an empirically robust hypothesis is proposed: deliberate symbolic artifacts of material culture preceded, triggered, and facilitated the origination of modern language and abstract thought. Of the three classic features of modern humans—art, language, and abstract thought—language has been an exclusive focus of many twentieth-century theorists, a focus attributed to advances in logic and computational modeling, with background assumptions from a simplistic epistemology. Hypotheses by Cassirer and Langer that art was prior to and facilitated the origination of language were dismissed by what we now know were false premises. It is proposed here that art enabled the critical features of displacement and communally shared semantic content required by human languages. The proposal also provides a causal answer to Harnad's "symbol grounding problem." A first-approximation sketch of the empirical model is given.

Keywords Decoupling • Mirror neurons • Shared imtentionality

R. Nolan (✉)
Department of Philosophy, Stony Brook University, Stony Brook, NY 11974, USA
e-mail: rita.nolan@stonybrook.edu

M. Pina and N. Gontier (eds.), *The Evolution of Social Communication in Primates*, Interdisciplinary Evolution Research 1, DOI: 10.1007/978-3-319-02669-5_13,

1 Introduction

Behaviorally modern humans display art, language, and abstract thought. Although the concepts of *art*, *language*, and *abstract thought* are inexact, we know a great deal about the behaviors they characterize and our exact knowledge of those behaviors has recently been significantly expanded by research results in different sciences. These new results reveal how considering this trio as a suite has significant implications for problems in the general theory of the evolution of modern humans that have so far resisted solution, implications in particular concerning the origination of language and, with language, abstract thought. Thus, in framing the title question in terms of behavioral modernity, no question of historical primacy versus interdependency between behavioral and anatomical modernity is insinuated (but see Tattersall, this volume, for discussion). Rather, this chapter proposes that there are benefits to the general theory in considering how the empirical evidence reveals that the same basic cognitive substructures are shared by all three traits and then sketches—to a first approximation—a causal account of their origin that is empirically testable both in individual brain structures, in principle, and in communal behavior as evidenced in archeological data on material culture as well as in current observation.

2 Twentieth-Century Approaches to the Question

2.1 *Modern Empiricism as Background for the Focus on Language*

Twentieth-century efforts to understand modern humans largely concentrated on language. Language provides humans with a source of unbounded information in contrast to what an individual can learn from first-hand experience and non-linguistic sources alone. Thus, a language is, among other things, an information storage and transmission system, so certainly language is central to explaining how humans came to excel other species in dominating their joint environment.

It is common for new explanations of phenomena to draw models from new technologies, and much theoretical interest in language in the last century was initially driven by two technological advances, those in logic (e.g., Frege 1879; Russell 1903; Russell and Whitehead 1910–1913; Wittgenstein 1922) and those in computational modeling (e.g., Turing 1937). Together, these results were mined to provide exact analyses in linguistics (Harris 1956, 1968; Chomsky 1957, 1965), in the development of formal linguistics. All these advances were routinely understood as cohering with modern empiricism and its conception of human cognition, which will be called here the "standard model" of human cognition—"standard model" meaning that the model is a widely accepted unquestioned assumption in normal science. Accordingly, modern empiricism provided background assumptions for the work on language that grew from those technological advances.

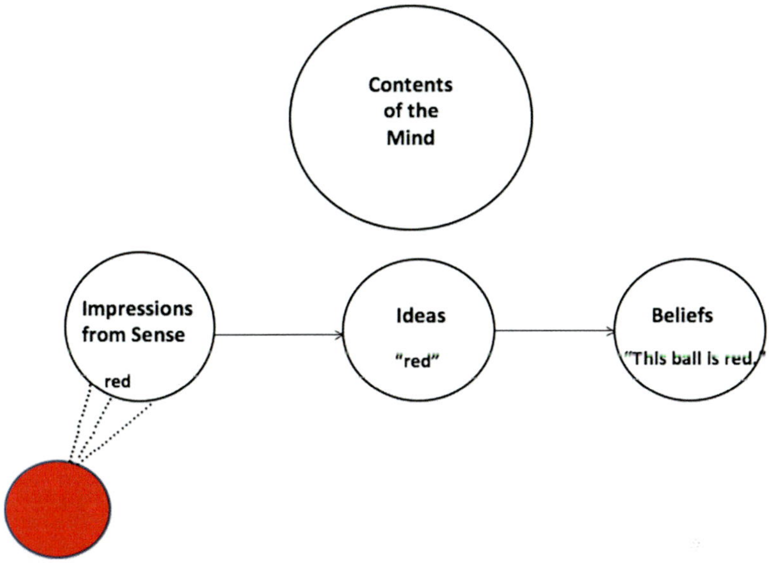

Fig. 1 Basic features of the standard model

2.1.1 Skeleton of the Model

The aspects of this model of human cognition relevant to our topic are simple and can be abstracted from the many debates and refinements it underwent. The model begins by analyzing the contents of the mind into three basic categories: impressions of sense such as the perception of red, ideas such as the idea of red, and beliefs such as the belief that the object referred to in "This ball is red" is red (Fig. 1).

In this model, beliefs are of two basic kinds. Beliefs may be about matters of fact such as the belief about some object referred to in "This ball is red" that it is red or they may be about relations among ideas such as, in the simple case, "Red is a color."

Different areas of study are assigned to these different types of beliefs: Science concerns beliefs about matters of fact, while beliefs about relations among ideas are the provenance of logic (Fig. 2).

Of course, every detail of this summary was opened to constant review, criticism, and debate throughout the history of modern empiricism. These fine points are not at issue here. The point of this review—or caricature—is to display the extent to which features of this model of human cognition themselves invite the conclusion that among the trio of features mentioned—art, language, abstract thought—it was language whose appearance was taken generally to mark the uniqueness of modern humans.

It is not surprising that applications of the standard model of human cognition to the question of behavioral modernity took human language to be the feature

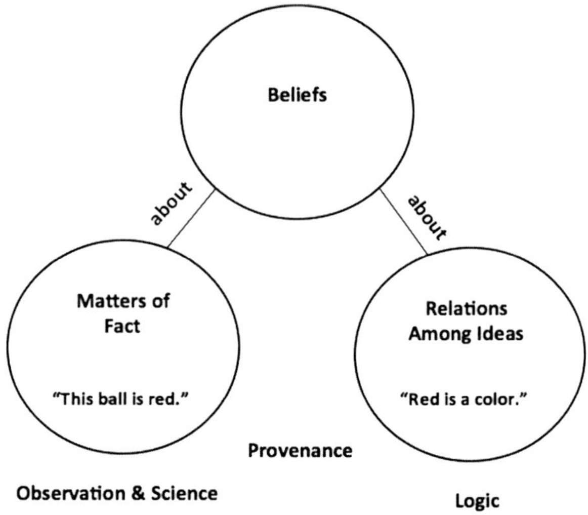

Fig. 2 Two types of belief—two areas of study

of modern humans that foremost needed theoretical explanation, for the model is itself *linguistically saturated*. That is to say, the model assumes and actively uses linguistic categories and structures in representing all the items included as "mental content" as well as in representing "external" items to which the mental items are understood to refer. Efforts to correct this fatal shortcoming while retaining the model (beginning perhaps with David Hume, one of its eighteenth-century founders) have been understood by many as unsuccessful. This fact has not prevented the model from having widespread influence on many research programs.

The ubiquity of this model as conceptual background, independent of all the intricate philosophical debates (cf., e.g., Carnap 1928 and Goodman 1951) concerning the naively realistic form of it that I have sketched, helps substantially to understand why human language was also widely assumed to be the uniquely singular trait that marks the evolutionary ascendance of modern humans and the one on whose origin alone rests the explanation of the evolutionary emergence of modern humans, anatomically and behaviorally. Indeed, Chomsky's ultimate selection of recursion as the only essential feature of the "Faculty of Language in the narrow sense" (FLN in Hauser et al. 2002) reflects, despite his avowed rationalism regarding nativism, his adherence to the mandate of the standard model that essential features of language must be purely syntactic and fully representable by formal structures alone since they must be relations among ideas and contain nothing about matters of fact. Both truths and falsehoods can be represented linguistically, so language itself is essentially neutral to matters of fact, on that way of looking at the matters.

Many critiques of using modern empiricism as a model of human cognition were given by distinguished philosophers and psychologists—by James (1890), Goodman (1951), Wittgenstein (1953), Grice (1957), Pribram et al. (1960), Austin (1962), and Bruner (1973), to name a few. These critiques went mostly unacknowledged by those using it as their standard model of human cognition.

2.2 An Alternative Speculation: Art as the Key Factor in the Emergence of Modern Humans

More tenuous speculations offered a different approach to the theory of modern humans. Cassirer (1923, 1923–1929) and Langer (1942) both proposed that modern humans are those who have the forms of symbolism required for the production of art. Specifically, Cassirer speculated in his *Philosophy of Symbolic Forms* (1923–1929) that myth, language, and scientific thought—including abstract thought—evolved historically in that order. Langer developed Cassirer's theme, proposing that art, language, and abstract thought all display the unique features of modern humans. I quote from the cover of her wildly popular book, *Philosophy in a New Key*, "The topics it deals with are many: language, sacrament, myth, music, abstraction, fact, knowledge—to name only the main ones. But through them all goes the principal theme, symbolic transformation as the essential activity of human minds." Later, Nelson Goodman in *Languages of Art* (1968), influenced by both Cassirer and Langer, added to the intuition of a primacy of art for humans by defending the claim that esthetics/philosophy of art is a branch of epistemology, understanding epistemology as the theory of knowledge. Goodman's rejection of the empiricist model of human cognition is implicit here (1968) as well as in his earlier work, *The Structure of Appearance* (1951).

2.2.1 The Alternative Proposal Dismissed

The proposal that art historically preceded the development of both language and abstract thought, clearer in Langer but nascent in Cassirer, was dismissed with this argument we now know to be faulty: The first record of art produced by humans was European cave art, dated around 30,000 BP. Humans who migrated to Europe from Africa before 30,000 BP navigated large bodies of water in groups. This navigation required a communication system as complicated as language. Therefore, art did not predate language.

But the more general criticism of this and all proposed answers to the title question in the twentieth century and earlier was the generally accepted one since the nineteenth century that they make claims that are not open to empirical confirmation. Pre-modern humans are not around to be observed.

2.3 A Note on the Terms "Art" and "Language" as Used Here

Some remarks are needed about how I am using the terms "art" and "language" herein. For simplicity, and for the most part, I want to avoid taking positions on ongoing debates within the special sciences and I adopt a minimalist approach.

2.3.1 Art

Those who pronounced European cave paintings to be the first examples of art likely held the imitation theory of art, first proposed in Plato: Art is an imitation of reality. Realistic pictorial representation is the noteworthy feature of art as it is understood on the imitation theory. The striking realism of the European cave paintings cannot be denied, but the imitation theory provides a highly restricted conception of art. Realistic pictorial representation does not capture the conception of art as we know it today, however inexact that conception is. In the present context, I use "Deliberate Symbolic Artifact of Material Culture" to capture the relevant notion of art, for simplicity and to avoid digressions for which there is no space here. (Contemporary proposals regarding what art is include noteworthy ones with relevance to our topic.) By "symbolic" I mean only that some significance is invested in the artifact independent of any technological functionality it may have, where that significance may include but is not limited to pictorial representation or other denotative dimension. For instance, the significance may be expressive rather than descriptive. In particular, by "symbolic" I do not mean referential, as in *"x is a symbol for y"*

2.3.2 Language

In this chapter, by a "language" I mean to refer to a complex of practices used by the members of a language community that characteristically displays this cluster of features: systematic semantic–syntactic structures, recursion, unbounded productivity, and displacement. This usage is consistent with the features of human languages that have been proposed by linguistic experts (Chomsky 1957, 1959; Hockett 1963 and earlier). By "recursion" in this context I understand the characteristic property that there are syntactic properties of a language that permit the construction of utterances that have never been made before. No syntactic limit is built into the syntax of the language, so the production of novel utterances (or other strings) is syntactically insured. My usage may seem to run counter to some recent usages, but note that I am not ascribing any essential properties to a language but only characteristic features, so my usage is not offered as an alternative to any essentialist usage.

This cluster of characteristic features of a language sustains it as scaffolding for abstract thought. There are two facts about languages so understood to take note of because they will figure later in this chapter. Individuals may participate in the complex of practices using different modalities: speech, signing, or inscriptions. But the complex of practices in which they are used, the language, is modality-neutral; this is the first fact to take note of. This important fact is neither mysterious nor magical. Schools, marriages, governments, and political parties are not identical to their membership or to their particular actions and practices. It is characteristic of humans to create these things, but it does not follow that these things are fictions. Although they owe their existence to socially contingent phenomena, they have causal effects on the rest of the world and that earns them the status of realities. A language, as characterized, is used by a community through its individual members. Thus, and this is the second fact to note, there must be some conformity and coordination in meaning and usage by the members of a language community. This second fact gives rise to, but is different from, the problem dubbed by Steven Harnad "The symbol grounding problem" (Harnad 1990). To put it simply here, how does more or less the same (abstract) meaning get attached more or less uniformly to a word by the different members of a language community? A word is not related to its meaning in the natural way that smoke is related to fire. What then "grounds" the meaning to a symbol such as a word or utterance?

3 Twenty-First-Century Approaches to the Question of Behavioral Modernity

The focus remains on language, but the structuralist grip of the twentieth century has been loosened by technological advances in areas such as the life and behavioral sciences. New avenues for research models within many fields appear and hypotheses about the origins of human language and modern humans abound. It is not possible to review any of them in depth here, so I give a representative sample for purposes only of illustration and apologize to their authors for my very brief comments, intended to highlight areas where the "art first" hypothesis might bring a welcome empirical or theoretical dimension to their proposals.

Six Competing Answers Currently, a myriad of recent contenders vie to explain the emergence of behaviorally modern humans or of their language. Here are some noteworthy ones:

1. The "enhanced working memory" hypothesis of Wynn and Coolidge (Balter 2010);
2. The "demographic" theory of Boyd and Shennan that language emerged from the congregation of migrating populations (Culotta 2010);
3. The hypothesis that language emerged as an elaboration of primate gesturing (Arbib 2012; Corballis 1999, 2002; Rizzollatti and Arbib 1998);
4. The hypothesis that language emerged as an elaboration of vocal signaling (MacNeilage 2008);
5. Andy Clark's "kludge" account, that an unconnected jumble of changes in all the above categories resulted in our distinctive features including language (Clark 1987);
6. Sterelny's "niche construction" hypothesis (Sterelny 2012).

Each of these hypotheses is important and undoubtedly contributes to the final, multiplex story, and those who focus on a single human feature do not maintain that it is the only precondition for the origination of language or of modern humans. Nevertheless, each has its own explanatory gap that is filled, if at all, in a highly speculative way. For example, (1) is an account that rests on individual biology, while (2) is a collective account based on communal activity, but the explanation needed must coordinate individual biology with group practices if it is to sustain a satisfactory account of human language origination. Regarding (3) and (4), vocal communicating and gesturing are different modalities of language use, but the same language can be used in either or any other modality. That is to say, a language is modality-neutral. Moreover, the bases of (3) and (4), gestures and vocalizations plus imitation, seem to open these accounts to the objections given long ago by Chomsky (1959) and others against Skinner's behaviorist account of individual language learning, in particular that, apart from highly conventionalized contexts, most utterances people make are ones they have never heard others make. The conception of language assumed fails to represent salient features of human language, notably, the cluster of features of human language that serves to sustain it as scaffolding for abstract thought. For example, as language is used, the semantic content of a linguistic message is normally "displaced"—it does not describe or refer to what is present to perception for the speaker–hearer and may not even refer to something that has a perceptual aspect. (5) and (6) each come close to a "no theory" theory, albeit in a principled way, claiming that many different factors contributed to the emergence of modern humans with none of them as far as we can know being of exceptional importance. Perhaps so. It is indisputable that a vast number of conditions occurring over a long time coalesced to result in behaviorally modern humans. But even acknowledging our limited understanding of the actual historical course of events that culminated in the emergence of modern humans, it nevertheless appears that their emergence happened so quickly in evolutionary time as to require an exceptional event as cataclysmic to the relevant transformation (Tomasello and Call 1997, p. 401), an event that perhaps qualified as a "tipping point."

4 The Blombos Conjecture: The Invention of Meaning

Cassirer and Langer proposed that the key to human uniqueness among species was their symbolic thought in all its behavioral manifestations rather than, say, a cleverness communicating with highly complex signals. Others have more recently continued this theme, notably Deacon (1998), Donald (2002), and Gardenfors (2003).

We now have empirical falsification of the first premise of the argument cited earlier, the premise that the proposals of Cassirer and Langer lack initial plausibility because there is no evidence of symbolic behavior by humans before the European migrations around 30,000 BP. We now know, owing to the archeological discovery at Blombos SA by Christopher Henshilwood (Henshilwood et al. 2002), that remarkably art-like artifacts were produced by humans at least as early as 75,000 BP and probably earlier since there is no evidence that the Blombos artifacts were the first such artifacts made (Fig. 3).

In addition, there are new theoretical results from different disciplines that can provide the framework for an empirically robust hypothesis that shows how an event in evolutionary history may explain the remarkable transition to full behavioral modernity.

Fig. 3 Engraved Ocher from Blombos Cave *Photo courtesy* of nsf and Chris Henshilwood

I focus now on three major results of recent scientific research: in comparative cognitive psychology, Tomasello's (2005 and 2008) "Shared Intentionality" hypothesis; in behavioral neuroscience, the discovery of mirror neuron systems (Rizzolatti and Arbib 1998; Rizzolatti and Craighero 2004; Arbib 2005, 2012); and in archeology, Henshilwood's discovery of the Blombos Artifacts (Henshilwood et al. 2002). Each is well supported by evidence, and I propose that the integration of these results warrants a renewed consideration of the proposal that art, understood as Deliberate Symbolic Artifacts of Material Culture, historically preceded and facilitated the origination of language and the subsequent growth of abstract thought, catapulting humans to their dominant status on the planet. I consider each research result in turn. Central to the positive proposal constructed is a concept of "decoupling" that I adapt from Sterelney (2012).

4.1 The Shared Intentionality Hypothesis

Tomasello's (2005) Shared Intentionality Hypothesis proposes that a significant difference between modern humans and closely related species is that members of other species can grasp the intentions of others, while humans are also capable of sharing intentions with one another, thereby enabling a massive degree of cooperative activity. Human language is one vehicle for sharing intentions, but evidence of shared intentionality occurs in early infancy and prior to language skills. This evidence leads Tomasello to conclude that shared intentionality and the cooperative behavior it underwrites were crucial contributions to language origination.

4.1.1 Shared Intentionality via Language

Human languages are vehicles of shared intentions in two ways, however. First, it is commonly understood that languages are vehicles for the expression and promulgation of belief. The English sentences "What he did was wrong" as well as "That's a chair," uttered in appropriate contexts, are expressions of culturally common beliefs, beliefs that are both learned and sanctioned within a cultural community. This appears to be the sense in which languages are vehicles of shared intentions in which Tomasello (2005) is interested, for his goal was to determine how such a capacity for shared intentionality may provide the basis for understanding cultural differences: They are promulgated by different shared intentions, understood as background cultural beliefs, within different cultures. Enculturation itself is, among other things, a process of coming to share common beliefs, often as a result of explicit learning of culturally condoned and reinforced utterances.

But there is another way in which a language is a vehicle of shared intentions. Grice (1957) proposed that the meanings of utterances themselves be understood in terms of the shared intentions of speakers of a language, and Tomasello derived his notion of shared intentions from Grice's analysis of linguistic meaning in terms of shared intentions. Grice proposed his analysis of linguistic meaning as an account of what he dubbed "non-natural" meaning—what previously had been called "conventional" meaning—to distinguish it from natural signs, signals, and symptoms. The other way in which a language is a vehicle of shared intentions is in its dependence for its continued existence on its users sharing the meanings of its utterances. Indeed, the enterprise of sharing beliefs itself requires assignment of shared meanings to utterances stating those beliefs. Thus, Grice builds recursion into his pragmatic account of linguistic meaning: A speaker's meaning something x by a linguistic utterance u is to be understood in terms of the speaker's intention that the hearer understands something x in virtue of that utterance u and that the hearer understands that the speaker intends the hearer to so understand that utterance. In case the reader is confused by this analysis, imagine yourself hearing the utterance *Pas de lieu Rhone que nous* without having any French proficiency. The utterance will sound like gibberish to you. But suppose someone then shows you the written sentence "Paddle your own canoe." Your perceptual experience of the understood English sentence is entirely different because you experience its meaning rather than just hearing sounds. This example is adapted from William James's (1890) chapter, "The Stream of Thought". Grice's proposal recognized that a human language is a complex, recursive, and intentional system whose functioning requires members of a language community to share the meanings of utterances, meanings understood in his analysis as speaker intentions. At the time it was proposed, Grice's pragmatic analysis of linguistic meaning was understood as a critical response to the empiricist model of cognition that underwrote behaviorist accounts of language such as that of Skinner (1957) as well as to the behaviorist accounts themselves. There was no implication by Grice that the intentions referred to could be found inside the "black box" nor any implication that there would be a biological representation of these shared intentions. No specific theory of mind is implied by Grice's analysis, neither a reduction of psychological states to brain states nor an identification of psychological states with states of any other stuff besides the stuff of which people are made—if people are "made of stuff". The device was a purely formal one, but one that attributed much more than observable behavior to speaker–hearers of linguistic utterances. It did not eliminate meanings so much as offer a new way of conceiving of them. It was definitely a way of conceiving of the meanings of utterances and of linguistic meaning in general, as a phenomenon that was not capable of being captured by reference only to physical stimuli.

Grice's proposed analysis of meaning in terms of shared intentions invites consideration of Harnad's "symbol grounding problem." In one form, this is the problem of accounting for how multiple individuals who speak the same language come, in the first place, to assign the same meaning to the utterances of their language. I have no interest here in recommending an old concept of meaning in any technical way and use the term only colloquially here to refer to the semantic content of our utterances. For language to work for us as it does, for example to convey information about ourselves and the world to one another, about the future, and the imaginable and unimaginable, requires that abstract semantic content be assigned uniformly across individual speakers of a language. The major problem is that no hypothesis on offer proposes how it could come about in the first place that the same abstract semantic content could be uniformly learned by different members of a language community.

While Tomasello derives his notion of shared intentionality from Grice, he does not seem to take up the second sense in which a language is a vehicle of shared intentions and that Grice employs, the sense in which a language requires that its users share the semantics of the language, more simply (and perhaps less theory laden), that they assign more or less the same meanings to its utterances. To quote from James (1890), "(N)o word in an understood sentence comes to consciousness as a mere noise. We feel its meaning as it passes…" (p. 271).

Despite this concern, Tomasello's proposal that the enlarged ability of modern humans to share intentions is an important part of the account of the massive growth in their cooperative activity is very persuasive. But whether language was a result of or a cause of either that enlarged ability to share intentions or that massive growth in cooperative activity remains unanswered. Perhaps it was both to some extent.

4.2 *Mirror Neuron Systems (MNS)*

The macaque monkey MNS is an "observation/execution matching system" for the action of grasping an object. The striking result of this research (Rizzolatti and Arbib 1998; Rizzolatti and Craighero 2004; Arbib 2005, 2012) is the discovery that the same MNS is activated when a monkey sees another monkey grasp an object such as a banana and when the monkey itself grasps the object. MNSs, then, provide an explanation of the neurophysiological bases for (some) imitative learning. On the assumption that human protolanguage was a gestural system of signs, it has been proposed that human protolanguage was learned by imitating the observed gestures of others. In this way, multiple extended MNSs endow humans with massive imitative abilities and the emergent system of gestural signs provides the scaffolding for the development of full human language.

It was mentioned briefly above, in discussing accounts of language as an elaboration of vocal or gestural communication, that a major problem with such accounts is that they are subject to the same objections that were detailed by

Chomsky (1959) against behaviorist accounts of language. In particular, Chomsky argued persuasively that no complex amount of "training up" through complex schedules of operant conditioning could alone result in fluency in a language because the mastery of the syntactic–semantic system of a language requires mastery of a rule-like system in which the rules cannot be extrapolated from a finite sample of utterances, let alone the small sample that small children are exposed to before they start speaking/signing their language. While I do not consider the answer "It's innate" to be a satisfactory explanatory strategy, I do not believe that linguistic fluency can be explained using imitation as the only basic explanatory function.

In his recent book, Arbib (2012) presents a complex account of language evolution with many other basic and advanced functions beyond imitation. The resulting account of language evolution bears little resemblance to a behaviorist account of language learning. Both a community with a shared semantics and displacement are central in Arbib's account. In this book, as in his other work, postulated features of protolanguage bear explanatory weight. Whether the postulated complex functions, some of them implicitly rule like, and the features attributed to protolanguage are found to be empirically sound is a matter for other scientists to judge.

There is a general problem in postulating processes in order to get a desired result that can be illustrated by computational modeling of the features that seem central to human language.

An account of language origination might be expected to provide a sound account of these pervasive features of full language: systematic syntactic–semantic structures, unbounded productivity (novelty), recursion, and displacement (Hockett 1963). If we see human language as playing a substantial role in the contrasts between modern humans and other hominids, then its critical feature of displacement is one, for example, that must be specifically accounted for in an adequate theory of its origination. This is the feature whereby things that are absent from the immediately perceived context of the utterance may nevertheless be talked about, in contrast to animal communication systems wherein messages are about things that are more or less part of the immediate perceptual context of the communicants. Counterfactual utterances, those that contemplate future plans or past disappointments, practical calculations, and moral deliberations, reflections on logic, as well as proofs of theorems all are about things that are inherently not perceptually present; some are about things that have no perceptual aspect. Indeed, the displacement problem and Harnad's symbol grounding problem coalesce to a large extent, for it is because our discourse is not usually "grounded" by references only to what is perceptually proximate that the question what "grounds" linguistic symbols arises. It is because our discourse is understood by members of our language community despite its being "displaced" that the symbol grounding problem arose.

Massive parallel processing models of multiple dynamic neural networks have been designed to circumvent the symbol grounding problem, and Harnad (1990) himself seems hesitantly to acquiesce that, perhaps, they do solve the problem. I have more reservations. That humans have solved the symbol grounding problem is evident in their successful use of language, without knowing exactly how we did it. Yet it seems to go without saying that computational models can be built successfully to model anything that is possible. In principle, many different computational models can be successfully built to model the very same thing. So I do not find it surprising that computational models have been built that produce, on some scale, replications of some or any aspects of human language use. Convincing evidence outside the model is necessary to determine whether any of them accurately represent how humans did in fact solve the symbol grounding problem. I locate the problem this chapter addresses within the latter frame, how did humans actually do it. This is a different question from whether there can be computational models that do it, to which the answer is obviously positive. Developing such models took intense work and must be appreciated. But they should not be misinterpreted.

The existence of MNSs in macaque monkeys empirically establishes that they possess action schemata in which there is recognition of the agent of an action, of the movement of the action (reaching and grasping), of the target of the action (the banana), and of the agent's goal or intention in so acting (to grasp the banana).

Apparently, MNSs do provide a neurophysiological basis for imitative learning in monkeys, but only on condition that the target of the action to be learned—that is, the object changed by the action (e.g., the banana)—is proximal, either within the learning monkey's perceptual field or highly cued by context (Rosenbaum 2009). In learning to imitate an action that it sees performed, the monkey generally must see the whole action performed, including its successful completion—the other monkey's achieved grasp of the banana.

It seems indubitable that early humans who were not behaviorally modern also had MNSs so that they could anticipate the goal of an action for some actions and when the target was proximal. At some point in natural history, humans changed from being able to grasp some intentions of others to being able to share the meanings of utterances, i.e., the wide-scale shared intentionality that Grice attributes to human language users. The question that emerges as key is how could a population traverse the vast divide between these two abilities?

4.3 The Blombos Artifact and the Invention of Meaning

The discovery of the Blombos SA ochers by Henshilwood et al. (2002), reliably dated at least 75,000 years BP, conclusively shows that Deliberate Symbolic Artifacts of Material Culture were produced long before migrations of behaviorally modern humans into Central Europe required communication systems comparable in complexity to languages. This discovery opens again the possibility that art preceded, facilitated, and triggered the emergence of language and with it the growth of abstract thought. But now, we have empirical resources that take that possibility out of the realm of speculative rumination and into the realm of the empirically testable. This is a possibility not to be taken lightly in view of the long, empirically unproductive history of the general question.

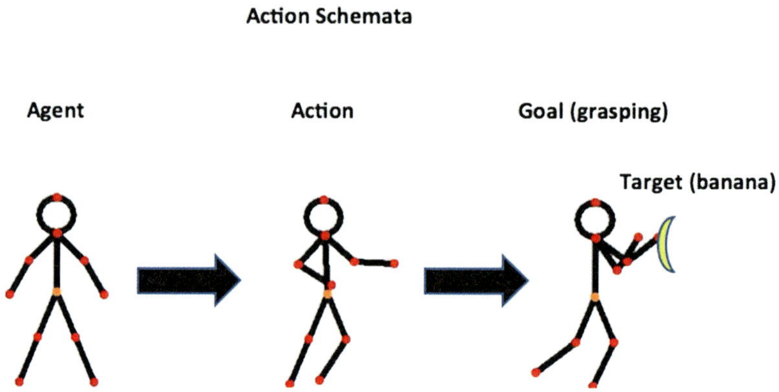

Fig. 4 An action schema for the mirror neuron system

The recent research results just discussed make it possible to construct a causal model for the emergence of behaviorally modernity in humans in which art precedes and triggers the emergence of language and the beginnings of rational thought. The causal model proposed draws from Tomasello's and Arbib's results, in a limited way, and on Grice's analysis of linguistic meaning in terms of shared intentionality. To these, I add the concept of "decoupling" borrowed from Sterelny (2012). It is a causal model for the emergence of shared intentionality as an approximate necessary condition for human language, for language origination, and for the growth of abstract thought made possible by language. In fact, all three of the classic features of behaviorally modern humans do share common cognitive features, as Cassirer, Langer, and Goodman were convinced, and the proposed model reflects that fact.

4.3.1 Preliminary Sketch of a Causal Model

Recall that the operation of MNS in monkeys requires that the monkey has an action schema by which it distinguishes the *agent* of the action, the *movements* that constitute the action, and the intended *goal* of the action, including its *target*. Contemporary humans also have MNSs, and both monkeys and humans are able to imitate some of the actions of others because they have MNSs (Figs. 4, 5).

For macaque monkeys—and we may suppose for pre-behaviorally modern humans—the target must be proximal, either within their perceptual field or highly cued, unless the complete action schema is a familiar one in a limited repertoire.

4.3.2 Tools, Ornaments, and Status Symbols

Certainly, making and using tools and status ornaments implicates MNS as well and their archeological appearance likely reflects the trajectory of cognitive development in humans. The objects themselves, the tool and the ornament, can be

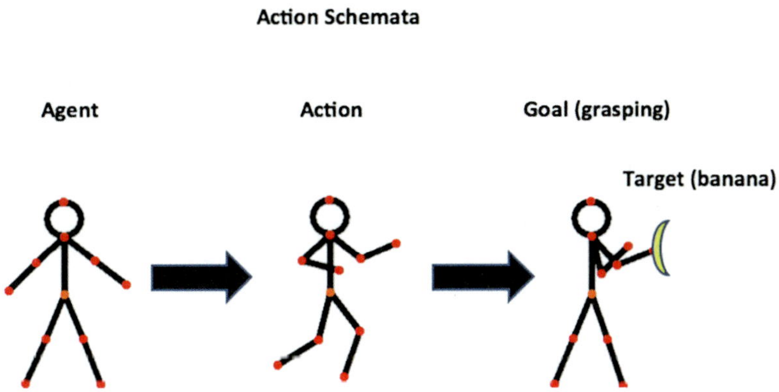

Fig. 5 The goal of the action is successful

understood as coding distinct action schemata associated with them, analogous to the monkey's action schema for grasping a banana. Since the targets for actions using tools and status ornaments—the prey or the people—are normally not present when they are being made for their single-purpose action, it may be thought that they require some degree of displacement, as when things not present are talked about. Possibly, shell beads, also found at Blombos, were calibrated, pairing size with status, but that too might be accounted for by simple association and coding for specific and discrete action schemata, without requiring language or other complex abstraction involving goals and targets (or speakers intentions) that are not perceivable. But human language use is distinctive from tools and status ornaments in many ways, significantly in its noted features of displacement, unbounded productivity/novelty, systematic syntactic–semantic structures, and recursion. Art and rational thought do, however, have analogs of these features. *Productivity* There is no intrinsic limit to the number of distinct and different art objects or rational thoughts. *Displacement* Both are characteristically "about" what is not simultaneously perceived. *Syntactic–semantic Structures* Within any particular art style, there are regularities between the execution, that is, the way the artist makes the work, and the artist's intention/goal in making the object, analogous to syntactic–semantic structures. Goodman (1968) provides an extensive analogy between art and language, perhaps straining the issue when disanalogies imposed by the art medium do not readily fit contemporary analyses of language. A significant difference from language regarding this feature is obvious: The prototype/stereotype of "art object" in contemporary human culture is a painting, and these are routinely considered to be analog products that assume, for example, continuities of color or shadings of one color into another as part of the intended made art object. Full language, on the other hand, is presumed to have discrete words and phonemes associated individually with meanings that seem to be experienced as analog items—inexact and shading into one another, contrary to formal semantic theory. In comparison with the relations between words and their meanings/semantic contents, the presumed relations between ornaments taken as indicative of status, while they are abstract, are simple relations learnable by simple association; both the ornament and the status—e.g., requiring deference, hierarchical position, wealth, beauty, etc.—have a single or small number of perceivable and relatively discrete "action goals," like the action of grasping the banana. Thus, tools and status ornaments seem to mark an important step in the evolution of modern human cognition, but the evidence does not require an account of their origination that is much beyond what can be explained on the basis of MNSs alone.

4.3.3 The Institutional Creation of Deliberate Symbolic Artifacts of Material Culture

What we do know is that at some time much earlier than 75,000 BP, some human attentively performed some actions resulting in permanently affecting an object. We may suppose the agent was observed while doing those actions, that the actions appeared to be deliberately done, and that the action schema was a novel one to those observers. Perhaps the agent's actions did not seem to be directed toward achieving any particular goal. Or perhaps there appeared to be no goal that the agent was perceived to accomplish with the object, in the way that an artist "does" nothing to the art work once the artist regards it as finished except possibly show it to others. Or perhaps the agent did things with the object that were noticed but not comprehended. It is also possible that no agent deliberately performed the hypothesized actions, but the object was assimilated by others to the products of deliberate action.

I am not suggesting the above description for the Blombos artifacts, for certainly there are other contexts possible for them than what I have just described, in particular that they were notational inscriptions of a communication system. It is apparent from looking at the ochers that they were products of deliberate action by their makers, and the fact that there is more than one such ocher, with similarities and differences between them, lends some initial plausibility to the conjecture that they were notational inscriptions. But it is not my intention to be making any proposal regarding their "true" interpretation, and fortunately, that does not matter to this chapter's project. Their existence around 75,000 BP suggests that much earlier one, or more, first object was deliberately affected similarly and that the agent's goal may not have been apparent. Likely, such objects attracted community attention because their artifactual features were noteworthy in some way, as is common with objects we call "art" today.

How the creation of such an artifact is to be taken or understood would be an opportunity for problem solving for our big-brained predecessors. We must suppose they had some communication skills even although not modern language and perhaps they could "consult" non-linguistically in some way with its producer. More likely, the group would look to its leaders to learn how to "take" it. Perhaps the alphas of the group could decide what to do with the artifact and how to take it, independent of its maker's goal, if any. Perhaps whoever created it would be automatically placed among the leaders and so could direct the community on how to "take it" or what to do with it. Perhaps, even, the leaders proposed that the community treat some object as the product of deliberate intentional action.

Among all these possibilities, once the material artifact is let us say made, given the right material conditions, the artifact survives its maker even though its being made by someone on purpose—and by hypothesis no clear immediate instrumental use could be read off the proximal perceptual horizon—remains obvious in the artifact's appearance. Even we can see from looking at it, despite our cultural remoteness, that the markings of the Blombos ocher have the look of being the result of a deliberate action sequence.

Borrowing a concept from Sterelny (2012), I propose that this imagined first artifact was *decoupled* from the action schema in which it was produced, leaving the persisting material artifact to be invested with an interpretation by the community, probably through its leaders. On the supposition that such an artifact could have been produced by a pre-linguistic human, this investment of significance could initially consist only in practices and actions taken in relation to it, perhaps, as Cassirer suggested, practices and actions that became ritualized. But as long as it persists, perhaps initially aided by witnesses to the initial action that produced it, other artifacts with variations in their properties from the original could also be produced, introducing the possibility of systematic variances in attributed significance as the initiation of systematic semantic–syntactic structures.

The key to how such material artifacts can account for displacement and symbol grounding is in their natural decoupling from their origin; that is, the end stage of the MNS of the action, the successful completion of the action, is already distinguished from the movements and the agent because the target is a distinct material object from the agent or the actions. Unlike vocalizations or gestures, the persistence of artifacts of material culture after they have been produced naturally decouples them from their agent and the actions that produced them. They are, as well, capable of novelty if not unbounded productivity, again displaying rudimentary linguistic properties. Likewise, recursion becomes possible in variations of the effects on like objects produced by other agents.

I have called the hypothesized first creation of a Deliberate Symbolic Artifact of Material Culture "the invention of meaning" to indicate its role as an object invested by a community in common significance that the artifact can carry independent of its physical and perceptual context, in a primitive analog to displacement. Its material persistence as an artifact with that significance is what grounds its meaning/significance. The significance invested in Deliberate Symbolic Artifacts of Material Culture is grounded naturally: It is grounded in the object so interpreted by the community.

5 Conclusion

So much of this account has involved matters of language; it is easy to lose sight of the fact that the proposal offered is the "art first" proposal: Art, as Deliberate Symbolic Artifacts of Material Culture, preceded, triggered, and facilitated the origin of human language and the subsequent growth of abstract thought. This proposal claims initial plausibility on the grounds that there are analogs of the same cognitive features in all three realms, art, language, and abstract thought. These have not been set out separately but brought up in the text where fitting. There are enough analogs of this sort to warrant calling each of the three realms a cognitive practice.

The account is empirically robust and is available to confirmatory empirical evidence on the level of the individual and on the communal level. It is required of any viable account of the origination of language that the account be able to provide the link between uniform communal assignment of meaning/semantic content and individual representation of meaning/semantic representation. First, the institutional creation of such artifacts implies that significance is assigned to them uniformly, more or less, throughout the community. Regarding evidence of such communal artifacts, we are directed to seek material archeological evidence of clear cases of such artifacts prior to 75,000 BP, and of course, that project is enthusiastically underway in South Africa. Additional empirical evidence may come in the form of evidence of communal ritual, whether or not there are related material artifacts.

Regarding empirical evidence of individual assignment of meaning/semantic content, ongoing research on brain imaging and mirror neurons may be able to definitively prove the ability of an individual to decouple the target of another's action from the agent and the agent's actions/movements, and perhaps this has already been done.

Although my preliminary sketch of this proposal needs refinement in several directions, it appears to be a candidate for further confirmatory evidence in individual neurophysiological research into MNSs with naturally decoupled completed goals of action schemata or targets of action that are naturally decoupled, completed artifacts. It also seems to invite archeological research into evidence of other early Deliberate Symbolic Artifacts of Material Culture, for example, evidence of artifacts in more ephemeral media which may not have survived environmental trauma or artifacts that do not resemble the types of material artifacts that are familiar to us.

Finally, the "art first" hypothesis is not intended as a rival of the many proposals that have been put forward for the origin of language or for the origin of modern humans. Instead, it is proposed as possibly complementing some of those proposals and able to be integrated with them in the search for the answer to our common questions.

References

Arbib M (2005) From monkey like action recognition to human language: an evolutionary framework for neurolinguistics. Behav Brain Sci 28:105–167. doi:10.1017/S0140525X05000038

Arbib M (2012) How the brain got language: the mirror system hypothesis (studies in the evolution of language). Oxford University Press, New York

Austin JL (1962) Sense and sensibilia. The Clarendon Press, Oxford

Balter M (2010) Did working memory spark creative culture? Science 328:160–163. doi:10.1126/science.328.5975.160

Bruner J (1973) Beyond the information given. WW Norton and Company, New York

Carnap R (1928) Der Logische Aufbau der Welt. Felix Meiner Verlag, Leipzig. English edition: Carnap R (1967) The logical structure of the world: pseudoproblems in philosophy (trans: George RA). University of California Press, CA

Cassirer E (1923) Language and myth. (trans: Langer S (1946)) Courier Dover Publication, Chelmsford

Cassirer E (1923–1929) Tr 1955. The philosophy of symbolic forms. Yale, New Haven

Chomsky N (1957) Syntactic structures. Mouton and Co, The Hague

Chomsky N (1959) A review of Skinner's verbal behavior. Language 35:26–58. doi:10.2307/411334

Chomsky N (1965) Aspects of the theory of syntax. MIT Press, Cambridge

Clark A (1987) The kludge in the machine. Mind Lang 2:277–300. doi:10.1111/j.1468-0017.1987.tb00123.x

Corballis M (1999) The gestural origins of language. Am Sci 87:138–146. doi:10.1511/1999.2.138

Corballis M (2002) From hand to mouth: the origins of language. Princeton University Press, Princeton

Culotta E (2010) Did modern humans get smart or just get together? Science 328:164. doi:10.1126/science.328.5975.164

Deacon T (1998) The symbolic species: the coevolution of language and the brain

Donald M (2002) A mind so rare: the evolution of human consciousness. WW Norton, New York

Frege G (1879) Begriiffsschrift. Verlag von Louis Nebert, Halle

Gardenfors P (2003) How homo became sapiens: on the evolution of thinking. Oxford University Press, Oxford

Goodman N (1951) The structure of appearance. Harvard Unversity Press, Cambridge

Goodman N (1968) Languages of art. Bobbs-Merrill, Indianapolis

Grice HP (1957) Meaning. Philos Rev 66:377–388. doi:10.2307/2182440

Harnad S (1990) The symbol grounding problem. Physica D 42:335–346. doi: 10.1016/0167-2789(90)90087-6

Harris Z (1956) Introduction to transformations and discourse analysis papers. University of Pennsylvania, Philadelphia

Harris Z (1968) Mathematical structures of language. Interscience tracts in pure and applied mathematics, New York

Hauser M, Chomsky N, Fitch W (2002) The faculty of language: what is it, who has it, and how did it evolve? Science 298:1569–1580. doi:10.1126/science.298.5598.1569

Henshilwood C, d'Errico F, Yates R, Jacobs Z, Tribolo C, Duller GAT et al (2002) Emergence of modern human behavior: middle stone age engravings from South Africa. Science 295:1271280. doi:10.1126/science.1067575

Hockett C (1963) The problem of universals of language. In: Greenberg J (ed) Universals of language. MIT Press, Cambridge

James W (1890) The principles of psychology. Henry Holt, Boston

Langer S (1942) Philosophy in a new key: a study in the symbolism of reason, rite, and art. New American Library, New York

MacNeilage PF (2008) The origin of speech. Oxford University Press, Oxford

Pribram C, Miller G, Galanter E (1960) Plans and the structure of behavior. Holt Rhinehart and Winston, NY

Rizzolatti G, Arbib A (1998) Language within our grasp. Trends Neurosci 21:188–194. doi:10.1016/S0166-2236(98)01260-0

Rizzolatti G, Craighero L (2004) The mirror-neuron system. Annu Rev Neurosci 27:169–192. doi:10.1146/annurev.neuro.27.070203.144230

Rosenbaum DA (2009) Putting thoughts into action. In: Presentation, conference on language, thought, and motor control. Stony Brook University, Stony Brook, NY

Russell B (1903) The principles of mathematics. Cambridge University Press, Cambidge

Russell B, Whitehead AN (1910, 1912, 1913) Principia Mathematica, 3 vols, Cambridge University Press, Cambridge

Skinner BF (1957) Verbal behavior. Copley Publishing Group, Boston
Sterelny K (2012) The evolved apprentice (Jean Nicod lectures). MIT Press, Cambridge
Tomasello M, Call J (1997) Primate cognition. Oxford University Press, USA
Tomasello M (2005) Understanding and sharing intentions: the origins of cultural cognition. Behav Brain Sci 28:675–735. doi:10.1017/S0140525X05000129
Tomasello M (2008) Origins of human communication. MIT Press, Cambridge
Turing AM (1937) On numbers, with an application to the *Entscheidungsproblem*. P Lond Math Soc 42:230–265. doi:10.1112/plms/s2-42.1.230
Wittgenstein L (1922) Tractatus logico-philosophicus. Harcourt, Brace and Co, Inc, NY; Kegan Paul, Trench, Trubner and Co, LTD, London
Wittgenstein L (1953) Philosophical investigations. Blackwell, London

Experiments and Simulations Can Inform Evolutionary Theories of the Cultural Evolution of Language

Mónica Tamariz

Abstract Language is a complex adaptive system supported by humans. With evidence and illustrative examples from recent computer simulation and experimental work, this chapter defends that much linguistic structure can be explained as emergent phenomena. Cultural processes such as social transmission to new generations or the patterns of speaker interaction, operating in large populations over many generations give rise to structure at different levels, from categories of phonemes to compositional structure. This position shifts the burden of explanation of linguistic structure from a biologically evolved mental organ to more tractable cultural processes, which are being investigated with a host of innovative empirical methods.

Keywords Cultural evolution of language • Experiments • Iterated learning • Replication • Emergence

1 The Cultural Evolution of Language

The phrase "language evolution" has several meanings. For some, it refers to the *genetic* innovations that appeared in the *Homo sapiens* lineage and that have allowed us to learn, use, produce and understand linguistic behaviour. Sociocognitive and neural capacities such as cooperation, conformity, symbolicity, shared intentionality, imitation or vocal control, which are heavily involved in language, are extraordinarily developed in our species compared to our closest relatives in the phylogenetic tree, namely other apes.

M. Tamariz (✉)
Language Evolution and Computation Research Unit, Department of Linguistics and English Language, School of Philosophy, Psychology and Language Sciences, The University of Edinburgh, Dugald Stewart Building, 3 Charles Street, Edinburgh EH8 1AD, UK
e-mail: monica@ling.ed.ac.uk

M. Pina and N. Gontier (eds.), *The Evolution of Social Communication in Primates*, Interdisciplinary Evolution Research 1, DOI: 10.1007/978-3-319-02669-5_14,

A second meaning of "language evolution" refers to the *cultural* evolution of linguistic structure. Features of languages like sounds, words or larger constructions can appear, change, move from one language into another, and disappear, giving rise to large-scale patterns of language birth, death and diversification. And all this happens in the historical timescale, through the cultural mechanisms involved in language use and communication in modern humans.

Traditionally, it has been assumed that changes in properties of individual languages such as sound, semantics, morphology or syntax were best explained by *cultural* mechanisms stemming from production and perception biases or population contact. It was concurrently assumed that another kind of properties of languages deemed to be more fundamental, perhaps universal—such as an arbitrary relationship between linguistic signals and meanings, having a closed repertoire of sound categories, coming to be shared by a community of speakers, or being structured in such a way that an open set of novel messages can be produced and understood by other members of the community—required explanations involving *genetic* evolution (e.g. Pinker and Bloom 1990).

One key assumption in this chapter that contrasts sharply with the assumptions outlined above is that cultural processes can explain fundamental properties of language. This perspective has shifted the explanatory emphasis from human genetic evolution towards human cultural transmission. This assumption is the subject of this chapter.

The extended synthesis expands the scope of evolutionary studies both within and beyond biological processes. Thus, the impact of culture on the evolution of humans is the subject of dual inheritance theories, where genetic and cultural information co-evolve (e.g. Cavalli-Sforza and Feldman 1981; Boyd and Richerson 1995; Richerson and Boyd 2005; Odling-Smee et al. 2003). The cultural evolution of language in the second sense proposed above adds a new dimension to evolution: purely cultural processes occur over historical time, sometimes within a few human generations, in a timescale where the biological evolution of humans is irrelevant. New communication systems emerge in the face of novel communicative needs, and those systems evolve through processes indistinguishable from natural selection, neutral evolution, mutation, or gene transfer (of course, applied to cultural, rather than genetic or epigenetic information). The first part of this chapter expounds the assumption that language, a communicative behaviour, is a cultural-evolutionary system (along the lines of Croft (2000), following Hull 1988; and Ritt 2004) and then goes on to describe a selection of recent empirical behavioural experiments and computer simulations whose results are interpreted in terms of that assumption. The second part of the chapter develops the interpretation to lay some foundations for a language-centred theory of language evolution based on previous frameworks [notably Croft (2000) and Ritt (2004)] and informed by the results of the experiments and simulations reviewed in the first part. In this theory, the sounds, words and grammatical constructions we produce undergo replication, variation and selection. Humans are simply a (complex) instrument that mediates replication and selection, while concepts constitute the niches that linguistic items compete for.

1.1 Language as a Complex Adaptive System

Many authors investigating cultural evolution have found it useful to consider that language is a complex adaptive system (Gell-Mann 1992; Beckner et al. 2009). A complex adaptive system is composed of many elements that interact with each other. As interactions unfold, their outcomes inform the ongoing interactions. As a result of this self-organization process, *emergent* properties may arise. A classic example of a complex adaptive system is a flock of birds. Each bird has a local rule that attempts to stay within a certain distance and follow the general direction of its close neighbours. The distance and direction can change from one moment to the next, and this change may be affected by the bird's own behaviour. Those are the local interactions. The emergent properties are the flock as a coherent unit and the typical flock motion patterns that look so mesmerizing from a distance. Each bird's individual behaviour is not intended to generate a flock; neither is the flock behaviour predictable in practice from the sum of local interactions—sensitivity to the precise initial conditions is one of the defining characteristics of complex adaptive systems. We know what kinds of patterns to expect, but in a particular instance we cannot predict with certainty what the next state of the flock will be. A good illustration of the notorious difficulty to predict the behaviour of a CAS is the weather, with countless air molecules interacting under local conditions of pressure and temperature from which the likes of storms, tornadoes or spells of dead calm emerge.

In language, we have a *local level* where individual instances of linguistic behaviour are produced typically in interactions between speakers for particular, usually communicative, purposes in a given context. A large number of such interactions give rise to *emergent properties* such as linguistic regularities and categories or coordination of conventions at the level of the population. The actors in a particular linguistic interaction normally do not intend to generate a population-wide system; rather, they just want to communicate about something, there and then. Any population-wide or language-wide patterns are unintended side effects, emergent properties or, in the terminology used by Keller (1994), results of the action of the "invisible hand". Moreover, predicting the long-term outcomes of language change at the population level is so complex that it has not even been attempted.

1.2 Evolutionary Processes

Over the course of this chapter, we will see evidence that cultural transmission can lead to self-organization and emergence, but also that it involves evolutionary processes such as inheritance, variation, neutral evolution and selection. The first part of this chapter discusses recent studies that focus on different aspects of language evolution. The model of cultural language evolution presented in the second part puts this discussion in the context of other frameworks of the cultural evolution of language and, crucially, relates processes to mechanisms, for instance, inheritance to babbling and language learning; variation to social mechanisms of generation and spread of variation in the sociolinguistics literature; and selection to social interactions and to the structure of meanings.

Languages do change, but they are also remarkably stable. Studies of language families (Dunn et al. 2005) and individual words (Pagel 2009) claim to have reconstructed lineages of language traits up to 10,000 years into the past. This stability is due to linguistic information being culturally inherited by new speakers, in other words, by infants very faithfully learning the language of their social group. This is why the mechanisms of language learning by infants are of interest to the study of language evolution. But we do not stop learning when childhood ends. Rather, learning continues throughout speakers' lives, as we are exposed to and create innovations during usage. Therefore, the mechanisms of language usage by adults are also of interest to cultural language evolution.

The low-level mechanisms that mediate the inheritance of linguistic information, both during learning and over usage, include imitation and conformity. An important aspect of imitation is copying behaviour irrespective of its function, or even of whether it has a function. This human capacity typically enables the accumulation of traits that leads to the cumulative complexification (Heyes 2013; Boyd and Richerson 1995; Tomasello 1999) of language structure.

Inheritance is not perfect, however, or there would be no evolution. Innovations can be introduced into languages through production and perception errors, as the result of speakers' efforts to express novel meanings, or through contact with other languages. And once we have different variants of the same linguistic items (a sound, a word, etc.), there may be evolutionary competition between them.

Neutral evolution, defined for biology (Kimura 1983), where variants spread following random dynamics, has been highlighted as an important mechanism of evolution in language (Nettle 1999) and culture (Bentley et al. 2004; Herzog et al. 2004). But selective pressures are also at work and a host of factors can affect the structure of languages as they are used and transmitted to new members of the population. Some examples of selective pressures are as follows:

- Cognitive biases mean that certain words, sounds or constructions are more likely to be used because they are easier to learn, process, produce or perceive than others. The preferred variants may end up being more frequent in the language than their competitors.
- Social constraints that make speakers prefer linguistic elements that are original, fashionable, conformist or complex and features that serve to identify speakers as part of a group, or to distinguish them from another group will also leave their mark at the language level.
- Patterns of connectivity in the population, influencing how many other speakers one interacts with and how often, or whether the patterns are homogeneous in the whole population or not can have an impact on the structure of emergent languages.
- The structure of the world and particularly any features of the meanings that speakers want to communicate about can also affect the ways those meanings are expressed.

These selection pressures may result in certain linguistic patterns being more likely to be learned by new individuals, more apt to spread in a population, or more efficient for the purposes of communication. The following section describes recent computer and behavioural models of the evolution of communicative systems, which are then analysed in terms of the evolutionary processes they reveal.

2 Computer Simulations and Experiments

Aside from having important theoretical implications, prioritizing cultural-evolutionary explanations over biological-evolutionary ones has opened new methodological avenues to explore the origins of linguistic structure, especially with experiments that model the transmission of linguistic information both during communicative usage and over learning by new generations of speakers. Mathematical and computer simulations have been applied, hand in hand with behavioural experiments, in many successful lines of research: experiments can verify simulation results, and simulations can be used to construct models based on experimental results. This section specifically reviews and discusses a selection of experiments and associated computer simulations looking at how cultural-evolutionary processes shape the structure of languages.

The following sections describe a selection of studies that explore the creation of conventions, or individual signals that have an agreed meaning for the interlocutors that use them; the spread of conventions through a population; the emergence of cultural systems such as vowel systems; and finally the cultural emergence of linguistic structure. I do not present an exhaustive literature review, but rather a sample of classic and new experiments and simulations with the aim of illustrating the evolutionary way of thinking about the language transmission and usage sketched above. Each study highlights an aspect of this approach and will be accompanied by a discussion of the methods and results in terms of evolutionary processes and elements. Along the way, I will point out further questions that could be tested empirically by extending or adapting the studies described. Finally, the last section brings those elements together to outline a theory of the cultural evolution of language.

2.1 The Emergence of Conventions

Imagine you are on holiday in a country where you have no common language with the locals. One morning, you go to reception to borrow a hair dryer. You put to work your best gesturing abilities to describe what you want, and the reception man tries to be as helpful as possible. In the midst of your gesturing, you say "Wet hair! Wet hair!" and at that precise moment, your interlocutor produces a hair dryer from a drawer. Upon seeing your happy face, the man says with a relieved

Fig. 1 One player's view of the game: *on the left*, the player's own board; *on the right*, the partner's board. Adapted from Scott-Phillips et al. (2009)

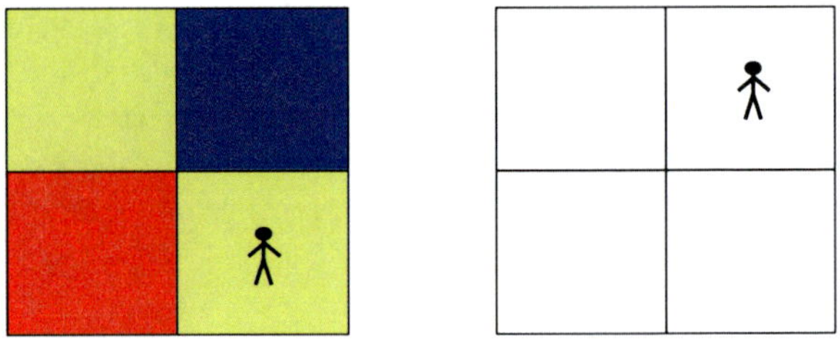

understanding smile "Wet-air! Wet-air!" The next morning, when you need the hair dryer, you ask the man for "wet-air" and immediately get what you want. The new word may be used by your friend when she wants to borrow the same item and you advise her to ask for "wet air". Similarly, the other receptionist may overhear and learn the new word and thus be able to help you or your friend with your requests in the future. You have created a new convention, and it has begun to spread beyond its creators, namely you and the receptionist.

Several things need to be in place for this to be able to happen. You need to have a shared communication channel, in this case gesture, that both you and the man at reception interpret as such signals.

For this to happen, the receptionist has to recognize your gesturing as communicative. This seems such obvious parts of the communicative interactions we are involved in every day that it is easy for us to take it for granted. However, it is not as obvious as it seems. Chimpanzees in the wild, for instance, do not interpret signals such as pointing as communicative without training, and tend to look at a pointing finger instead of at the direction where it is pointing (Povinelli et al. 2003). Humans, in contrast, have the inclination to interpret signals as communicative and also have a dedicated channel, most commonly speech, through which we expect to receive communicative signals. Even if one of these conditions is not met, we can still get by with the others, as in the hair dryer example, where there was not a shared language, but the gestures were readily understood as communicative.

But if there were no dedicated medium, shared signals or pre-established signaller and receiver roles, would we still be able to create an effective communication system? Two studies by Scott-Phillips et al. (2009) and Galantucci (2005) address this precise question. Scott-Phillips and colleagues designed an experiment where two people sat at two connected computer terminals and played a cooperative game. Each player had a playing board, composed of four coloured quadrants (Fig. 1) and a little character that could be moved from the centre of one quadrant to the centre of an adjacent quadrant by pressing the arrow keys. The quadrants could be red, green, yellow or green. The aim of the game was for both players to place their characters, each on their board, on the same colour.

Each player saw on the screen both their own and their partner's playing boards and characters, but they could not see the colours in the partner's board. The players could not see or hear each other, they were simply told about the task. In

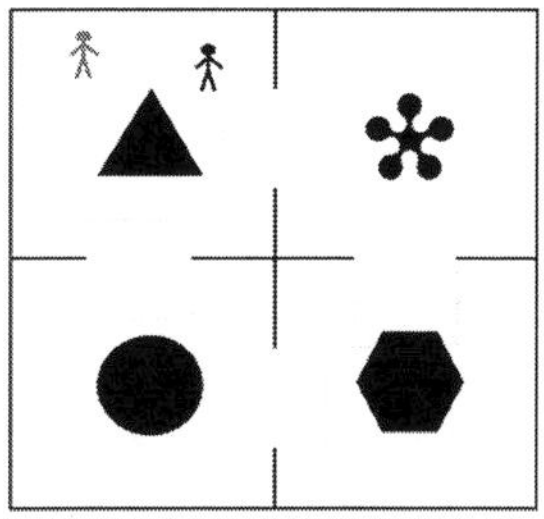

Fig. 2 Galantucci's game map: four rooms connected by doors. Adapted from Galantucci (2005)

these conditions, agreeing on a colour seems impossible—can you come up with a solution? After playing some games, however, most pairs managed to create a communication system, and in all cases, the process was the same. The first step to success consisted on the two players implicitly deciding on a default colour, so they would both land on that colour and thus score a point. The default colour tended to be red, perhaps the most salient one. But this strategy had a problem: sometimes, red was not present in one or both boards. The next step to success happened as a reaction to this situation. When a player did not have red on her board, she would attempt to let her partner know by moving her character for instance from left to right and back to the left, repeatedly. Then, she would land on another colour, say blue. This situation prompted the association of blue with left–right motion. At this point, the first communicative signal was created, at the same time as a communication medium, the character's motions, was discovered. After this insight, the other two colours were soon associated with other character motion patterns, and the game task, supported by communication, became trivial.

This experiment highlights the versatility of communication systems and the resources of our drive to communicate: any information pattern can become a signal. The characters' movements, which in principle served simply to move between quadrants, are co-opted for a communicative function. Also, the whole system is bootstrapped from an initial heuristic, namely always landing on the same colour and a situation where the heuristic fails.

A somewhat related study addressing the creation of a communication system de novo is that of Galantucci's. Here, two people also played a cooperative computer game; in this case, the task was for the two characters to meet in the same quadrant; quadrants were now identified not by colour, but by a shape (Fig. 2). During the game, each player only saw the room he or she was in and had to infer the whole map from experiencing moving through the doors.

There was an additional important difference between this experiment and the one described above: here, there was a dedicated communication channel. Beside each computer were a digital pad and pen, and the players could see on their screen what they and their partner wrote or drew. The experiment instructions did not mention this pad, and if the players attempted to use it, they would discover that what they wrote was heavily distorted, as if they were writing on a moving tape and faded rapidly from view.

As before, it is difficult to score points consistently in this game without communication and, here again, not all pairs of players found a solution. Those who succeeded, however, developed signals that were adapted to the communication channel and to the meaning structure. The distorting-pad prevented players from writing letters or numbers, or from drawing the figures found in the rooms; the signals they employed were those most immune to the distorting effects of the pad like vertical lines or small marks. As for the meanings employed to identify the rooms, several strategies were apparent. Some pairs numbered the rooms by drawing one, two, three or four small marks on the pad. Others tried to represent the triangle, circle, etc. iconically, for instance by relating to the number of vertices in each figure: three for the triangle, five for the star, one for the circle and six for the hexagon. A third group drew a line on either side of the pad to indicate whether they were in a room on the left or on the right. The last solution is ambiguous, since there are two signals for four different rooms, but it was complemented by the following strategy: at the beginning of a game, each player would tell the other which side of the board they were in. Then, one of the players would always move first. This combination of ambiguous signal plus conventionalized turn-taking allowed success at each game.

In two continuations of this experiment, Galantucci progressively increased the task's difficulty. The successful pairs from experiment 1 went on to the next level, which had a 3×3 board, and then to a 4×4 one. The drawings in the new rooms included an umbrella, a bird, a star, a hash or a crown. Performance levels in the complex environments turned out to be dependent on the type of strategy employed at the easier level: numerical systems were easily adapted to the extended boards, but iconic and side-based systems were not. The constraints of the drawing pad did not allow to keep drawing new sufficiently distinct icons for the figures. And the ambiguity of the side-based system grew exponentially with each increase in board size.

These differences can be understood in terms of adaptiveness of the signal systems to the game tasks. The main function of a communicative signal is unambiguously to point to one meaning among several possible ones (in this case, one room out of the four, nine or sixteen in the board). One of the traits that make signals useful, or likely to be reused, and therefore "fit", is being distinct from each other. In the initial, 2×2 board, the iconic, number-based and position-based-plus-turn-taking solutions were all adaptive. But in an extending meaning space, only the numerical one proved to be adaptive. In the former case, we can talk of a number of adaptive *independent* signals, each distinctly pointing to one room. However, in the latter, it is the *system* that is adaptive, as it allows for the repertoire of signals to be extended in a way that preserves and expands the communicative function. We will see more of adaptiveness at the system level when we talk about experiments with languages below.

In a further study dealing with the early evolution of signals (Garrod et al. 2007), pairs of participants created graphical conventions to represent a series of concepts in a pictionary-like task. In each game, the "director" and the "guesser" each saw a list of sixteen concepts. One of these concepts was selected and given to the game director, who would draw something for the guesser. If the latter guessed correctly what the target concept was, the pair scored a point. Garrod and colleagues were specifically interested in the role of feedback between the players in the emergence of arbitrary signals, and their experiments involved two kinds of feedback: one manipulated whether the two participants exchanged the roles of drawer and guesser in the games and the other whether the guesser could interact with the drawer, for example by asking for clarification.

Their results revealed that the players obtained higher scores when both types of feedback were allowed. Moreover, the drawings became simpler and more arbitrary (less likely to be identified by onlookers) only if at least one kind of feedback was available.

Brown (2012) has criticized the conclusion that the final simple drawings in these experiments are truly arbitrary, arguing that the even if onlookers could not identify the drawers accurately, the players themselves could still see traces of the initial iconic relationship between drawing and concept. For instance, two inverted V's would be impossible for an onlooker to recognize as a representation of the concept "cartoon". But the players and creators of the signal would probably still interpret the V's as stylized versions of a cartoon rabbit's ears. However, after a few generational transmission events, new users of the signal would not know about its origin and, therefore, for them, it would be truly arbitrary.

In Garrod et al.'s (2007) experiments, two-way interaction between the players allowed them to know that they were "on the same wavelength": they knew what a drawing meant for both of them, in the context of the game (e.g. a rabbit had been enough to make the guesser select "cartoon" from the sixteen possible concepts). They knew how and why the drawings changed over the course of the game (e.g. one of them drew the rabbit's ears and, before he had time to finish the drawing, the guesser gave an answer; from then on, the ears would be enough to identify the concept "cartoon"). This neatly demonstrates how a purely cultural process, social interaction, can contribute to the emergence of arbitrary communicative signals.

A cultural-evolutionary analysis of the three studies described can be framed in terms of adaptation of the emerging systems to constraints. A pressure inherent to these communicative tasks, indeed, the function of any communicative system, is *expressivity*: if we are to avoid ambiguity, a distinct signal is required for each meaning. The expressivity of a system depends in turn on the flexibility of the *communication medium*, which is decreasingly constrained in the three experiments. The main difficulty in Scott-Phillips et al. (2009) experiment was that the players needed to *exapt,* or co-opt, the characters' movements, whose original function was simply to move from one quadrant to another, for the novel function of providing a communication medium. Galantucci's (2005) and Garrod et al.'s (2007) experiments had a dedicated graphical communication channel, but in the case of the former, rapid-fading and linear motion prevented drawing normally. Nevertheless, some simple signals arose that were easy to produce within these limitations and sufficiently distinct from each other. As the expressivity pressure increased as new rooms were added to the board, the contrived communication medium made it impossible for some of the emergent systems to meet the expressivity requirements, and the participants failed to solve the task. Garrod et al.'s participants needed to disambiguate among sixteen referents, the same number as in Galantucci's third experiment, but in this case, they used drawing, a familiar medium both for game directors and guessers. This made all the difference, and all players succeeded in the communicative task, which suggests that given more time and practice Galantucci's participants would have the opportunity to explore the vast space of possibilities offered by their limited communication channels and evolve complex, structured systems. I say vast because, after all, rapid-fading and linearity are also characteristics of speech, which is short-lived and does not allow going back to revise what was produced earlier.

2.2 The Spread of Conventions

The players in Scott-Phillips et al. (2009), Galantucci (2005) and Garrod et al. (2007) managed to create new communicative conventions and to use them successfully over and over to achieve a goal. In real languages, innovations like new words and expressions are created all the time, usually spurred by the context, by knowledge shared by the interlocutors, by the need to express a new meaning, or by the desire to express something in a new way. Most of these innovations are never used again or, as in the games above, are only ever used by their creators, but some may catch on among new interlocutors of the creators, like the intercultural solution "wet air" for hair dryer in the story spread to a friend and to other people working in the hotel. A subset of all innovations will be adopted by more and more speakers and a few among them may spread to a whole linguistic community via the connections of the social network.

Coordination, or convergence on the same solution to express a meaning by the whole population can be viewed as an emergent property of languages. At each interaction, the interlocutors just wish to communicate with one another; but complex patterns of interactions involving many interlocutors may result in coordination at the population level. This hypothesis that shared conventions emerge through cultural processes of self-organization was tested in series of influential computer simulation studies carried out by Steels (1996, 1998, 2003, 2006).

The basic skeleton of these simulations, laid out in Steels (1996), includes a population of agents who are able to learn associations between signals and meanings, and who interact in pairs playing "language games". In each game, a speaker and a hearer are selected from the population and a set of objects are chosen as the context of the game. One of these objects is marked as the topic of the game. The speaker names a distinguishing feature of the topic (e.g. the colour and shape, that singles it out from the rest of the context), and subsequently the hearer points at the object that he thinks the speaker was referring to. With this information, both agents update their vocabulary in order to align to each other. Over many such interactions, involving many different player pairs, the agents' vocabularies, initially empty and subsequently idiosyncratic to each pair, end up being shared by the whole population.

This dynamics is mirrored in another experiment where eight people play pictionary games in pairs (Fay et al. 2010). The pairs change several times so that in the end, every player has played with everyone else. Within each pair, the game procedure was the same as in Garrod et al. (2007) where there was role swap and feedback. The initial pairs of players normally developed different representations for each concept, for example, the concept "cartoon" could be represented as the drawing of a rabbit for a pair, a Simpson character for another and Mickey Mouse ears for yet another. But often, after several partner changes, the whole population converged onto the same representations. In some cases, however, two or more variant representations remained.

How did this happen? The patterns of variant spread observed in Fay et al.'s population could have resulted from neutral evolutionary dynamics whereby variants that happen to have a higher initial frequency have a higher probability of spreading to the whole population through random processes. (In biology, this dynamics is instantiated as genetic drift.) In fact, Steels characterizes the spread of information in the population in his 1996 simulation described above and in other associated studies as neutral evolution, where all the possible ways to name the objects are equally likely to spread. Several studies have highlighted the power of neutral evolution to explain cultural evolution, for example, the names given to babies (Bentley et al. 2004) and the breeds of dog that people tended to buy (Herzog et al. 2004) changed according to the neutral model of evolution. Others have focused on the role of selection pressures on the spread of cultural variants (Richerson and Boyd 2005). Linguistic innovations are believed to spread by a mixture of neutral and biased transmission (Nettle 1999; Trudgill 2004; Blythe and Croft 2009).

The nature of variant spread in Fay et al.'s (2010) experiment has been explored in a recent simulation study (Tamariz et al. in preparation). Mirroring the experimental design, the simulations had eight agents arranged into pairs playing communication games about sixteen concepts, and undergoing the same number of interactions and partner exchanges as in the experiment. The spread patterns obtained in the simulations were compared to those recorded in the experiment.

The study considered several possible mechanisms of spread: first, a model of neutral evolution where each agent chose which variant to produce randomly from the two variants he saw in the previous round—his own and his previous partner's. Some empirical data points, for instance those for concepts where convergence had not been achieved, were similar to the results of the neutral evolution simulations, but the majority converged faster than predicted by this model. The second model included a bias for conformity that increased the chances that the two players in a pair would coordinate by agreeing on the same representation variant. Some of the remaining data points were captured by this model, indicating that local coordination could also be at work. Indeed, the goal of each interaction is to communicate successfully with the current partner, and using the same variant as him or her is a good strategy to achieve this. The third and final model tested included a selective pressure, namely content bias: certain variants were intrinsically more likely to be copied than others. This model captured all the remaining data points, and in fact the models that best fitted most empirical spread patterns included a degree of content bias.

These results further support that multiple processes (neutral evolution, content biases, social heuristics) may operate on the evolution of culturally transmitted variants. Additionally, they provide evidence for replicator dynamics operating at the cultural level (Blythe and Croft 2009): the computer models in Tamariz et al. (in preparation), assumed that each representation variant was, in general, faithfully replicated (when it was reproduced by a player at each round in the game) but could also mutate (when it was modified by a player).

This study reveals how some variants spread while others do not, but it does not address *what* makes some variants more likely to be copied than others. Are the winning variants easier to produce, process or perceive than others? Are they particularly clever or ingenious? Are they simple? Or iconic? Future manipulations of the type of cultural information transmitted in experiments like Fay et al. (2010) informed by studies on cognitive salience combined with computer simulations such as those in Tamariz et al. (in preparation) could help us understand precisely what kinds of content biases operate on cultural transmission.

2.3 The Cultural Emergence of Systems

In the work reviewed thus far, new conventions emerged and spread in a population. Participants in different experimental games started to use their own behaviours as signals, with communicative intent or to designate concepts; but these signals were largely independent from each other, there were no categories, rules or interactions that justified treating them as a unified system. This contrasts with many cultural realms, notably language, which behave, as we have seen, as complex adaptive systems. Next, I describe some studies where properties typical of systems emerge out of cultural processes of interaction and transmission, and then, I discuss the evolutionary processes they reveal.

De Boer (1999, 2000, 2001) observed that languages have a fixed repertoire of vowels whose organization is strongly constrained. In particular, vowel categories are widely spread out in the acoustic-articulatory space so they maximize occupation of the available space and minimize overlapping. This results in vowels that are easiest to distinguish from each other by hearers. de Boer designed a simulation with two agents who could produce, perceive, and remember speech sounds characterized by three realistic parameters: tongue position, height and lip rounding. The two agents played repeated imitation games (swapping roles at every interaction) as follows: the initiator would select a vowel prototype from memory and produce a version of it (with noise). The imitator would hear it, interpret it as one of the prototypes in his memory, and produce a version of this prototype (again, with noise). The initiator would then interpret this as a prototype; if this was the same as the initial prototype, the game was successful. Both agents could then update their memories with the information from the game by merging close vowels, throwing away seldom used vowels, adding new random vowels or moving produced vowels closer to the perceived ones within a prototype. After many iterations of the game, the vowel categories spread over the whole articulatory space and maximized the distance between them, optimizing distinguishability like natural vowel systems. This final arrangement is an emergent property at the level of the whole system.

Wedel (2006) offers an interesting evolutionary approach based on exemplar models to explain the mechanisms for the formation and stabilization of vowel (or other sound) systems such as those modelled in de Boer (2001). Each sound category exists as a distribution of "exemplars", or individual instances of production in a linguistic community. We learn by hearing and producing many such sound exemplars, which all leave a memory trace in our minds. Each individual's knowledge of a sound is based on a different sample and therefore is slightly different from every other speaker's knowledge. With such pervasive individual differences, how come sounds do not change constantly over time and across a population? Wedel suggests the answer is *blending inheritance*, a mechanism for the transmission of features with continuous values that is capable of maintaining stable categories over time. Each exemplar of, say, vowel *oo*, we produce is based not on a single exemplar in our memory, but on the all the exemplars in that vowel category. The production target may have the average tongue position, height and lip rounding values of all the *oo* exemplas in our memory. It is harder to argue for replication in this case, as each production is clearly not necessarily causally linked or similar to a particular previous exemplar, and establishing lineages is therefore not obvious. Nevertheless, Wedel persuasively argues that the categories behave as discrete replicators. The stable categories achieved with blending inheritance translate into a multimodal distribution of acoustic values in the output data produced by speakers. New learners exposed to this input will go on to produce (by blending inheritance) sounds with the same underlying distribution properties as the one they learned, so, over the generations, both the statistical properties of the distributions of exemplars produced and the categories are maintained. As Pierrehumbert puts it, "phonological representation (is) an error-correcting code" (2012: 175).

Recently, Verhoef and colleagues devised a behavioural experiment somewhat related to de Boer's vowel category simulations, modelling the emergence of a system of signals (Verhoef et al. 2011, 2012). In this case, the cultural mechanism at work was not interaction in a pair (*closed-group method*) in Mesoudi and Whiten's (2008) terminology, but transmission to new learners (*transmission chain method*). Verhoef and colleagues had a human participant learn how to play a set of twelve different, random whistles with a slide whistle—a kind of flute with an embolus inside that can be pushed in or out to produce higher or lower sounds. He or she would subsequently attempt to play back the twelve sounds, no repetitions allowed. These twelve new whistles were used to train the next participant in the diffusion-chain experiment (see next section for more details on this paradigm).

Fig. 3 A transmission chain of drawings involving nine participants. The *top row* shows the eight original drawings, and *subsequent rows*, the drawings produced by each of the participants

The second participant's output would be the input to the next, and so on for ten "generations". Over repeated transmission, the sounds changed dramatically and the final whistles were typically a *system* composed of a small number of sub-whistles recombined in different ways. At each episode of transmission, the structure of the information transmitted changed a little, but this happened without any intentionality on the part of the participants, who were merely trying to reproduce what they had learned. Over the generations, some of the more difficult-to-remember patterns would not be reproduced in the output, while novel whistles would be produced; the pressure to produce twelve different sounds encouraged reuse of remembered sub-parts in those novel whistles produced. In addition, the sub-parts were sometimes reduplicated or reversed. The final systems were, consequently, simpler in the information-theory sense—they had much lower entropy, i.e. were more compressible than the initial ones—and easier to learn—later-generation participants only needed to memorize a few units and a few ways to recombine them. Out of the initially continuous sound space, a small set of discrete patterns emerged. These are usually very different from each other, as was the case with the vowel categories that emerged in de Boer's simulations and therefore easy to distinguish, memorize and produce.

Figure 3 shows some results from a recent small study carried out by the author of this chapter as a teaching exercise. The design of this study was the same as that in Verhoef et al.'s (2011, 2012), but here, eight drawings were transmitted instead of twelve whistles. The ten generations of drawings show visually the same kind of recombination and simplification processes attested in the results of Verhoef et al. (2011, 2012). The processes involved here, again, were reduplication reversal and recombination. For example, the sixth participant in this chain (7th row of drawings in Fig. 3) reversed the sixth drawing (or perhaps recombined the shape of the rightmost drawing and the lines at the end of the Z-shapes from the third drawing from the right at the previous generation) and also added a horizontal line at the bottom of drawings one and four. It is also apparent that three categories of drawings become increasingly obvious over the generations: the three Z-like drawings on the right; the three drawings with steps (drawings two, three and five) and the two closed drawings (one and four). A short description of the final set of drawings, then, would probably involve the three categories plus the details that differentiate the items within each category, for instance the direction of the end lines of the "stairs" or the direction and presence of endlines in the Z-like drawings.

The drawings, initially unrelated to each other, evolve to form a *system* of related categories. I argue that the emergent units (the categories and the details) are replicators, as they fulfil the necessary criteria of similarity (copying fidelity), causality and information transfer between model and copy, longevity and fecundity (Dawkins 1976; Ritt 2004; Sperber 2000; Godfrey-Smith 2000). Once the sub-units in the whistles in Verhoef et al. (2011, 2012) and in the small drawing experiments have stabilized, after a few generations, they are faithfully and reliably copied in such a way that we can trace their lineages; they are causally connected to previous productions of the same units; and information about their structure is transferred from originals to copies.

Additionally, the small number and systematic nature of the replicator set at the final generations is an adaptation to the elements involved in replication. First, limited exposure time to the original whistles or drawings only provides limited opportunity to memorize long complex patterns. Second, a cognitive preference for regularity leads the participants to reuse patterns and processes they have memorized and extend them to the whistles and drawings they cannot remember well.

2.4 The Evolution of Regular Linguistic Structure: Systematicity Between Signals and Meanings

So far we have seen evidence that individual *communicative signals* on the one hand and structured *systems* of replicable units such as whistles and drawings on the other can emerge and stabilize over repeated use and transmission. Next, we will look at simulations and experiments showing that structured *systems of signals*, such as artificial languages, can also emerge from the same cultural processes.

All natural languages are compositional: the meaning of a complex linguistic utterance is a function of the meanings of its elements and the ways in which they are put together. For instance, the meaning of a sentence such as "man bites dog" depends on the meaning of its components and their order (compare "dog bites man"). Compositionality is thus the key property of languages that allows us to recombine words and constructions in infinite ways to express new meanings. Kirby (2001) published a simulation study that demonstrated that compositionality can emerge from cultural transmission dynamics alone. This was the first of an ongoing family of simulations and experiments applying the iterated learning model to explore the role of transmission in language. Iterated learning is the "process in which the behaviour of one individual is the product of observation of similar behaviour in another individual who acquired the behaviour in the same way" (Scott-Phillips and Kirby 2010: 411). Kirby's (2001) seminal simulation involved chains of agents learning artificial languages composed of a set of meanings, and their names. Every generation in the chain included three steps: first, an "adult" agent was given some meanings and it must name them using the signals in its memory, or inventing; next, a learner agent learns the language (the associations between meanings and signals) produced by the adult; finally, the learner becomes the adult for the next generation and the old adult is discarded. Learning involved agents adding the new associations to their memory and streamlining redundant information; here is an illustration of the streamlining process: if an agent had the associations {john, eats}$\Longrightarrow$"johneats" and {tiger, eats}$\Longrightarrow$ "tigereats", the streamlining process will replace them with the more general {x, eats}$\Longrightarrow$"*xeats*", {john}$\Longrightarrow$"john" and {tiger}$\Longrightarrow$"tiger".

The initial signals were invented random letter strings produced by the first adults, but the languages kept changing over the generations until eventually they stabilized. The crucial result was that the final, stable languages resulting from the simple iterated learning dynamics were compositional: each value of each meaning dimension had an associated letter string; and letter strings were combined according to rules to form the complex meanings. Kirby's insight was that the language adapted to the transmission process: unlike the initial ones, the final, compositional languages were stable and did not change over transmission. Even if the input only included part of the language, its compositional structure allowed the agents to reconstruct the complete language. If compositionality, a fundamental property of languages, can be explained by cultural processes of learning and transmission, reasoned Kirby, perhaps we should focus more on how languages have adapted to humans through cultural mechanisms and less on how humans have adapted to language through biological-cognitive evolution.

This and related computer models inspired a line of experimental work that confirmed and complemented the simulation results. The first modern experiments on artificial language iterated learning are those described in Kirby et al. (2008). The dynamics closely mirrored those in Kirby (2001): here, a human participant was trained on (a half of) an artificial miniature language: words referring

Fig. 4 Drawings from representative chains in the two conditions in Tamariz and Kirby (in press). Generations 0, 1, 4, 7, 10, 13, 16, 19 and 22 are shown from a chain in the memory condition (*top*) and one in the copy-from-view condition (*bottom*)

to objects; then, she was asked to name the full set of meanings. From her output, half of the object–word pairs were selected and given to the next learner as training input. The learner then was asked to name *all* the objects, and so forth for ten generations. Each language consisted of 27 words referring to as many meanings: all the possible objects combining three shapes (triangle, circle and square), three colours (blue, black and red) and three motions (spiral, horizontal and bounce). While in the simulation the first agent (generation) in a chain invented its own signals, in the experiment the first participant was given randomly constructed words. The results revealed that the initial languages where each object was associated to a random word became structured. For instance, in one transmission chain, all the objects moving horizontally ended up being called "tuge" and all the objects with a spiral motion, "poi".

The languages in this experiment and in Kirby's (2001) simulations became increasingly stable over generations. This can be interpreted, again, as the emergence of stable replicators. In the absence of any communicative requirements, the only task given to participants was to learn and reproduce the language as faithfully as possible, and the languages readily adapted to the task. Moreover, the fact that the word-replicators referred to categories of objects (those moving horizontally or spirally) indicates adaptation of the words to the structure of the meanings. The few remaining words in the final language were associated not to random collections of objects, but to meaningful categories.

These results were different from those in Kirby's (2001) simulation in an important respect: there were a lot fewer words than meanings. The simulations had an implicit bias for unambiguous mappings where a distinct signal was associated to each meaning. In a second experiment, Kirby et al. (2008) also introduced a bias for diversity, or expressivity when they had to select the items from one participant's output to construct the training set for the next one, instead of doing it randomly as in the first experiment, they removed as many items with duplicate words as possible. The languages emerging in this condition did not stabilize to the same degree as in the first experiment, but reproduction fidelity kept increasing. But the most dramatic effect of this subtle manipulation was the emergence, in some languages, of *compositional* structure over the generations. In the languages where compositionality emerged, different parts of the words (the beginnings, middles and ends) became associated with colour, shape and motion categories. In one of the chains, whose final-generation language is represented in Fig. 4, word beginnings were reliably associated with colour, and word-endings with motion. In a perfectly compositional system, word-middles would be associated with shape; and there would be no variability such as the one found in the example in Fig. 4 for colour blue (which is expressed as either *ku* or *hu*). This variability, incidentally, does not introduce ambiguity for comprehension, but it creates a degree of uncertainty in production—should I use *hu* or *ku*?

In Kirby et al.'s (2008) second experiment, replicators also emerged. In this case, they were not whole signals like the "tuge" or "poi" words from the first experiment, but elements in the signals, and their positions, which also became increasingly stable over the generations—in other words, evolved to become easy to replicate.

In contrast with the signal-only systems (whistles, drawings), in systems where signals are paired with meaning, certain letter strings (words or parts of words) become increasingly consistently associated to features of the drawings and thus acquire a symbolic or referential function. The "meaningful" strings, in turn, become easier to remember—to replicate—both because they "mean" something and because of their higher frequency. When a participant needed to name an object that she had not seen during training, say, a red horizontally moving triangle,[1] she would be likely to use letter strings that had been associated with red colour, horizontal motion or triangular shapes in her training set. The adaptive solution that emerged over the generations under the double constraint to be easy to replicate and functional was a compositional system: only nine segments, combined in different ways, could express 27 meanings. Compositionality is, in fact, an efficient solution to the problem faced by languages unambiguously naming a structured set of items (Brighton et al. 2005). Compositional languages have three important interrelated properties: first, they are *expressive*, as they allow one-to-one unambiguous mappings between signals and meanings, which are good for communication. Second, they are *compressible*, as many meanings can be expressed with few signals; this increases the replicability of the system, since only a few items have to be memorized. And third, as a consequence of the first two properties, the languages are extendable; in other words, they allow the expression and understanding of novel meanings.

The difference between experiments one and two that allowed the emergence of compositionality only in the second was the removal duplicate signals from the training input. In a real language, this bias should arise from the need of words to refer to meanings unambiguously if the communicative interaction is to be successful. This last point was tested in a recent series of experiments, where Kirby et al. (in preparation) combined the iterated artificial language learning paradigm with communicative tasks. The main question they addressed was: what is the relative contribution of transmission to new learners (transmission chains, as in Kirby 2001; Verhoef et al. 2011, 2012; Kirby et al. 2008) and usage [closed-group

[1] For the most part, participants were unaware that there were asked to name novel objects.

interactions, as in the communicative experiments by Scott-Phillips et al. (2009), Galantucci (2005), Garrod et al. (2007) or Fay et al. (2010)] to the emergence of structured systems? In their experiments, Kirby et al. (in preparation) had pairs of participants play a game on two computers. Following Kirby et al. (2008), they were first trained on a language; but then, instead of simply reproducing the words, they *used* the language to play a cooperative communicative game together. The last words produced for each of the objects at the end of their game were used to train the participant pair in the next round in the game, and so on for six rounds. The key manipulation was whether the next round was played by the same pair of players (closed group) or a new pair (transmission chain).

The communicative interactions between the pairs were inspired by the "language games" in Steels' (1996) computer models and involved a speaker and a guesser. The speaker was shown a target meaning, which she had to name. The guesser then saw on his screen the signal typed by the speaker and an array of possible meanings, and had to choose which of the meanings corresponded to the signal. If the chosen meaning was the same as the target, they jointly scored a point and the players exchanged roles at every game.

Overall, the final languages were, as in previous experiments, easier to learn (more stable across rounds) and more structured (systematic and compositional) than earlier ones. As far as the role of the communicative task is concerned, the hypothesis was held: the final languages contained hardly any homonyms, so they were fully expressive and effective for communication (more points were scored at the final than earlier rounds). In other words, the communicative task achieved the same effect as the expressivity pressure introduced in experiment 2 in Kirby et al. (2008). So, as long as we are not interested in other aspects of communicative interaction [such as feedback, as in Garrod et al.'s (2007) pictionary studies], the simple iterated learning-and-recall with anti-homonymy filter is a valid design. If, however, we wish to explore aspects of communicative interaction, we can choose the iterated *communication* design.

As for the roles of closed group versus transmission chains, the languages in the closed-group condition—where the same two people went through repeated rounds of training and play—obtained higher scores and were more stable than those in the chain conditions, where the pair was replaced by a fresh one at each round. Crucially, the chain languages showed a strong increase in compositionality, but those in the horizontal condition did not change in that respect and remained a set of distinct, but idiosyncratic and unrelated signals.

It seems, then, that transmission to *new learners* is the key element for structure to emerge in these experiments. In evolutionary terms, they represent the generation turnover that renovates the pool of linguistics variants. New learners bring to the dynamics the necessity for inheritance of information between generations, a crucial element in any Darwinian system. In the iterated communication experiments, having several consecutive learners exerts pressure for stable, faithfully replicable languages; and communication selects for expressive languages. The adaptive solution to this double constraint is, again, compositionality: a simple system with few elements to memorize and reproduce, and a few combinatorial rules that make the languages robust against memory failure, because it is possible to generalize from a few items to the whole set of meanings.

For compositionality to be at all possible, an additional requirement is that the meanings are structured. The repetition of meaning features (colour, motion and shape) in different objects requires that several meanings share the same features. I have spoken of this match between words and object categories in terms of adaptation of forms to meanings. This hypothesis is upheld by an experiment by Perfors and Navarro (2011) that shows the impact of the structure of the meaning space on the final language. They run iterated language chains like those of Kirby et al. (2008), but used an especially designed set of meanings: squares of six different sizes and six different levels of brightness.

They manipulated the structure of the meaning space to make one dimension more salient than the other. In the control condition, the values of size and brightness were evenly distributed. In the *size condition*, there were three smaller sizes and three clearly larger sizes. In the *brightness condition*, there were three lighter and three clearly darker shades.

Their words were consonant–vowel–consonant syllables, and they did not apply a filter to eliminate homonyms, so they expected ambiguous languages to emerge, similar to those of Kirby et al. (2008, experiment 1). However, while Kirby et al. had no prediction as to which meaning would come to be expressed (e.g. motion in the case above, or colour, or shape), Perfors and Navarro's (2011) manipulation of the meanings predicted that each unevenly distributed meaning space would favour a particular categorization. This is precisely what they found: in the size condition, the words tended to categorize the objects by size, for instance a word for larger shapes and another for smaller shapes, while in the brightness condition, the emerging categories were aligned with the darker and lighter shapes.

The idea that spurred Perfors and Navarro's study was the Bayesian prediction that the outcomes of an iterated learning chain could be explained by the learners' prior biases. This is most clearly shown in Kalish et al. (2007). These authors run chains of participants who had to implicitly learn mathematical functions: each participant was shown a horizontal bar of a certain length on the screen and had to respond to it by producing a vertical bar of some length. After they had produced their vertical length, they were given feedback as to what the response should have been. The horizontal and vertical magnitudes were, in fact, the x and y values of a function. For instance, in the case of the simplest linear function, $y = x$, they learned to respond so that the longer the horizontal bar, the longer the vertical one should be. They initiated a total of eight transmission chains with different mathematical functions including the above-mentioned positive linear function $y = x$; the inverse negative function $y = 1 - x$; a nonlinear function; and a random correspondence of y values to x values. After nine generations, in all but one of the chains, the function had turned into a positive linear $y = x$, and the remaining one into the negative linear function $y = 1 - x$. These results were interpreted in Bayesian terms, with the posterior distribution of linear functions (the final seven positive linear and one negative linear functions) reflecting the prior (the cognitive

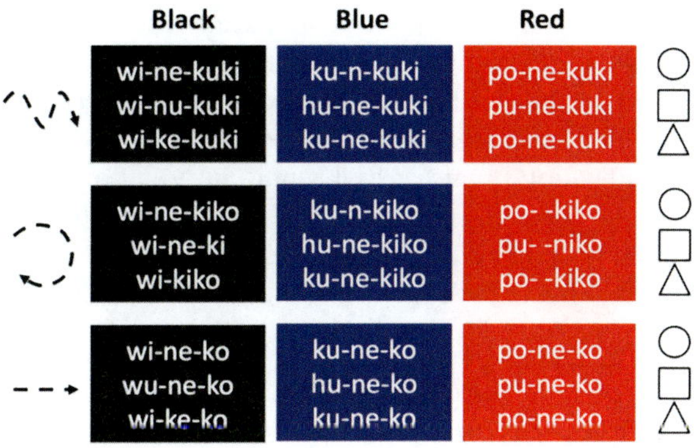

Fig. 5 The words produced by a fourth generation participant in one of the chains Kirby et al. (2008) (Exp. 1). Hyphens have been added for clarity's sake

preference for linear functions, especially the positive linear one, Kalish et al. 2004).

Perfors and Navarro's posterior (final) distribution of languages could not simply be a reflection of (prior) cognitive biases, since this prior was presumably the same for the participants in the size and the brightness conditions. With their ingenious experiment, they showed that cognitive priors are not fixed. Instead, they can be affected by external factors, in this case by the structure of the world, which suggested alternative partitions or categorizations of the meaning space. This experiment is a transparent demonstration of a cultural system—a set of words that are transmitted generation after generation—adaptively responding to the structure of its environment, the meanings that the words refer to.

Iterated learning experiments are designed to establish the role of repeated transmission on language structure and to highlight how cultural information adapts to its own transmission (Kirby et al. 2008). They assume a tight learning bottleneck, with very little training and a pressure to generalize the language learned to express novel, unseen meanings. This may be a valid assumption for language learning, but in many other realms of culture, learning involves extensive teaching and feedback (think of the years of formal education or learning to play an instrument). The differential effect of these two types of learning—without and with feedback—is tested in a study by Tamariz and Kirby (in press). This is, again, an iterated learning experiment, where participants had to look at a drawing for ten seconds and then reproduce it as accurately as they could. The drawing produced by one participant would be the original for the next one, and the initial drawing in all chains was a meaningless doodle (Generation 0 in Fig. 5). In half of the transmission chains, the original drawings were removed from view after the ten seconds (modelling unsupervised, limited training), but in the other half, they remained in full view while the participants copied them (feedback). In this way, the memory element of the transmission could be explored.

In the memory condition, the drawings, as in the previously described whistle and drawing studies, became simpler over the generations (Fig. 5, top). They did so by turning smaller and more streamlined, but also by transforming into conventional numbers of letters—memorizing “capital R” is much more economical than memorizing the description of a complicated doodle. The drawings in the copy-from-view condition, however, although changing, retained the initial level of complexity and remained meaningless (Fig. 5, bottom). This result provides a clear indication that simplification is caused by keeping information in memory—even if only for a few seconds as in this experiment. Conversely, it shows how different aspects of transmission (or inheritance), such as learning, keeping in memory and reproducing a pattern, affects the structure of the pattern in distinct ways.

2.5 Conclusions from Experiments and Simulations

This sample of empirical studies has shown a variety of evolutionary processes in action: inheritance of information as it is passed on from interlocutor to interlocutor and from experienced user to learner; and selection of information patterns that are best adapted to environmental factors such as other patterns, people’s biases and the structure of the world.

We have learned that social communicative interactions are required for the emergence and spread of communicative systems and conventions. We have also seen that when many patterns evolve together, they are influenced by each other and become a system. Finally, language has peculiarities that make it special in several respects: linguistic forms compete for meanings, and they need to be flexible enough to express endless novel meanings during usage; one efficient adaptative solution to this is compositional structure.

The next section integrates this knowledge into a theoretical framework of language evolution.

3 Elements for a Theory of the Cultural Evolution of Language

The first part of this chapter reviewed a selection of recent experiments and computer simulations focusing on how they implement cultural-evolutionary processes such as inheritance, variation and selection. This second part outlines a theoretical framework for cultural evolution, centred on communicative systems. The main elements discussed in this framework are, first, *inheritance* of linguistic information: What are the mechanisms of language transmission? Can we talk of replication of linguistic patterns? And second, *selection*: What is the environment where linguistic patterns evolve? What effects do environmental factors have on the patterns?

I will start off by highlighting some high-level commonalities between some of the processes in language evolution and in the origin of life. The beginning of life, before DNA and other complex molecules existed, was characterized by cyclical chemical reactions involving autonomous replication, or "continued growth and division which is reliant on input of small molecules and energy only" (Szostak et al. 2001). Replication occurred whenever new similar molecules were produced. Variation was brought about by random changes in the molecular structure and by recombination of different molecular parts by horizontal transfer. The feature of this early life system that is relevant to the present discussion is that it did not include translation: the molecules did not code for anything in the way genes today code for proteins. The only "function" of these molecules was self-replication, and the dynamics of the system selected the best replicators: molecular structures that replicated more faithfully increased in frequency and therefore produced even more (faithful) copies of themselves. In the long run, the best replicators would come to prevail. This contrasts with present-day genes, stretches of DNA, which, by virtue of coding for proteins, have functions contributing to the success of the organism that carries them.

The transition between "selection for replicability" only (or the evolution of replicators) and the addition of "selection for function" (or natural selection) came about when the replicating molecules began to code for proteins which, in turn, altered the environment where the molecules replicated. This transition is the third of the major evolutionary transitions proposed by Maynard Smith and Szathmáry (1995). For Woese (1998), it is *the* major transition of life; he calls it "the Darwinian threshold" because it marks the beginning of genes defined by their functions, which constitute the units of natural selection. After the Darwinian threshold, vertical transfer of genetic information leads to an increasingly permanent organismal phylogenetic trace (Woese 1998).

An analogy of these processes in the origin and evolution of language would be the view that humans began to produce vocalizations that carried no symbolic or referential meaning, perhaps similar to birdsong. The "musical protolanguage" hypothesis of language origin (e.g. Darwin 1871; Okanoya 2002; Fitch 2010) does just this. Versions of this hypothesis share the assumption that our hominin ancestors evolved the capacity for *vocal learning*, that is, for faithfully imitating vocal patterns—or, more widely, motor patterns including rhythmical, gestural or vocal sequences. Among our closest relatives, we are the only species capable of (and indeed prone to) imitating behaviour even if it has no apparent function.[2]

The stage of the origin of life that Woese (1998) would call pre-Darwinian corresponds in the musical protolanguage hypothesis to sounds that are transmitted socially, but which have no communicative function—maybe tunes or dance patterns, hence the "musical" name of this hypothesis. The sounds that were faithfully copied would persist over time, the rest would not: this is selection for replicability. The Darwinian threshold would be crossed when sounds and their combinations began to be produced and understood as meaningful. Now, as well as sounds being selected for replication, certain sound combinations would be

[2] Vocal learning has been attested in several other distant branches of the tree of life including some species of cetaceans, elephants, bats, song birds, parrots and hummingbirds (Jarvis 2006).

selected for because they conveyed useful meanings, or because they conveyed them well. This is selection for function.

However, the two levels of selection, for faithful replicators and for function do not need to have occurred sequentially either in life or in language (as is proposed to have occurred in the musical protolanguage hypothesis). In fact, a co-evolutionary scenario involving the human capacity for imitative vocal (or more generally, motor) learning, the presence of increasingly complex vocalizations in the environment and co-opting those vocalizations for communicative purposes, in the style of Dor and Jablonka (2010) is equally if not more plausible and has the advantage that it does not require to posit a non-communicative function for the early vocalizations.

In the following sections, I argue for a theoretical approach to language evolution that involves these two levels of evolution in language: on one hand, our species has evolved imitative skills that ensure the faithful replication of the stuff of language—mainly, sounds. On the other hand, combinations of those sounds have functions: we use them to communicate meanings to each other. Communication involves many interrelated factors such as the concepts we entertain and their structure, the alignment of concepts in interlocutors, patterns of social interaction, which may pose selection pressures on the evolution of linguistic items.

3.1 What Evolves in Language Evolution?

The opening paragraphs in this chapter characterized language as behaviour, and what follows is based on this view: I will talk about linguistic patterns, and by that I mean patterns we produce: speech sounds, words, constructions and structural patterns. Linguistic patterns, therefore, do not include the functions of words, constructions, etc. The functions of linguistic items, or "meanings" in an extended sense, include semantic meaning but also all the nuances a word produced in context may convey—the identity and status of the speaker, the degree of formality or informality of the context, the nuances of meaning perceived in the particular context, etc. Those factors, in the current approach, together with speakers and their intentions, constitute an important part of the *environment* where linguistic behavioural patterns evolve. This contrast with Croft's (2000) model of the evolution of language, where linguistic replicators or "linguemes", include not only the sounds uttered by speakers but also the meanings of those utterances. Linguemes are linguistic conventions (sound patterns plus their shared meaning) that are replicated each time they are used and are passed on to new generations through usage and learning. The current approach is closer to that of Ritt's (2004), who concedes that the replication of meaning together with the form is highly problematic and gives a nuanced definition for replicators from which meaning has all but disappeared, leaving only the sounds uttered (or, more specifically, the neural activation patterns that lead to the sounds being uttered). To reiterate, in the present model, linguistic replicators are, more in accordance with Ritt's proposal, purely behavioural, while meanings are part of the environment where they evolve.

The main argument for putting behaviour at the centre of this theoretical framework is that meanings, the mental representations corresponding to linguistic forms, are much more variable between speakers than the linguistic behavioural form. The number of possible combinations of percepts and concepts must be vastly greater than the number of linguistic patterns found in any one language, if only because the same patterns are reused in multiple and diverse occasions and contexts. Naturally, there is something in common among all the occasions and contexts where the same form is used, but there is also much conceptual information in the brain that is under- or non-specified in linguistic forms. If experience incrementally contributes to the function of each linguistic construction, and if we assume that individual experiential histories are unique, then the individual differences in meaning for each linguistic pattern will be orders of magnitude greater than differences in the corresponding linguistic pattern. We may say that, for each word, there are as many meanings as there are speakers in the language.

The set of all meanings in an individual yields an overall meaning space constituting "a complicated network of similarities overlapping and criss-crossing" (Wittgenstein 1953: 66e). These complex entities cannot be the same for two speakers and may even be different for the same speaker on different occasions. Behavioural linguistic patterns, on the other hand, have similar linguistically relevant features between speakers. It is true that even leaving aside non-linguistic differences such as voice timbre, quality or volume, there is still variation in the forms of words or sounds within the same language. (But "correct speech perception irrespective of the acoustic variation between the different speakers and word context" can be explained by "the existence of such neuronal populations in the human brain that can encode acoustic invariances specific to each speech sound" (Näätänen 2001:1.)

In his model of cultural evolution, Sperber (1996) defends that meanings (mental representations) "are more basic than public ones" (ibid: 78). And yet, he maintains that the cultural transmission of mental information is fundamentally transformational and those transformations are not inheritable. Any stability in cultural (mental) representations across individuals and over time is explained by "attractors" or "points or regions in the space of possibilities, towards which transformations tend to go".[3]

Further arguments and evidence for the higher stability and fidelity of transmission of public productions can be found, paradoxically, in the midst of expositions about the primacy of mental culture. In a critique to Sperber (2000), Dennett (2006) claims that public cultural items such as recipes, wheels or renditions of a musical piece can be faithfully transmitted "thanks to the shared norms for (…) analog processes already inculcated in the apprentice". In Dennett's argument, however, fidelity relies on the apprentice already being enculturated. Tomasello, Kruger and Ratner (1993) look precisely into enculturation and assert that "human beings 'transmit' ontogenetically acquired behaviour and information, both within and across generations, with a much higher degree of fidelity than other animal species". Richerson

[3] An epidemiology of representations: A talk with Dan Sperber. *Edge*, 164, 27 June 2005. http://www.edge.org/documents/archive/edge164.html#sperber. Accessed 6 June 2011.

and Boyd, while arguing for mentalistic culture, observe that: "[I]nformation in one person's brain generates some behaviour—some words, the act of tying a knot, or the knot itself—that gives rise to information in a second person's brain that generates a similar behaviour. If we could look inside people's heads, we might find out that *different individuals have different mental representations* of a bowline, even when *they tie it exactly the same way*" (2005: 63–64; my italics). On the same vein, Hodgson and Knudsen note, in their evolutionary approach to economics, that "with habits, replicative similarity is necessarily present at the behavioural level, but unlikely at the neural or genotypic level" (2004: 288). Shennan points out that "the resemblance between the inputs and the public outputs is often very striking" (2002: 47), as illustrated by the continuity observed in many prehistoric pottery traditions (ibid: 47). Another remarkable example of this continuity is the persistence of the same designs and manufacture processes in the Oldowan and Acheulean stone tools in the archaeological record for *over one million years* with negligible modification (although no assumptions can be made about the (lack of) stability of the mental representations of the producers of those tools, since they were hardly human). Not quite as long but equally impressive are the timescales of linguistic items proposed by Pagel (2009), who argues that the origin of the oldest words may be traced back over tens of thousands of years. In the same chapter, he proposes that words, phonemes or syntax constitute "discrete heritable units" (Pagel 2009: 406, Table 1), which are transmitted through "teaching, learning and imitation" (ibid). All these arguments together point to the cultural inheritance of public behaviours and indeed characterize this inheritance as *replication*.

Public cultural manifestations are caused by mental activity at the individual level. The use of "activity", as opposed to "representations", is not accidental. Mental representations (Sperber 1996) and Cognitive Causal Chains (Sperber 2006) involve semantic content and relationships. Behaviours and artefacts are the product of the implementation of neural motor instructions, which are in turn caused by other neural activity, perceptual and associative in nature. At this level of analysis, the causality pathways can in principle be established (for instance, with priming experiments that can reveal associations or with methods that provide windows into brain activity). Mental cultural representations, in contrast, are *emergent* from public manifestations. This is notably the case at the individual level, when the patterns of brain connectivity change in response to experience over life-long learning. Like all emergent or complex phenomena, mental representations are sensitive to local conditions and therefore unique and unpredictable—in other words, non-replicable.

3.2 Selection for Replicability in Language

Linguistic (behavioural) patterns can persist over long periods of time, therefore, because they are reproduced faithfully generations after generation. In most of the models reviewed in the first part of this chapter, reproduction is assumed (e.g. a skill given to the agents in a simulation) or expected (e.g. when human participants are expected to learn and reproduce a typed string of letters). A couple of the experiments, however, acknowledge that learning to produce the signals themselves is not trivial. When the participants in Galantucci (2005) and Verhoef et al. (2011, 2012) were confronted with their props (the distorting writing pad and the slide whistle, respectively), they had to learn the relationship between their movements and the output signals. During the course of the experiments, they became increasingly proficient with their props, and consequently more in control of the structure of the drawings and whistles they produced.

Similarly, human infants need to learn how to control their vocalizations (or signed gestures). Early in their development, they construct a perceptual-motor machinery that allows them to faithfully reproduce specifically the sounds of their language. This machinery develops during babbling. From 5 to 7 months of age, infants tune their motor-articulatory and auditory-perceptual capacities to accurately match the patterns (phonemic categories, intonation patterns) of their ambient language (Braine 1994; Vihman et al. 2009), at the same time as they imitate other motor skills (Iverson et al. 2007; Thelen 1981, cited in Vihman et al. 2009). One proposed mechanism underlying faithful imitation of sounds is an "articulatory filter" (Vihman 1993) whereby sound patterns that the child has already produced during babbling become more perceptually salient. This allows infants to notice frequent patterns in the input speech stream, which prompts further repetition (Vihman et al. 2009). Patterns produced in babbling that are not reinforced by the external input are repeated to a lesser extent, resulting in a repertoire of sounds that resembles that of the ambient language. De Boer's model, where distinct vowel categories emerged out of feedback between agents, models some aspects of the learning dynamics that goes on during babbling.

As we have seen, the replication of phonemic categories in new speakers can be modelled as blending inheritance (Wedel 2006). The resulting stable categories[4] translate into a multimodal distribution of acoustic values in the output data produced by speakers. Maye et al. (2002) elegantly demonstrated how distributions of acoustic values in the input translate into functional categories through statistical learning. When they exposed 6- and 8-month-old infants to sounds from a phonetic continuum with a bimodal distribution, the infants were able to discriminate between sounds from both ends of the continuum. Then, the distribution of the sounds was unimodal; however, the infants would not discriminate between the same two sounds. This indicates that the infants inferred two categories from the bimodal distribution and a single category from the unimodal one. It is this sensitivity to the input's surface statistics that sculpts the fuzzy but distinct, functionally discrete, sound categories. Infants will go on to produce (by blending inheritance) sounds with the same underlying distribution properties as the one they learned, so, over the generations, the statistical properties of phonemic categories are

[4] They are stable only relatively speaking since, like any replicator in an evolving system, they are subject to mutation.

maintained. Extending this dynamics to a population leads to the emergence of shared systems of phonemes, as modelled in de Boer (2001) and others (Oudeyer 2006; de Boer and Zuidema 2010).

In established languages, selection for replicability may be difficult to detect because an optimal stable state has been reached, but even then it still would act as a stabilizing mechanism, tending to maintain things as they are. In emergent systems, it should lead to the appearance of replicator lineages. An emergent language, Nicaraguan Sign Language, which was spontaneously created by a community of deaf children brought together to a school for the deaf in Managua only a few decades ago, provides a window into the genesis of linguistic systems (e.g. Senghas and Coppola 2001; Senghas et al. 2004; Sandler et al. 2005; Aronoff et al. 2008) and thus gives us the opportunity to examine the forces that operate on the origin of phonemic categories.[5] In Nicaraguan Sign Language, selection for replicability was at work during the emergence of phonemes and continues to stabilize the existing phonemes. Al-Sayyid Bedouin Sign Language (Aronoff et al. 2008) is another recent sign language, but here, stable phonemic categories are not attested. This may be due to its different population circumstances, though. Nicaraguan Sign Language appeared around 1977 and has now hundreds of Signers. Al-Sayyid Bedouin Sign Language, although it has been around since the 1940s, is used along with spoken language in a smaller mixed community where deaf people are a (sizeable) minority. Perhaps, phonemes will emerge in this language in the future.

During the first year of the life of an infant, the emphasis of language-related learning seems to be directed towards constructing the segmentation of her acoustic-articulatory space that allows her to produce faithful copies of the phonemic categories of the ambient language. As was the case with early life self-replicating molecules, the main "function" of these phonemic categories is self-replication, and the dynamics of the system selects, over generations, for the easiest-to-produce or easiest-to-perceive sounds. Ontogenetically, an infant's initial babblings have no meaningful content, so individual sounds or their combinations are not selected because of their functions. At this very early stage, Darwinian dynamics with respect to meanings does not exist.

3.3 Selection for Function in Language

The accurate replication of phonemic categories is pivotal for linguistic replication, as the exemplars in these categories are the discrete, replicable and combinatorial units which recombine to form larger linguistic patterns—strings of

[5] For Kirby (in press), the emergence of duality of pattern, and also of compositionality and the split between content and function words, represents instances of major evolutionary transitions within language.

phonemes, ways to organize strings of phonemes, intonation patterns, etc. These larger patterns are recurrently produced across similar contexts for similar objectives; in other words, they have functions. Functions are defined by the association of a pattern with the contexts where it is produced and the effects that it is perceived to achieve. With each production, the produced pattern—but neither the context nor the function—is replicated. The same pattern can be produced for more or less similar—but rarely identical—functions, which, in turn, will be required in more or less similar—but rarely identical—contexts. Language learning includes the process by which we become able to use the same patterns as fellow speakers for similar functions (in appropriate contexts); but in that process, the only thing that is replicated (or copied with similarity, transfer of information and causality between original and copy) are sounds and sound combinations, nothing else.[6] Sounds, as we have seen, are replicated thanks to the perceptual and motor learning that takes place early on in an infant's life. (Meaningful) sound combinations are replicated when they are produced in communicative contexts.

The moment sounds or combinations of sounds become symbolically associated with meanings, the system crosses the Darwinian threshold. From that point onwards, as well as sounds being selected for their replicability, sound combinations are selected for because they convey certain meanings, and lineages of words and other constructions can be traced. When linguistic forms compete for meaning niches, we can talk about selection for function.

The mechanism for replication of construction-form replicators involves symbolicity—our capacity to arbitrarily associate a pattern to a meaning (Deacon 1997), which is subsumed in *arbitrary imitation*—our capacity to reproduce of arbitrary symbols observed in others. The main foundation of arbitrary imitation is a kind of imitation variously referred to as "imitation learning" (Tomasello 1996), "true imitation" (Zentall 2006), "observational learning" (Carroll and Bandura 1982), "blind imitation" (Gergely and Csibra 2006) or "complex imitation" (Heyes 2013). In arbitrary imitation, copies of the forms (or means) rather than the functions (or ends) are produced. This is opposed to emulation, where the observer focuses on the goal and employs any means to achieve it, not necessarily the ones used by the model.

Arbitrary imitation is hugely developed in humans, particularly in human children, but not quite there in non-human primates (Tomasello 1996; Whiten et al. 2004). Chimpanzees, for instance, like human children, can imitate complex behaviour sequences in order to achieve a particular goal (Horner et al. 2006). However, if a chimpanzee discovers that an element in the sequence is unnecessary for the goal, or has no function, it tends to stop producing it. In contrast, in the same circumstances, four-year-old children tend to stick to the complete, partially pointless sequence (Horner and Whiten 2005). It is not clear whether

[6] Of course, in sign languages, it is not sound combinations, but bodily signals that are replicated. And both in spoken and signed languages, some conventional or quasi-conventional co-speech gestures and sounds can also be said to replicate.

children do not analyse the sequence into units, or they do not look for sub-goals, or, even if they do analyse the sequence and realize that an element is unnecessary they nevertheless continue to reproduce it; whatever the exact nature of the process, the result is that the focus of imitation in children are sequences of behaviour. Children's behaviour, therefore, is less rational than that of chimpanzees, but it is more conformist and it implies a high degree of confidence, or trust, that useful information is out there at their disposal. Arbitrary imitation is especially well developed in human children, attested by their tendency to engage in pretend play, as well as by a number of experiments showing that they will imitate intricate actions even when they are obviously over-elaborated for the intended goal (Meltzoff 1988; Horner and Whiten 2005; Lyons et al. 2007; Flynn and Whiten 2008; Whiten et al. 2009). Arbitrary imitation learning is also the mechanism behind the "ratchet effect", which makes cumulative evolution possible in human culture (Tomasello et al. 1993). These studies, in sum, show that humans focus on *means* as they unquestioningly imitate observed arbitrary behavioural patterns, whether they have a utility function or not, other primates focus on *ends* when they only reproduce the actions that (they are persuaded) are functional.

For arbitrary imitation to be possible, two types of abstraction are necessary: abstraction of form from function and abstraction of the signal from the producer of the signal. The first type of abstraction concerns the "arbitrary" part of arbitrary imitation. Arbitrariness is a property of the symbolic associations that link linguistic forms with their meanings (de Saussure [1916] 1983). Forms and meanings are not transparently related to each other, but rather, we simply learn and accept that they are conventionally linked. Apart from language, cultural institutions such as money, democracy or rituals also rely on arbitrary conventions (a bank note or a voting ballot have the value they have because everyone behaves as if they do), while others, like technology, cannot do so.[7] In order to dissociate a signal—a behaviour—from its function, an arbitrary imitator must be able to *abstract form from function*—or means from ends—that is, decouple an action from its iconic or primary utility function.

The attribution of a novel function to an existing behaviour is not trivial, in fact human adults and children over six find it difficult and display what is called *functional fixedness*: solving a task by using a tool for a novel function is slower if they already had associated the tool with its known utility function (Adamson 1952; Defeyter and German 2003; German and Defeyter 2000). Chimpanzees, incidentally, show extreme functional fixedness (Hanus et al. 2011), while human children under six do not, and are happy to assign new functions to tools that already had a known function. In the experimental game where participants co-opted their movements around the board to communicate the colour of the box where

[7] The function of a piece of technology is constrained by the properties of its formal structure (German et al. 2007). A hammer is good for hammering and a fishhook is good for catching fish, and they are not interchangeable (and never will be, no matter how much people agree on the opposite).

they would land (Scott-Phillips et al. 2009), functional flexibility happened, not without difficulty, because the players had been given a riddle they had to solve cooperatively and they may have been actively looking for any useful informative cues. The motions around the board were sufficient to meet the limited expressive requirements of the task, namely to distinguish between only four colours. Functional fixedness, therefore, may hinder learning new arbitrary associations. But a cognitive mechanism that may favour arbitrary imitation is pattern completion (Tamariz 2011), which brings about a feeling of relief when an incomplete pattern is completed. This relief is called secondary reinforcement (Miller and Dollard 1941; Osgood 1953) and is noticeable for example while listening to music (Keller and Schoenfeld 1950), when patterns that confirm our expectations bring about pleasure, but patterns that contradict our expectations produce unease. Secondary reinforcement is exacerbated in certain conditions like Tourette syndrome (Prado et al. 2008) or obsessive-compulsive disorder (Rasmussen and Eisen 1992; Summerfeldt 2004). Pattern completion is closely related to the automatization of motor productions, which has also been proposed to have evolved as a facilitator of language production (Deacon 2007). Pattern completion, thus, does away with the necessity of a utility function. A learned pattern is completed for the sake of completing it.

The second type of abstraction required for arbitrary imitation concerns the "imitation" aspect. Arbitrary communication requires that individuals are able to copy behaviour that they have observed in another individual; in other words, they must be able to assume the role of both receiver and producer of behaviour. For this to be possible, they must be able to *abstract the signal from the producer of the signal*. This capacity is called role-reversal imitation (Tomasello 1999) and is much more developed in humans than in other closely related species. One of the most striking examples of a communication system created and learned through interaction by chimpanzees is ontogenetic ritualization (Tomasello and Call 1997). An example of this kind of ritualization is a baby chimp raising its arms and trying to climb on an adult's back. After this has happened a few times, it is sufficient for the baby to slightly raise its arms for the adult to understand her request and act accordingly. But these learned rituals have limitations: each of the participants has its role and those roles do not change. And the ritual is restricted to this particular pair of individuals who share first-hand experience of the history of the interactions. In ontogenetic ritualization, the behaviour of each of the participants in the interaction is indivisibly attached to its performer. Humans, on the other hand, do role-reversal imitation spontaneously.

It is possible that the two forms of abstraction are manifestations of a single cognitive adaptation to be less rational, overcome logical expectations, accept any sort of incoming information and flexibly integrate patterns in the input even if their cause is not understood—a part of the process of self-domestication proposed by Deacon involving the "de-differentiation of innate predispositions and an increase in the contribution by a learning mechanism" (Deacon 2007: 92).

In the following section, this theoretical framework is supported by evidence from the empirical studies reviewed in the first part of the chapter.

3.4 Cultural-Evolutionary Dynamics in Language

The structure of languages is affected by many and varied pressures. Most of them can be categorized as related either to transmission/learning (e.g. cognitive bias relating to production, perception and processing) or to communication/usage (e.g. alignment of concepts in interlocutors, patterns of social interaction or meaning structure). Transmission and communication are intricately intertwined, as the normal way to learn language involves using it communicatively, except in one respect, highlighted above: the sounds of the ambient language are learned thanks to through a unique combination of perceptual-motor and statistical learning in early infancy. The outcome of babbling is the capacity to reproduce the sounds of a language accurately, in other words, a *mechanism of replication* for linguistic sounds. Once this mechanism is in place, the sounds can be used for communication.[8]

Selection for replicability and selection for function may be easy to tell apart in the example of the origin of life, because the replication of "functionless" molecules may be explained by chemical processes, which are distinct from the selective pressures deriving from genetic function. However, in the case of language, it is difficult to find a plausible explanation for the analogous process, namely the repeated imitative production of gestures or vocalizations that do not have a function at the origin of language. But, as pointed out above, selection for replicability and for function do not have to operate (or have evolved) sequentially—they may do so simultaneously in an interactive way. The first experiment described in this chapter, by Scott-Phillips et al. (2009), may be a model of the interaction between the two types of selection in the origin of communication systems. Here, remember, the two players' characters had to land in rooms of the same colour. At the beginning of a game, when the players were exploring the game, the movements around the rooms were random, and typically they were not copied—although it is not impossible that the fact that one participant started to move gave his partner the idea of moving around too. When they acquired a communicative function, that is, as soon as they were produced and understood as communicative, they stabilized. They began to be faithfully replicated by both players *because* they had a function. In other words, functionality drove replicability.

In many of the experiments described above, the initial state of the system was usually a random set of signals—letter strings, drawings and whistles. This is probably not a good reflection of what the early stages of language were like—unless we accept the musical theory of protolanguage, where we would have a large set or even a system of non-communicative vocal or gestural signals before they took on a communicative role. More likely, signals became communicative

[8] Notwithstanding the fact that the processes of perfecting the production of the sounds of a language and using vocalizations communicatively usually happen simultaneously for some months or years.

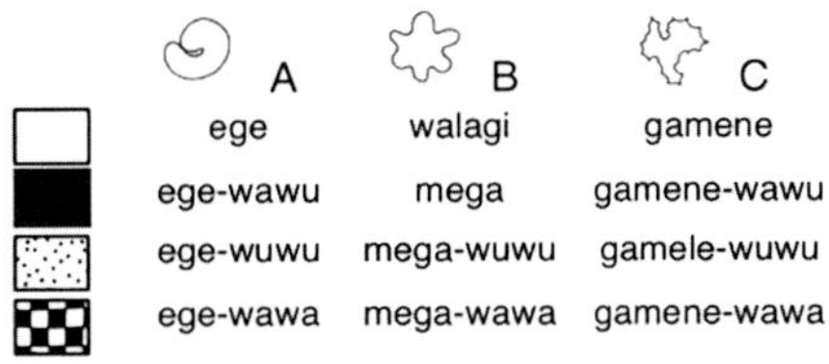

Fig. 6 The sixth and last generation from one of the languages in the Vertical transmission condition in Kirby et al. (in preparation). Hyphens have been added to make the compositional structure more visible. In this language, the first segment refers to the shape (ege means *shape A*; mega *shape B*; and gamene, *shape C*) and the second segment, to the texture (no ending means *white*, wawu means *black*, wuwu means *dotted* and wawa means *checked*). There is one irregular word: walagi, for *shape B*, *white*; and one irregularity: gamele, instead of gamene, for *shape C*, when it has a *dotted texture*

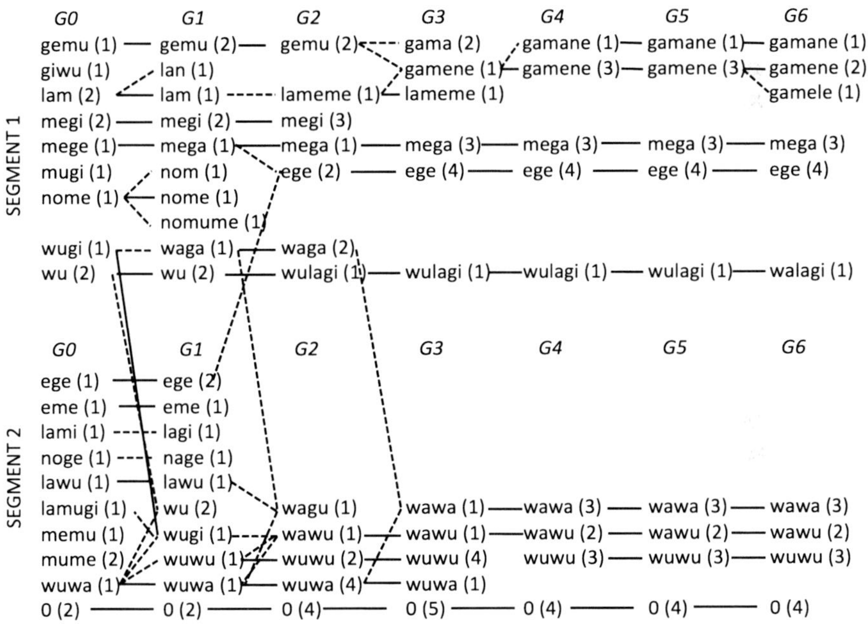

Fig. 7 Coalescent trees of the first and second segments in the words produced at generations G0 to G6 from one of the languages in Kirby et al. (in preparation). The trees were generated following the methods described in Cornish et al. (2009). *Black lines* indicate perfect replication and *dotted lines* indicate recombination or probable descent with modification. The frequency of each segment type at each generation is shown in *brackets*

very early on, as modelled in Scott-Phillips et al. (2009), or in the pictionary games (Garrod et al. 2007; Fay et al. 2010), where the drawings were communicative from the start. Nevertheless, the experimental transmission chains initialized with random signals show how selection for replicability and selection

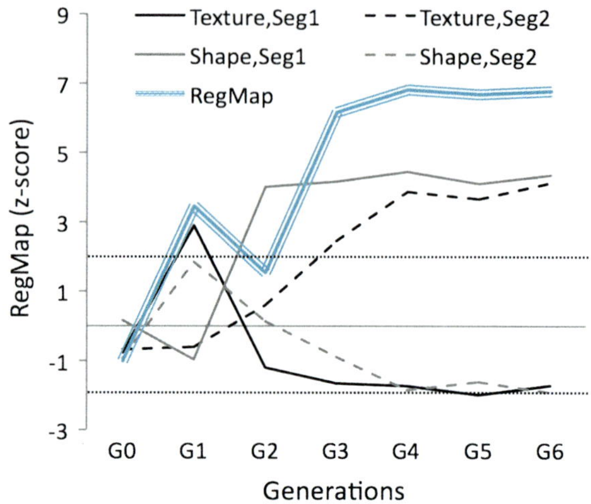

Fig. 8 *RegMap* (*double line*) and partial *RegMap* values for generations G0 to G6 from the same language in Kirby et al. (in preparation) as in Fig. X. The partial *RegMap* values were calculated following the methods described in Cornish et al. (2009). *RegMap* was calculated by running the same method on the partial *RegMaps*. Z-scores calculated with a Monte Carlo simulation ($N = 5000$). Values near 0 indicate random or irregular mappings. Values above 1.96 indicate the mappings are significantly more regular than expected by chance, while values below -1.96 indicate they are significantly less regular than expected by chance

for function transform randomness into structure. Thus, in languages like those in emerging in Kirby et al. (2008, in preparation), we see the increasing prevalence of more reproducible and increasingly meaningful signal units. The following analysis on one of the artificial language families generated in Kirby et al. (in preparation) clearly illustrates the simultaneous action of selection for replicability and selection for function. Figure 6 shows the final generation of this particular language chain, which began with a random language and was learned used communicatively by six consecutive pairs of players.

The coalescent tree in Fig. 7 shows the evolution of the first and second segments in the language from the initial random language, to the language produced by the sixth generation, and illustrates the emergence of stable replicators. The initial languages contain many different segments, both in first and second position. As the languages evolve, the segments mutate, blend and move from one position to another, while they decrease in number. The surviving segments have higher frequencies (they are reused in several words) and become increasingly stable towards the latter generations, where mutations and position changes are almost non-existent. It is also interesting to note that wild mutations do not occur. The players in the experimental game, even when they could not remember the words for an object, did not invent a totally new word, or introduce new letters; they behaved in a very conformist way (even though they were not required to) and produce only words similar to the set they had been trained on.

The effects of selection for function are clear in Fig. 8, which plots a measure of the level of compositional structure between word segments and meanings (Fig. 6). Partial *RegMap* quantifies the confidence that a segment is consistently associated with a meaning. The graph shows how, from generation G2 onwards, the first segment is clearly, and increasingly stably, associated with shape and the second segment, with texture. The overall *RegMap* value measures the overall confidence that each segment is consistently associated with a meaning in a one-to-one, unambiguous relationship.

In this particular language, which referred to meanings with two features—shape and texture—the signals have split into two meaningful units, with the first one adapted to conveying shape and the second one to conveying texture. In the languages from the experiments in Kirby et al. (2008), where there were three dimensions of meaning (shape, colour and motion), signals split into three meaningful units, each adapted to one dimension (Cornish et al. 2009).

The fitness of a letter-string replicator in these languages (the likelihood that it would be reproduced by the next generation) was determined by how memorable it was, which in turn depended on (a) replication factors, e.g. how easy it was to produce, or how similar it was to letter strings in the native language of the players; and (b) functional factors, e.g. how meaningful it was, or how reliably it was associated with a meaning dimension. The effects of replication factors are apparent in Verhoef et al.'s (2011, 2012) whistle experiments, where the final, evolved whistles were easier to produce than the initial ones. Functional factors are apparent in all the iterated learning of miniature artificial language experiments, with adaptation of forms to meaning space structure being most obvious in Perfors and Navarro's (2011) study where word categories aligned with either the size or the colour of the square objects they had to denote, depending the salience of the differences in square size of colour.

The (functional) fitness of linguistic patterns is also affected by their being associated with certain social values or social identities (Labov 2001; Croft 2000; Richerson and Boyd 2005) or having an iconic relationship with a meaning, as illustrated by the paradigmatic case of words that are cross-culturally preferred to designate a rounded figure (like "bouba" or "maluma") or a spiky figure (like "kiki" and "takete") (Kohler 1929; Ramachandran and Hubbard 2001). Some of the fitter drawing systems produced in the distorting-pad to denote the different rooms in Galantucci's (2005) experiments were iconically related to their meanings: lines on either side of the pad referred to rooms on the left or the right of the board. The fitness of a linguistic pattern is also influenced by how often speakers need or wish to refer to their meaning. For instance, the English lexical form "oil-lamp" is not very fit nowadays, as its referent ceased to be frequent in the homes of English speakers. Conversely, the appearance of the Internet has selected for the corresponding word form "Internet", which is now infinitely more frequent than only two or three decades ago. The form "gay", on the other hand, used to be selected for through its association with the meaning "happy", whereas now it is probably even fitter because of its connection with the commonly expressed concept of "homosexual". Finally, a trait that enhances fitness specifically in communication systems is being distinct from other linguistic pattern replicators: the drawings produced in the distorting pads in Galantucci (2005) or the words for the object in the communicative games in Kirby et al. (in preparation) were functional and therefore had higher chances of being reused, only if they were distinct from each other. Each drawing, or each word, in a system adapted specifically to one of the meaning niches available.

Natural languages are culturally transmitted institutions and therefore have to be continuously learned by new speakers. Human learners are able to faithfully learn the sounds of their language during babbling and subsequently reproduce them accurately. The most adaptive sounds are those that are easiest to learn and reproduce by the mechanisms involved in babbling. Languages persist because speakers use them for their communicative purposes. The most adaptive linguistic (sound) patterns are those that best convey relevant meanings. The meanings, the speakers and their cognitive capacities and communicative needs are the environment where linguistic replication, innovation and selection take place.

4 Conclusion

This chapter has presented a model of the cultural evolution of language based on mechanisms that are attested in experiments. The main argument for replication stems from the claim that functions help stabilize arbitrary forms. Theories of cultural evolution have not managed to find a consensual framework and I believe this is because they were focusing on the most interesting part of culture: shared social constructs and values, etc. that exist in people's minds. Such mental representations are not faithfully replicated, they do not "leap across brains", and these have constituted serious problems with theories such as memetics. Ideas, values and cultural institutions continue to exist and evolve because the behaviours that give rise to them are faithfully replicated by generation after generation of humans. But because mental representations are emergent from individual experiential paths, they cannot replicate. They may be similar in the same way as the precise paths of two birds in a flock or the noses of grandfather, father and grandson are similar; they belong to the same *kind* of paths and noses, but each is unique. In this chapter, I have described mechanisms for the replication of the public manifestations of linguistic information—and the same mechanisms could well be at work in other cultural institutions, from greetings to money or justice. Culture, including language, exists because of human brains and the knowledge, beliefs and values that emerge in them. But culture, including language, would not exist without human bodies—hands, mouths, arms—that replicate the public behavioural and material expressions of mental constructs.

References

Adamson RE (1952) Functional fixedness as related to problem solving: a repetition of three experiments. J Exp Psychol 44:288–291

Aronoff M, Meir I, Padden C, Sandler W (2008) Roots of linguistic organization in a new language. In: Arbib M, Bickerton D (eds) Special Issue on Holophrasis, Compositionality and Protolanguage. Interact Stud 9(1):131–150

Beckner C, Blythe R, Bybee J, Christiansen MH, Croft W, Ellis NC, Holland J, Ke J, Larsen-Freeman D, Schoenemann T (2009) Language is a complex adaptive system: position paper. Lang Learn 59(S1):1–26

Bentley RA, Hahn MW, Shennan SJ (2004) Random drift and culture change. P Roy Soc Lond B Bio 271(1547):1443–1450

Blythe R, Croft W (2009) The speech community in evolutionary language dynamics. Lang Learn 59(S1):47–63

Boyd R, Richerson PJ (1995) Culture and the evolutionary process. University of Chicago Press, Chicago

Braine MDS (1994) Is nativism sufficient? J Child Lang 21:9–31

Brighton H, Smith K, Kirby S (2005) Language as an evolutionary system. Phys Life Rev 2:177–226

Brown JE (2012) The evolution of symbolic communication: an embodied perspective. Unpublished PhD thesis. The University of Edinburgh

Carroll WR, Bandura A (1982) The role of visual monitoring in observational learning of action patterns. J Motor Behav 14:153–167

Cavalli-Sforza LL, Feldman MW (1981) Cultural transmission and evolution: a quantitative approach. Princeton University Press, Princeton (Monographs in Pop Biol 16)

Cornish H, Tamariz M, Kirby S (2009) Complex adaptive systems and the origins of adaptive structure: what experiments can tell us. Special issue on language as a complex adaptive system. Lang Learn 59(4S1):187–205

Croft W (2000) Explaining language change: an evolutionary approach. Longman, Harlow

Darwin C (1871) The descent of man and selection in relation to sex. John Murray, London

Dawkins R (1976) The selfish gene. Oxford University Press, Oxford

Deacon TW (1997) The symbolic species: the co-evolution of language and the brain. Norton, New York

Deacon TW (2007) Multilevel selection and language evolution. In: BH Weber BH, Depew DJ (eds) Evolution and learning: the Baldwin effect reconsidered. MIT Press, Cambridge, MA

de Boer B (2001) The origins of vowel systems. Oxford University Press, Oxford

de Boer B (2000) Self-organisation in vowel systems. J Phon 28(4):441–465

de Boer B (1999) Evolution and self-organisation in vowel systems. Evol Communication 3(1):79–103

de Boer B, Zuidema W (2010) Multi-agent simulations of the evolution of combinatorial phonology. Adapt Behav 18(2):141–154

Dennett D (2006) From typo to thinko: when evolution graduated to semantic norms. In: Levinson S, Jaisson P (eds) Evolution and culture, a Fyssen Foundation symposium. MIT Press, Cambridge

de Saussure F ([1916] 1983) Course in general linguistics. Duckworth, London

Defeyter MA, German TP (2003) Acquiring an understanding of design: evidence from children's insight problem solving. Cognition 89:133–155

Dor D, Jablonka E (2010) Plasticity and canalization in the evolution of linguistic communication: An evolutionary developmental approach. In: Larson L, Deprez V, Yamakido H (eds) The evolution of human language. Cambridge University Press, Cambridge

Dunn M, Terrill A, Reesink G, Foley RA, Levinson SC (2005) Structural phylogenetics and the reconstruction of ancient language history. Science 309(5743):2072–2075

Fay N, Garrod S, Roberts L, Swoboda N (2010) The interactive evolution of human communicative systems. Cognitive Sci 34:351–386

Fitch WT (2010) The evolution of language. Cambridge University Press, Cambridge

Flynn E, Whiten A (2008) Cultural transmission of tool use in young children: a diffusion chain study. Soc Dev 17(3):699–718
Galantucci B (2005) An experimental study of the emergence of human communication systems. Cognitive Sci 29(5):737–767
Garrod S, Fay N, Lee J, Oberlander J, MacLeod T (2007) Foundations of representation: where might graphical symbol systems come from? Cognitive Sci 31:961–987
Gell-Mann M (1992) Complexity and complex adaptive systems. In: Hawkins JA, Gell-Mann M (eds) The evolution of human languages. Addison-Wesley, Reading
Gergely G, Csibra G (2006) Sylvia's recipe: the role of imitation and pedagogy in the transmission of human culture. In: Enfield NJ, Levinson SC (eds) Roots of human sociality: culture, cognition, and human interaction. Berg Publishers, Oxford
German TP, Defeyter MA (2000) Immunity to functional fixedness in young children. Psychon Bull Rev 7:707–712
German TP, Truxaw D, Defeyter MA (2007) The role of information about "convention", "design", and "goal" in representing artificial kinds. New Dir Child Adolesc Dev 115:69–81
Godfrey-Smith P (2000) The replicator in retrospect. Biol Philos 15(3):403–423
Hanus D, Mendes N, Tennie C, Call J (2011) Comparing the performances of apes (Gorilla gorilla, Pan troglodytes, Pongo pygmaeus) and human children (Homo sapiens) in the floating peanut task. PLoS ONE 6(6):e19555
Heyes CM (2013) What can imitation do for cooperation? In: Calcott B, Joyce R, Sterelny K (eds) Signalling, commitment and emotion. MIT Press, Harvard
Herzog HA, Bentley RA, Hahn MW (2004) Random drift and large shifts in popularity of dog breeds. P Roy Soc Lond B Bio 271:S353–S356
Hodgson GM, Knudsen T (2004) The Firm as an interactor: firms as vehicles for habits and routines. J Evol Econ 14(3):281–307
Horner V, Whiten A (2005) Causal knowledge and imitation/emulation switching in chimpanzees (Pan troglodytes) and children (Homo sapiens). Anim Cogn 8:164–181
Horner V, Whiten A, Flynn E, de Waal FBM (2006) Faithful copying of foraging techniques along cultural transmission chains by chimpanzees and children. PNAS 103:13878–13883
Hull DL (1988) Science as a process: an evolutionary account of the social and conceptual development of science. University of Chicago Press, Chicago
Iverson JM, Hall AJ, Nickel L, Wozniak RH (2007) The relationship between onset of reduplicated babble and laterality biases in infant rhythmic arm movements. Brain Lang 101:198–207
Jarvis ED (2006) Evolution for and against vocal learning in birds and mammals. Ornithol Sci 5:5–14
Kalish ML, Griffiths TL, Lewandowsky S (2007) Iterated learning: intergenerational knowledge transmission reveals inductive biases. Psychon Bull Rev 14:288–294
Kalish M, Lewandowsky S, Kruschke J (2004) Population of linear experts: knowledge partitioning and function learning. Psychol Rev 111:1072–1099
Keller R (1994) On language change the invisible hand in language. Routledge, London
Keller FS, Schoenfeld WN (1950) Principles of psychology: a systematic text in the science of behavior. Appleton-Century-Crofts, New York
Kimura M (1983) The neutral theory of molecular evolution. Cambridge University Press, Cambridge
Kirby S (2001) Spontaneous evolution of linguistic structure: an iterated learning model of the emergence of regularity and irregularity. IEEE J Evol Comput 5:102–110
Kirby S, Cornish H, Smith K (2008) Cumulative cultural evolution in the laboratory: an experimental approach to the origins of structure in human language. PNAS 105(31):10681–10686
Kirby S, Tamariz M Cornish H, Smith K, Compression and communication drive the evolution of language (in prep)
Kohler W (1929) Gestalt psychology. Liveright, New York
Labov W (2001) Principles of linguistic change, volume II: social factors. Blackwell, Oxford
Lyons DE, Young AG, Keil FC (2007) The hidden structure of overimitation. PNAS 104:19751–19756

Maye J, Werker JF, Gerken L (2002) Infant sensitivity to distributional information can affect phonetic discrimination. Cognition 82(3):B101–B111
Maynard Smith J, Szathmáry E (1995) The major transitions in evolution. Freeman, Oxford
Whiten A, McGuigan N, Marshall-Pescini S, Hopper LM (2009) Emulation, imitation, over-imitation and the scope of culture for child and chimpanzee. Philos Trans Roy Soc [Biol] 364(1528):2417–2428
Meltzoff AN (1988) Infant imitation after a 1-week delay: long-term memory for novel acts and multiple stimuli. Dev Psych 24:470–476
Mesoudi A, Whiten A (2008) The multiple roles of cultural transmission experiments in understanding human cultural evolution. Philos Trans Roy Soc [Biol] 363:3489–3501
Miller NE, Dollard J (1941) Social learning and imitation. Yale University Press, New Haven
Näätänen R (2001) The perception of speech sounds by the human brain as reflected by the mismatch negativity (MMN) and its magnetic equivalent (MMNm). Psychophysiology 38:1–21
Nettle D (1999) Language diversity. Oxford University Press, Oxford
Odling-Smee FJ, Laland KN, Feldman MW (2003) Niche construction: the neglected process in evolution. Princeton University Press, Princeton and Oxford
Okanoya K (2002) Sexual display as a syntactical vehicle: the evolution of syntax in birdsong and human language through sexual selection In: Wray A (ed) The transition to language. Oxford University Press, Oxford
Osgood CE (1953) Method and theory in experimental psychology. Oxford University Press, New York
Oudeyer PY (2006) Self-organization in the evolution of speech. Oxford University Press, Oxford
Pagel M (2009) Human language as a culturally transmitted replicator. Nature Rev Genet 10:405–415
Perfors A, Navarro DJ (2011) Language evolution is shaped by the structure of the world: an iterated learning analysis. In: Carlson L, Hoelscher C, Shipley TF (eds) Proceedings of the 33rd annual conference of the cognitive science society. Cognitive Science Society, Austin
Pierrehumbert JB (2012) The dynamic lexicon In Cohn A, Huffman M, Fougeron C (eds) Handbook of Laboratory Phonology. Oxford U Press
Pinker S, Bloom P (1990) Natural-language and natural-selection. Behav Brain Sci 13(4):707–726
Povinelli D, Bering J, Giambrone S (2003) Chimpanzee 'pointing': another error of the argument by analogy? In: Kita S (ed) Pointing: where language, culture, and cognition meet. Erlbaum, Hillsdale
Prado HS, Rosario MC, Lee J, Hounie AG, Shavitt RG, Miguel EC (2008) Sensory phenomena in obsessive-compulsive disorder and tic disorders: a review of the literature. CNS Spectr 13(5):425–432
Ramachandran V, Hubbard E (2001) Synaesthesia: a window into perception, thought and language. J Consciousness Stud 8(1):3–34
Rasmussen SA, Eisen JL (1992) The epidemiology and differential diagnosis of obsessive compulsive disorder. J Clin Psychiat 53:4–10
Richerson PJ, Boyd R (2005) Not by genes alone: how culture transformed human evolution. University of Chicago press, Chicago
Ritt N (2004) Selfish sounds: a darwinian approach to language change. Cambridge University Press, Cambridge
Sandler W, Meir I, Padden C, Aronoff M (2005) The emergence of grammar: systematic structure in a new language. PNAS 102(7):2661–2665
Scott-Phillips TC, Kirby S (2010) Language evolution in the laboratory. Trends Cogn Sci 14(9):411–417
Scott-Phillips TC, Kirby S, Ritchie GRS (2009) Signalling signalhood and the emergence of communication. Cognition 113(2):226–233
Senghas A, Coppola M (2001) Children creating language: how Nicaraguan Sign Language acquired a spatial grammar. Psych Sci 12(4):323–328
Senghas A, Kita S, Özyürek A (2004) Children creating core properties of language: evidence from an emerging sign language in Nicaragua. Science 305(5691):1779–1782

Shennan S (2002) Genes, memes and human history: darwinian archaeology and cultural evolution. Thames and Hudson, London
Sperber D (1996) Explaining culture: a naturalistic approach. Blackwell, Oxford
Sperber D (2000) An objection to the memetic approach to culture In Aunger R (ed) Darwinizing culture: the status of memetics as a science. Oxford University Press, Oxford
Sperber D (2006) Conceptual tools for a naturalistic approach to cultural evolution In: Levinson SC, Jaisson P (eds) Evolution and culture: a Fyssen Foundation symposium. MIT Press, Cambridge, pp 147–165
Steels L (1996) Emergent adaptive lexicons. In: Maes P (ed) From animals to animals 4, proceedings of the fourth international conference on simulation of adaptive behavior. MIT Press, Cambridge
Steels L (1998) The origins of syntax in visually grounded robotic agents. Artif Intell 103:1–24
Steels L (2003) Evolving grounded communication for robots. Trends Cogn Sci 7(7):308–312
Steels L (2006) Experiments on the emergence of human communication. Trends Cogn Sci 10(8):347–349
Summerfeldt LJ (2004) Understanding and treating incompleteness in obsessive compulsive disorder. J Clin Psychol 60:1155–1168
Szostak J, Bartel D, Luisi P (2001) Synthesizing life. Nature 409:383–390
Tamariz M (2011) Could arbitrary imitation and pattern completion have bootstrapped human linguistic communication? Interact. Stud. 12(1):36–62
Tamariz M, Kirby S, Culture: copying, compression and conventionality. Cognitive Sci (in press)
Tamariz M, Ellison ME, Barr D, Fay N, Evidence for selection in the cultural evolution of human communication systems (in prep)
Thelen E (1981) Rhythmical behavior in infancy: an ethological perspective. Dev Psychol 17(3):237–257
Tomasello M (1996) Do apes ape? In: Heyes CM, Galef BG Jr (eds) Social learning in animals: the roots of culture. Academic Press, London
Tomasello M (1999) The cultural origins of human cognition. Harvard University Press, Cambridge
Tomasello M, Call J (1997) Primate cognition. Oxford University Press, Oxford
Tomasello M, Kruger A, Ratner H (1993) Cultural learning. Behav Brain Sci 16:495–552
Trudgill P (2004) New-dialect formation: the inevitability of colonial Englishes. Edinburgh University Press, Edinburgh
Verhoef T, Kirby S, Padden C (2011) Cultural emergence of combinatorial structure in an artificial whistled language. In: Carlson L, Holscher C, Shipley T (eds) Proceedings of the 33rd annual conference of the cognitive science society. Cognitive Science, Austin, pp 483–488
Verhoef T, de Boer B, Kirby S (2012) Holistic or synthetic protolanguage: evidence from iterated learning of whistled signals. In: Scott-Phillips T, Tamariz M, Cartmill EA, Hurford JR (eds) The evolution of language: proceedings of the 9th international conference. World Scientific, Hackensack
Vihman MM, DePaolis RA, Keren-Portnoy T (2009) Babbling and word: a dynamic systems perspective on phonological development. In: Bavin EL (ed) The cambridge handbook of child language. Cambridge University Press, Cambridge
Vihman MM (1993) Variable paths to early word production. J Phonet 21:61–82
Wedel A (2006) Exemplar models, evolution and language change. Linguist Rev 23:247–274
Whiten A, Horner V, Litchfield C, Marshall-Pescini S (2004) How do apes ape? Learn Behav 32:36–52
Wittgenstein L (1953) Philosophical investigations. Blackwell, Oxford
Woese CR (1998) The universal ancestor. PNAS 95:6854–6859
Zentall TR (2006) Imitation: definitions, evidence and mechanisms. Anim Cogn 9:1435–1448

The Emergence of Modern Communication in Primates: A Computational Approach

Antonio Benítez-Burraco, Ana Mineiro and Alexandre Castro-Caldas

Abstract It is biological structures (and their activities), and not the diverse functions they contribute to (i.e., forms of behavior), that evolve. We believe that the long-lasting controversy around when modern language appeared would benefit from a shift of focus, from "communication" to "computation." Computation is the activity performed by specific neural devices. Computational devices (and their neurobiological correlates), but not communication devices, have a common evolutionary history. We further expect that computational devices are functionally coupled to different interface systems, thus rendering diverse kinds of outputs and eventually contributing to different functions (forms of behaviors). Multiple evidence (genetic, neurobiological, clinical, archeological, fossil, and ethological) suggest that the computational device of human language (the faculty of language in the narrow sense, after Chomsky) is an evolutionary novelty that appeared along with anatomically modern humans. Importantly, this does not preclude that other extinct hominins had "language." It is just that the strings of symbols they were plausibly able to produce lacked certain structural properties that we can only find in extant oral or sign languages. Hominin oral "languages" (or better perhaps, "protolanguages") could have replaced signed "languages" at some early period during hominin evolution. Nonetheless, the gestural "languages" (or better, "protolanguages") hypothetically employed by other extinct hominids would have been less structurally complex than extant human languages are.

Keywords Computation • Hominin • Language evolution • Language modalities • Syntax

---

A. Benítez-Burraco (✉)
Department of Spanish Philology and Teaching,
University of Huelva, Huelva, Spain
e-mail: antonio.benitez@dfesp.uhu.es

A. Mineiro · A. Castro-Caldas
Institute of Health Sciences, Portuguese Catholic University, Lisbon, Portugal

M. Pina and N. Gontier (eds.), *The Evolution of Social Communication in Primates*, Interdisciplinary Evolution Research 1, DOI: 10.1007/978-3-319-02669-5_15,

1 Introduction

Not surprisingly, when it comes to the origin of social communication systems within primates, it is human language that has spilled the greatest amount of ink. Initially, the analysis of language evolution mostly relied on linguistic evidence, as provided by language contact studies, historical linguistics, and the like. Nowadays, this has also become an important concern for researchers with very different backgrounds (anthropologists, primatologists, or archeologists, but also molecular biologists and geneticists). The field has greatly benefited from this multidisciplinary approach. At the same time, the need for such an approach clearly reflects the complexity of the task.

There is an intense dispute around when and how modern language has evolved, and above all, what language has evolved from (and what for). To begin with, it is hotly disputed whether language has evolved by descent with modification (Brandon and Hornstein 1986; Pinker and Bloom 1990; Hurford 1992; Dunbar 1993; Newmeyer 1998; Donald 1999) or is instead an exaptation (Calvin and Bickerton 2000; Chomsky 1982, 1988; Gould 1991; Lieberman 1984; Piatelli-Palmarini 1989; Wilkins and Wakefield 1995). The latter possibility could imply that most components of language evolved to fulfill other functions (i.e., are preadaptations), but also that some of them (or even language itself) could be afunctional, non-specific, or dysfunctional by nature (Chomsky 1988; Piatelli-Palmarini 1990; Lightfoot 1999). As a consequence, it is also disputed whether language has evolved in a gradual fashion (from oral or even manual communication devices) (Newmeyer 1998; Corballis 2002; see also Calvin and Bickerton 2000), or if it instead suddenly appeared (Berwick 1998; Chomsky 2010). Another topic of interest is whether language exhibits continuity with other primate communication systems or instead derives from ape cognition (or from both) (Aitchison 1998; Ulbaek 1998; Bickerton 1990, 2009). A last concern is what language evolved for (assuming that it is an adaptative trait). Diverse evolutionary advantages have been suggested for language: (1) the optimization of hunter-gatherer abilities (Cziko 1995); (2) technological development (Gibson 1990; Ambrose 2001); (3) sexual selection (i.e., courtship and mating optimization) (Aiello 1998; Miller 1999; Wildgen 2004); (4) the improvement of child care and teaching (King 1996); or (5) the enhancement of social interactions within larger groups (Dunbar 1996, 1998; Dessalles 2000; Tomasello et al. 2005). Conversely, for researchers who argue that the (computational) device needed for language processing can be detached from the functions language ultimately fulfills (e.g., Chomsky), language could have arisen in neutral conditions.

2 The Evolutionary Puzzle

The analysis of language evolution primarily relies on two kinds of complementary evidence: comparative (that is, homologous faculties existing in other extant species) and fossil (that is, intermediate stages in the evolution of the faculty).

When it comes to living species, we run into the discontinuity problem. In a nutshell, no extant species has a communicative system that is endowed with the same combinatorial complexity and the symbolic nature as that of humans. In fact, we find more symbolic complexity in phylogenetically distant species, at least in natural conditions. For instance, non-combinable symbolic elements seem to be more common among *Cercopithecidae*. Hence, vervet monkeys or Diana monkeys make use of them to point to some of their predators (Cheney and Seyfarth 1990). Similarly, we find more complex (sound) strings in phylogenetically distant species, at least in natural conditions. Paradigmatically, some birdsongs can be described in terms of syllables, motifs, and song bouts, which suggest that some kind of sequential rules (or syntax) regulate their arrangement and combination (Marler 1970; Todt and Hultsch 1998; Okanoya 2002). Among extant primates, it is among *Hylobatidae* (Geissmann 2000) and *Cercopithecidae* (Ouattara et al. 2009) where we usually find sounds combined in different fashion. It is not entirely evident which animal faculties are homologous to human language. It could be either animal cognition (e.g., symbolic behavior) or oral communication [see Bickerton (1990, 2009) for a comprehensive discussion].

With regard to fossil evidences of language, they could also be essentially uninformative regarding the emergence of complex language. Fossil evidence is basically of two kinds: evidence of the auditive/vocal systems and evidence of symbolic behavior. Concerning the first kind of evidence, it could be problematic because of what we elsewhere have called the "form-function" problem [see for instance Balari et al. (2011, 2013)]. This problem has two sides. On the one hand, modern functions cannot be automatically inferred from human-like, language-related biological structures. A classic example is the descended larynx, which is also present in other, non-linguistic mammals (Fitch and Reby 2001; Fitch 2002). Moreover, linguistic units are not physical, but cognitive by nature. That is, being able to categorically perceive two different sounds [as most primates do (Zayan and Vauclair 1998; Thompson and Oden 2000)] does not entail being able to distinguish them as two different phonemes, that is, as sounds with contrastive meaning in a word. On the other hand, modern functions can exist even when some human-like, language-related biological structure is absent. Paradigmatically, in our species, sign languages are as grammatically complex as oral languages, although they use different reception and transmission channels (Sandler 2006; Sandler and Lillo-Martin 2006).

Concerning evidence of symbolism, there is an ample consensus in paleoanthropology in the sense that complex symbolic behavior implies language (Mellars 1996a, b; McBrearty and Brooks 2000). But, by *language,* we are referring here to a language endowed with semantic compositionality and productivity. It is this kind of language that enables to go beyond the "here and now" and to create displaced conceptual representations, both in time and space (Jerison 1985; Bickerton 1990; Dennett 1996). Evidence of symbolism per se in other extinct species are controversial, even among Neanderthals (Mithen 1996, 2006; Mellars 1996a, b; Tattersall 1998). Actually, other extant primates can learn and use symbols (Premack 1971; Savage-Rumbaugh 1986; Gardner et al. 1989). In truth, the hallmark of human language is not *symbolism* per se, but the fact that symbols are arranged in strings exhibiting certain formal properties. If we could eventually prove that some extinct hominins had a symbolic culture, we could not automatically infer that they were endowed with a modern faculty of language. Symbolic cultures are opaque by nature (Eco 1976), while linguistic meaning is open and productive by nature.

3 What Evolves in Language Evolution

It is not always clear which entity we are referring to when we talk about language evolution. Here, caution is in order. In fact, this is an important concern if we want to properly address this evolutionary conundrum. Firstly, we should not conflate *language* with one of the functions it fulfills, namely *communication*. Human language is a tool subservient to many other functions besides communicating (e.g., thinking and reasoning, marking social identity, expressing emotions, playing, etc.). As a consequence, when discussing language evolution, it seems more appropriate to focus on structural properties of language [to some extent some structural properties of language are said to be motivated by the functions it fulfills (Croft 1995)]. Secondly, we should not equate *language* with a *code* either. From a semiotic perspective, human languages are indeed codes (Morris 1946). But, natural languages are not merely sets of symbols that are arbitrarily associated with a plurality of meanings to allow a communicative interaction between partners. As we pointed out at the end of Sect. 2, these symbols are arranged in very specific ways. Hence, it is compositionality, productivity, recursion, binding, and the like that are the idiosyncratic features of human languages at the structural level. As a consequence, we further contend that it is not just structural properties of language that matter, but specifically how linguistic structures are generated. In other words, when asking about how language evolved, we should give priority to the questions of why modern humans (and only they) are able to generate such linguistic structures and how this capacity has evolved in our lineage.

In our opinion, this shift of focus is supported by current theoretical paradigms in biology. Modern biology claims that only biological structures evolve, but not the uses they are ultimately given as a consequence of their connections with other structures and the relationships existing between the organism and the environment [see Love (2007) for a discussion]. Ultimately, the evolution of biological structures results from the evolution of the developmental systems that control their growth. As noted by Raff (2000: 78), "(e)volution is biased by development."

4 The Computational Hypothesis: Overview

In Sect. 3, we argued that functional rationales for language ("communication," "symbolic behavior," and the like) are not entirely illuminating with regards to language evolution. In fact, neither "communication" nor "symbolic behavior" are natural classes with a common evolutionary history [see for instance Hauser et al. (2002)]. In Sect. 3, we concluded as well that it would be perhaps more informative to explore the way in which linguistic structures are generated. In fact, some linguists (e.g., Chomsky) have characterized *language* as a computational device: "A person's language is a computational system of the mind/brain that generates an infinite array of hierarchically structured expressions" (Chomsky 2005: 45). This computational system interfaces with a conceptual system (a set of symbols for concepts) to produce sentences, which are in turn exteriorized either acoustically (speech) or visually (signs). Notice that to compute is just to sequence and to relate groups of elements. Importantly, *computation* is the activity performed by specific sets of neurons. That means that computational systems are biological structures belonging to a natural class with a common evolutionary history. We thus arrive at our computational hypothesis: "language evolution is explained above all by the evolution of the neural substrate of the computational system employed for language processing." Eventually, this implies that it is neither animal communication nor animal cognition, but computational abilities that language has evolved from. Similarly, when analyzing the fossil register, it is not evidence of auditive or vocal systems, or of modern (i.e., "symbolic") behavior that matters, but of computational abilities (see below).

In order to implement our hypothesis, we will rely on a classic depiction of computational systems posited by Chomsky in the 1950s (Chomsky 1956, 1959). According to his Hierarchy, different formal languages are needed to handle different sorts of computational problems. Specifically, linguistic structures involving cross-serial dependencies suggest that natural languages should be characterized as Type 1, or context-sensitive languages within this hierarchy. It has been argued that formal grammars cannot satisfactorily apprehend all the complexities inherent in natural languages (Rogers and Pullum 2011) and specifically, that Chomsky's Hierarchy may be "too weak and too strong" when applied to human languages (Berwick et al. 2002: 2). Nonetheless, we think that they can illuminate their most basic properties. More importantly, this Hierarchy has a neurobiological correlate. Our ultimate contention will be that it is this neural architecture that can actually illuminate how human language evolved. In a nutshell, the automaton in Chomsky's Hierarchy equates to a computational device integrated by a pattern generator (or sequencer) and a memory "stack." Simply put, more memory resources allow the automaton to generate more complex structures. According to Ullman (2001) or Lieberman (2002), the neural substrate of the sequencer is the basal ganglia, although this patterning activity plausibly emerges from the coordinated activity of diverse subcortical and cortical areas, perhaps involving the cerebellum and the thalamus as well (Murdoch 2010; Barbas et al. 2012). In turn, the memory "stack" may be equated to the working memory that plausibly relies on the activity performed by diverse cortical areas. Given that systems of computation are a natural class with a common evolutionary history, the question of how language evolved can be reformulated in terms of how this neural substrate has evolved. It is evident that speaking (or signing) entails much more than just computing. For language to exist, this computational system must be coupled to a dictionary of symbols (aka *lexicon*) and to some device that allows to exteriorize and to interiorize the strings of symbols it ultimately generates (speech organs or signing mechanisms).

However, these interfaces are secondary to language. Once again, if language is to be merely characterized as a gestural or oral symbolic communication system, then ape strings of signs or lexigrams, such as those generated by some individuals reared in captivity (Gardner et al. 1989; Savage-Rumbaugh and Lewin 1994; Savage-Rumbaugh et al. 1998), should be regarded as *sentences*. However, these strings lack key properties of human sentences, namely, recursion or bound anaphora (i.e., binding or control relations). In the same vein, if it is exteriorizing mechanisms that matter, sign languages as employed by deaf people could not be regarded as natural languages. Nonetheless, their central properties parallel those of spoken languages [see Sandler and Lillo-Martin (2006) for a comprehensive characterization]. Hence, signs in signed languages also belong to lexical categories like noun, verb, adjective, adverb, pronoun, and determiner. Sign languages possess as well a lexicon of sign forms and a system for creating new signs in which meaningful elements (morphemes) are combined. Certainly, the combinatory system results more from non-concatenative processes (in which a sign stem is nested within various movement contours and planes in space). But, similarly to spoken languages, signed languages are also governed by constraints on morpheme combination and application of rules to particular forms. At the syntactic level, sign combination to form sentences is governed as well by phrase structure rules and syntactic principles. Sign languages follow universal constraints on syntactic form that have been proposed on the basis of data from spoken languages (for instance, constraints on co-reference and embedded structures). All signers can distinguish grammatical from ungrammatical combinations of signs in their language. Moreover, signs are not holistic gestures, but are composed of a finite inventory of contrastive meaningless units that resemble the phonemes of spoken languages (handshape, location [or place of articulation], and movement [orientation of the handshape]). Ultimately, sign languages are acquired by the child in the same fashion than oral languages (Newport and Meier 1985; Mayberry and Squires 2006), and similar brain areas are involved in processing both modalities (see Sect. 5 below).

We think that all these circumstances qualify our computational hypothesis.

5 The Computational Hypothesis: Comparative Evidence

When we look at "animal communication" with the Chomsky's Hierarchy lens, we only find T3 systems (i.e., regular grammars) among extant species. Some species [as Gentner's starlings (Gentner et al. 2006)] could perhaps access T2 systems, but only in experimental conditions [but see van Heijningen et al. (2009) and Berwick et al. (2011) for a critical view]. Conversely, we actually find T1 systems in nature, but they do not interface with conceptual capacities or with sound exteriorising devices. For instance, weaving is a motor activity that seemingly demands a context-sensitive computational system [see Lorenzo (2012) and for a comprehensive characterization]. What if this ability is more informative than "communication" per se for unraveling language evolution? In fact, comparative evidence ultimately suggests that the interface between a T1 computational system, a conceptual system, and a vocal–auditory system in our species is a contingent fact. For instance, in weaver birds, this enhanced, T1-type computational capacity has plausibly interfaced with a dictionary of "movemes" [i.e., motor primitives, after Del Vecchio et al. 2003] and with some mechanisms controlling movements. Of course, this possibility is nicely illustrated by sign languages too, in which an enhanced, specifically human T1 system has interacted with a gestural-visual exteriorizing/interiorizing device.

Not surprisingly, the comparative analysis of the neural substrate of the computational system of language in diverse species allows to find evolutionary continuity and real language homologs. This substrate seems to be a quite ancient neural circuitry similarly designed in birds and mammals (Teramitsu et al. 2004; Bolhius et al. 2010; Berwick et al. 2012). However, while the sequencer component is highly preserved, working memory exhibits a lower degree of evolutionary continuity, ultimately suggesting that cortex evolution is the key step for computational systems to evolve (Balari and Lorenzo 2013; Balari et al. 2013). Moreover, although they are peripheral with regards to the emergence of modern language (but not of language per se), both the conceptual system and the externalization/internalization systems exhibit a long evolutionary history too, also linked to the evolution of cortical areas.

It is not only comparative evidence that seemingly corroborates the functional unspecificity of the computational system of language (in the sense that it can interface with different systems rendering different outputs), and ultimately, the suitability of our computational approach to language evolution. In our own species, two lines of evidence also confirm this possibility. On the one hand, it is not just language, but motor or cognitive disorders that can be conceptualized as *constructional* disturbances. For instance, drawing disorders do not only entail a disturbance in single motor movements, but also in the planning of motor sequences per se. In fact, as pointed out above, motor sequences seem to be decomposable into primitives that are arranged according to combinatorial or syntactic rules (Flash and Hochner 2005, and references cited therein). Importantly, neuroimaging analyses suggest that brain areas involved in language processing greatly match brain areas involved in motor processing (Makuuchi et al. 2003; Makuuchi 2010). Actually, the brain seems to rely on basic neural "binding mechanisms" to generate any kind of composite objects at the representational level (Flash and Hochner 2005), thus eventually explaining why the same areas are activated. Importantly also, just as we considered the lexicon and the exteriorizing/interiorizing devices as peripheral regarding language, convergent evidence similarly suggests that biomechanical factors play a subsidiary role in movement, with movements being controlled by a "central" device (Dipietro et al. 2009). On the whole, this probably explains the comorbidity frequently observed between motor and language disorders. Hence, whenever one of these brain areas performing some basic computation (relevant for both language and motor planning) is affected, both motor and linguistic disturbances simultaneously appear. For example, dyslexia is sometimes comorbid with drawing deficits, to the extent that a visuoconstructive deficit has been claimed to occur in dyslexics (Eden et al. 2003; Lipowska et al. 2011). Moreover, it has been suggested that dyslexia could be caused by a general deficit in the rule abstraction mechanism inherent to sequential learning, which would simultaneously impair both visuomotor tasks and syntax (Vicari et al. 2005; Pavlidou et al. 2010). Similarly, linguistic and motor deficits co-occur in Huntington's disease, a neurodegenerative condition caused by the selective atrophy of the basal ganglia (Teichmann et al. 2005, 2008; Robins Wahlin et al. 2010).

On the other hand, recent research has revealed that (spoken) language areas within the left hemisphere are also recruited for sign language production and comprehension. Concerning production, the involved neural systems seem to be quite the same as in oral languages. So, despite the considerable differences between articulators (vocal tract vs. hands), the functional specialization of the neural system does not depend on the nature of the motor system involved (Emmorey 2002). It is true that some dissociation seems to exist at the neural level between signs and pantomime gestures. For instance, Corina et al. (1992) reported left-hemisphere dominance for linguistic signs (of American Sign Language [henceforth, ASL]) but no lateral effect when subjects had to produce gestures like waving good bye. At the same time, the systems involved in the motor planning needed for signing and the systems involved in producing non-linguistic signs may not be completely autonomous (Corina et al. 1999). Hence, recent research also conducted by Corina et al. (2000) with PET technology suggested that cortical regions involved in everyday perception of human actions is also recruited in linguistic processing by deaf native signers. Concerning comprehension, Neville et al. (1998) found that Broca's and Wernicke's areas, but also the superior temporal sulcus and the angular gyrus (the latter usually implicated in reading in hearing people) are activated in hearing and deaf native signers when asked to watch ASL sentences (the activation of critical left-hemisphere structures during sign language processing in both groups further suggests that this cannot be the result of a neural reorganization due to a "lack of auditory input"). Conversely, some differences can be observed regarding the localization of the lexicon. Hence, deaf and hearing native signers showed distinct patterns for open (in essence, derivational and constructional "productive" words such as nouns, verbs, or adjectives) and for closed (in essence, "dead words" in terms of lexical productivity such as articles, conjunctions, auxiliaries) ASL sign classes when compared to English speakers (Neville et al. 1997). Regarding English speakers, the event-related potential (ERP) response to closed class words was left lateralized, and for the native ASL signers, it was bilateral. In sum, the brain areas involved in language computation are conceived to welcome the information within the linguistic processing, independently of the modality (oral vs. signed). In other words, the brain seems to respect function rather than form (Emmorey 2002). In the same vein, Hickok et al. (1996) have claimed that left-hemisphere specialization for language is due to the "linguistic nature of the systems" rather than the sensorial features of the linguistic signal or the motor aspects of language production. Probably also, it is not the motor system or the perceptual mechanisms (audition vs. visuospatial processing) that guides the brain organization for language. It is possible then to consider the left hemisphere as specialised in the control of complex motor movement independently of whether those movements are linguistic or not.

In fact, the hypothesis that neural devices performing specific activities can be functionally coupled to different interface devices (thus contributing to diverse functions) is the mainstream approach to brain function in current neurobiology. For instance, Poeppel and Embick (2005) have claimed that "(t)he natural assumption is that the differently structured cortical areas are specialized for performing different types of computations, and that some of these computations are necessary for language but also for other cognitive functions" (p. 112) and that "(o)perations of a specific type have uniform computational properties, and it might be expected that certain brain regions are specialized to perform this type of computation" (p. 116).

6 The Computational Hypothesis: Fossil Evidence

This computationally oriented view of language evolution supports our previous criticism of the fossil evidences commonly used for inferring the presence of modern language in other hominin species (see Sect. 2). Accordingly, if they are essentially uninformative, it is basically because they are related to the so-called external systems [after Chomsky (Hauser et al. 2002)], that is, the externalization/internalization systems (speech/signing organs) and the conceptual system (dictionary of symbols). Again, this does not entail that they are uninformative with regard to language evolution (or the emergence of communication) per se among primates.

As the very existence of other full-fledged linguistic modalities besides spoken languages (paradigmatically, sign languages) already suggested, fossil evidence also confirms that the interface between a T1, human-like computational system, a conceptual system, and a vocal–auditory system (as observed in our species) is a contingent fact. On the one hand, this circumstance supports the view that some hominin species could have relied on other mechanisms (alternative to speech) for interiorizing/exteriorizing the "linguistic" (or rather perhaps, protolinguistic) sequences they were able to generate (see Sect. 8 below for a more detailed discussion about this possibility). On the other hand, it also validates the search for other, "non-linguistic" fossil evidence of language evolution, namely, evidence related to the computational system of language itself. In doing so, one ought to, of course, leave in second place the functions this system ultimately fulfills when it is attached to other devices. As we discussed in previous sections, it is neuronal structures (and their activities) that matter in evolution; moreover, it is the achievement of an enhanced computational system what would ultimately explains the emergence of modern language.

We have suggested two different evidences of this type (but we expect many others) (Balari et al. 2013). Knots are the first one. Remember that weaving or knotting demands a context-sensitive computational system (see Sect. 5). It seems plausible that whoever made knots could think in context-sensitive terms as well (and plausibly could also externalize that kind of thought), thus having modern language [see Camps and Uriagereka (2006), or Balari et al. (2011) for more comprehensive defenses of this view]. In the fossil register, knots are attested only from 27 Kya (=kiloyears) BP (=before present) (Soffer et al. 2000), although they can be inferred from about 75–90 Kya BP. Crucially, they are only associated with anatomically modern humans (henceforth, AMH) (d'Errico et al. 2005; Henshilwood and Dubreuil 2009). Lithic industries are the second kind of such evidence. Contrary to the static nature of prior hominin techno-complexes, including the Mousterian industries associated with Neanderthals, the AMH register shows a very quick succession of lithic industries that are more and more complex in time (Mithen 1996; Tattersall 1998; Mellars 2002, 2005; Wynn and Coolidge 2004). Complex language seems a key requirement for non-static cultures. (Non-static) culture can be construed as a continuous, self-feedbacked process of change (Dennett 1995; Dunbar et al. 1999; Levinson and Jaisson 2006). Above all, it is modern, syntactically complex language that fuels this endless change, given that it allows to explore virtually (and to transmit efficiently) new possibilities (Dennett 1995, 1996). For example, only modern language allows for achieving mental representations that can be displaced both spatially and temporally or that lack a real correlate (Jerison 1985; Bickerton 1990; Dennett 1996). Perhaps more importantly, non-static cultures demand an enhanced working memory (Coolidge and Wynn 2005; Wynn and Coolidge 2007). As we discussed in Sect. 4, the more memory resources available to the sequencer, the more complex strings the computational system is able to generate. If it is the case that only AMHs are endowed with an enhanced working memory, it is plausible as well that only they have had a T1 computational system. This ultimately implies that some important cortical reconfiguration occurred in our species, but not in other extinct hominins, allowing modern syntax to emerge (see Sect. 7 below).

On the whole, these two kinds of different non-linguistic evidence support the view that only AMHs among the hominins have complex language. Again, this does not entail that other hominins lacked language. It is just that their (proto)languages had been less structurally complex than AMH languages. Quite probably, they were endowed with regular, ape-like grammars (see below Sect. 8 for a more detailed discussion).

7 The Computational Hypothesis: Neurobiological Evidence

The discussion above implicitly entails a specific model of brain evolution within our clade. This mode of change had prompted the emergence of modern language only in our species. Under our view, some cortical reconfiguration occurred in our species that increased our working memory capacity and eventually allowed the resulting (and enhanced) computational system to interface with the "external systems." We think that this model is supported by diverse paleoneurobiological and genetic evidence.

In the last few years, language evolution has also been discussed from a neurobiological perspective. However, the search for homologs of the neuronal substrate of language has been focused on the attribution of functional equivalences to some of the "classical" language areas, particularly, to Broca's area. However, this can be problematic. As we discussed above, modern functions cannot be automatically inferred from human-like, language-related structures (we called this "the form-function problem"). Hence, although Broca's areas in monkeys and humans perform the same *activity* (i.e., they are *homologs*), they are not *functionally* equivalent, given that in apes it controls grasping and manipulation (interestingly, it also discharges when the monkey observes a conspecific making similar actions) (Rizzolatti and Arbib 1998).

Another common proxy for (modern) language is *lateralization*. A growing corpus of evidence suggests that the left hemisphere plays an important role in producing and understanding linguistic utterances. Lesions occurred in the perisylvian area of the left hemisphere provoke different types of language disorders, namely aphasias. In Paleoanthropology, left-lateralization patterns are usually inferred from handedness ratios. On the whole, the full inference is as follows: if some hominin species had a right-handedness ratio similar to living people, then it would have also had a human-like pattern of brain lateralization and, ultimately, have been endowed with linguistic abilities similar to AMHs [see for instance Frayer et al. (2010) on Neanderthals]. Many circumstances make this inference problematic [see Benítez-Burraco and Longa (2012) on this extinct species]. Briefly, the link between right-handedness, (structural and functional) brain lateralization, and language is not as straightforward as assumed, even within our own species. On the one hand, the correlation between handedness and verbal skills is weak to say the least (Natsopoulos et al. 2002; Nettle 2003), even at the brain level (Szaflarski et al. 2002; Selnes and Whitaker 2006). On the other hand, non-left lateralized brain configurations (as those exhibited by some left-handed people or some hemispherectomized subjects) do not compulsorily entail an impaired or disordered faculty of language (Foundas et al. 1994; Liégeois et al. 2008). Lastly, in "crossed aphasia," language deficit can occur in right-handed people after right-brain damage (Castro-Caldas et al. 1987; Falchook et al. 2013). In truth, the "linguistic brain" might not be as left lateralized as currently assumed (even in right-handed people), given that some key areas seem to be right lateralized [i.e., the caudate (Ifthikharuddin et al. 2000; Watkins et al. 2001)] and that different cortical areas of the right hemisphere are regularly recruited for language processing (Just et al. 1996). Additionally, from a phylogenetic perspective, both right-handedness and brain lateralization seemingly predate the evidence for (modern) language. For example, according to Holloway (1996), *Homo erectus* already showed a modern, right-handed pattern. In fact, it has been suggested that both human and ape handedness patterns are similarly influenced by task complexity (Uomini 2009). Moreover, structural and functional brain asymmetries predate the evidence for (modern) right-handedness (and of course, for language) in our clade.

Again, this is true both for extinct (Holloway 1981; Kyriacou and Bruner 2011) and extant primates (Holloway and De La Costelareymondie 1982; Cantalupo and Hopkins 2001). Eventually, it is possible that brain lateralization is only indirectly related to language. As Cochet and Byrne (2013) have pointed out, "there is some evidence that tool use served as a preadaptation for left-hemisphere specialization for language." In the same vein, "a growing body of work suggests that features of intentionality and hierarchical structure may explain the functional origin of cerebral and manual asymmetries." Incidentally, this possibility reinforces the convenience, when discussing language evolution, of detaching neural structures and their activities from the functions they ultimately contribute to. On the whole, language at the brain level seems to depend more on a particular connectivity pattern between different areas (performing basic types of computations) than on a specific pattern of structural and functional lateralization of the brain. Laterality might be primarily related to speech (i.e., the externalization of linguistic expressions), as Broca himself pointed out in his seminal work (Broca 1861: 334). After all, the articulators must perform symmetrically, and we cannot use them independently. But, as we have sufficiently argued, speech is a peripheral component of the human faculty of language.

On the contrary, it has been usually assumed that brain size constitutes too rough a proxy for language evolution [see Falk (1990) or Ayala and Cela Conde (2006), among many others]. However, when brains grow, structural changes (in the form of internal reorganization) occur that presumably give rise to functional changes with crucial consequences for language evolution. On the one hand, brain allometry changes, because late-maturing brain areas usually grow larger (Finlay and Darlington 1995). On the other hand, a more areas or activity nuclei appear, and they show a higher degree of lamination (Ebbesson 1980; Strausfeld et al. 2000). Finally, the mutual invasion of these areas and nuclei of activity (aka *connectional invasion*) is favoured (Deacon 1990a, b). We have hypothesized that some crucial steps for the emergence of a modern computational system (and ultimately, of modern language) can be a by-product of the increment of brain size occurred during our speciation (see Balari et al. 2013 for details). Hence, the more cortical resources, the more working memory available for computing (a key step for achieving a full-fledged computational system). Similarly, the more cortical resources, the more long-term memory available for storing information and ultimately, for having larger lexicons). Additionally, the connectional invasion of disjoint areas would have allowed different systems to interface (this is crucial for constructing a functional module from different computational devices, otherwise functionally non-specific by nature). In a similar vein, Boeckx (2012) has argued that the more globular configuration of (adult) AMH brains (compared to that of other extant hominins), with the thalamus located in a more central position and with neurons establishing more long-distance connections, created a new neuronal workspace that allowed for more efficient connections and information exchanges to take place and, eventually, for an enhanced computational efficiency. According to him, this new mode of combination, which allows for the formation of potentially unbounded hierarchical structures (Boeckx has called it *unbounded merge*), is not radically new. What is new is the fact that it is not constrained. As a consequence, it can combine any sort of elements. Of course, we do not rule out the possibility that some AMH-specific interconnection patterns have also contributed to these changes and, ultimately, to the emergence of modern language. For instance, the temporal lobe projection of the human arcuate fasciculus seems to be absent (or to be much smaller) in non-human primates (Rilling et al. 2008). Obviously, it is very difficult to infer brain connectivity from hominin endocasts, given that brain nerve tracts do not fossilize.

In turn, we expect that these changes in brain size (and connectivity) resulted from the modification of genes controlling neural proliferation. In fact, some of these genes have been positively selected in our clade and some substitutions or insertions/deletions have specifically occurred after our split from Neanderthals. It is plausible then that these changes could account for (some of) the observed differences in brain ontogeny between both species (Gunz et al. 2012). These dissimilar ontogenetic trajectories plausibly entail different patterns of brain connectivity and of the interface between processing devices, and ultimately, different faculties of language. For instance, *MCPH1* encodes a protein involved in DNA repair and in chromosome condensation during the cellular cycle (Trimborn et al. 2004; Xu et al. 2004). Its mutation gives rise to microcephaly, an atavistic condition characterised by a reduced cortical volume resulting from a decrease in the number of neurons (Woods 2004). Although the strongest signals of positive selection on this gene predate the split between Old World monkeys and great apes (Evans et al. 2004; Wang and Su 2004), AMHs bear a derived allele in some positions of the gene compared to Neanderthals (Green et al. 2010). In a similar vein, Boeckx (2012) has linked the emergence of his unbounded merge to some change in a gene (or genes) controlling the development of the brain (and the skull).

We cannot rule out the possibility that some of the changes that brought about modern language affected to genes involved in other processes, such as neural differentiation, migration, or interconnection. We will refer here to these genes as "language genes" in the plain sense that their mutation usually gives rise to clinical conditions in which language is impaired (of course, there is no such thing as language genes stricto sensu: genes are not blueprints!). Under our computational hypothesis, these genes would have stabilised and consolidated the reorganizational processes concomitant with brain growth. One of these "language genes" is *FOXP2*. This gene encodes a transcription factor that promotes the neuronal differentiation necessary for the development and the activity of cortico-thalamic-striatal circuits involved in motor planning, sequential tasks, and procedural learning (Vargha-Khadem et al. 2005; Fisher and Scharff 2009). In the human lineage, the protein FOXP2 has undergone two key changes (Enard et al. 2002), although these changes predate the split between Neanderthals and AMHs, about 300–400 Kya BP (Krause et al. 2007). However, an AMH-specific single nucleotide change (SNC) has been recently found within a regulatory region of the gene. This change may have modified its expression pattern in our species (Maricic et al. 2012). In the same vein, the Denisovan (the hominin species closest to Neanderthals) *CNTNAP2* shows a fixed ancestral SNC compared to the AMH gene (Meyer et al. 2012). *CNTNAP2* is one of FOXP2's targets and also a candidate gene for specific language impairment and autism (Alarcón et al. 2008; Bakkaloglu et al. 2008; Vernes et al. 2008). However, *Foxp2* mutations impair auditory-motor association learning in mice (Kurt et al. 2012). Therefore, this gene could actually be related to the externalization of language (i.e., speech), and not (at least directly) to the development of the computational system of language. Consequently, it seems necessary to still look for other substitutions and insertions/deletions that have occurred within the human lineage that have affected other different genes involved in neural proliferation, migration, specialization, or interconnection. Notice, however, that phenotypic innovations can arise in neutral conditions too (that is, without involving gene mutations), due to the very dynamics and the generative properties of developmental systems (Müller and Newman 2005; West-Eberhard 2005; Walsh 2007). That means that two species could eventually exhibit different faculties of language even if they are endowed with the same "linguistic genotype," particularly if environmental conditions bring about divergent brain ontogenies that give rise to different patterns of brain connectivity and ultimately, of interface between processing devices.

8 The Computational Hypothesis: Constraints on Language Evolution

Our last concern is how our hypothesis, if correct, constrains previous (and future) models of language evolution.

As we have suggested elsewhere in this paper, our view that a modern faculty of language (this primarily meaning a modern system of computation) is an evolutionary novelty of AMHs does not preclude the possibility that other extinct hominins had "language," i.e., that were able to productively combine symbolic elements into strings that were ultimately uttered or signed. It is just that these strings probably lacked certain structural properties that we can only find in current oral or sign languages (specifically, dependencies acting at an arbitrarily long distance). At the same time, language "sophistication," as posited by Corballis (2002), did not probably increase in a smooth way. As we discussed in Sect. 4, although syntax is not an all-or-nothing matter, it is not a continuum either. If our hypothesis is correct, the "languages" (or protolanguages) of other hominins would have been endowed with regular grammars (as ape "languages" are) or perhaps with context-free grammars.

The modality of these hominin "languages" or protolanguages (that is, the way in which linguistic sequences were ultimately transmitted, either orally or gesturally) is also a controversial point. For example, according to Corballis (2002: 123), "the facts of primate evolution favor an origin (of the language instinct) in manual gestures." Our evolutionary model is compatible with Corballis's "from hand-to-mouth" hypothesis, because of the functional independence of the computational system of language, the role played by "language areas" in extant primates (paradigmatically, by Broca's area), and the very existence (and properties) of sign languages, which suggest that orality is not a prerequisite for language. In fact, speech organs seem to have evolved quite slowly. Hence, modern vocal tracts and modern auditory systems are already attested in *Homo heidelbergensis* (Martínez et al. 2004; Martínez and Arsuaga 2009). This suggests that oral "language" predates the emergence of modern syntax within our lineage. Maybe this modern, enhanced-for-speech audio–vocal device already interacted with a conceptual system (responsible for thought) in other hominin species (it surely did in Neanderthals). In other words, the closest species to us would have probably been endowed with oral, symbolic communication systems, although less complex grammatically. Concerning more distant species, they could have been endowed with gestural "languages" or protolanguages, as Corballis suggests, but very probably, these communication devices would have been less complex than human languages. According to Boeckx (2012), the conceptual systems of other hominins (and even the putative "words" or signs they used) would not have been identical to ours, given that our enhanced syntactic capacity very likely reshaped them. Putting it differently, hominin semantics and phonology were not exactly the same as ours.

Finally, the hypothesis that orality replaced gestuality sometime after the split of our lineage from extant apes is also compatible with our model. However, co-evolution and co-intervention between both modalities is also plausible. After all, sign languages and oral languages have a common neural substrate (see Sect. 5). Probably, in the latest extinct hominins sign "languages" and oral "languages" would have coexisted as they do in modern humans.

9 Conclusions

To summarize, we contend that if we want to improve our research on communicative abilities in primates, we should redraw ongoing studies on language evolution by giving pre-eminence to computational issues and by diminishing the importance given to communication (and in fact, to any other function of language). In particular, we should acknowledge that:

- the faculty of language is, above all, a natural system of computation
- language functions are orthologous to this faculty
- in our species, it is the case that this system of computation interacts with a conceptual system and a vocal–auditory system (and occasionally, with a gestural-visual system)
- language (in a broad sense) arises from the interface between these three components
- each component can functionally interface with other devices (at the same time, the output of this interaction, even being non-linguistic by nature, can illuminate their properties)
- these components have a long-lasting evolutionary trajectory, with homologs (and perhaps analogs) in other extant species

- human-like language is not possible without an enhanced (i.e., human-like) computational system
- within our lineage this enhanced computational system is an AMH autapomorphy (that is, a derived trait), but has probably evolved in other vertebrate lineages too.

If we proceed in this way, we will be able to find real (that is, informative) language homologs in other extant species and also real (again, informative) language fossils from extinct species. Ultimately, we will obtain a more accurate, biologically grounded view of how communication evolved in our clade.

Acknowledgments Antonio Benítez-Burraco's research was funded by the Spanish Ministerio de Ciencia e Innovación and FEDER under the Project "Biolinguistics: evolution, development, and fossils of language" (FFI2010-14955). Ana Mineiro and Alexandre Castro-Caldas' research was funded by Fundação para a Ciência e a Tecnologia under the project with reference PTDC/LIN/111889/2009.

References

Aiello LC (1998) The foundations of human language. In: Jablonski NG, Aiello LC (eds) The origin and diversification of language. California Academy of Sciences, San Francisco

Aitchison J (1998) On discontinuing the continuity–discontinuity debate. In: Hurford JR, Studdert-Kenned M, Knight C (eds) Approaches to the evolution of language. Cambridge University Press, Cambridge

Alarcón M, Abrahams BS, Stone JL, Duvall JA, Perederiy JV, Bomar JM, Sebat J, Wigler M, Martin CL, Ledbetter DH, Nelson SF, Cantor RM, Geschwind DH (2008) Linkage, association, and gene-expression analyses identify *CNTNAP2* as an autism-susceptibility gene. Am J Hum Genet 82:150–159

Ambrose SH (2001) Paleolithic technology and human evolution. Science 291:1748–1753

Ayala F, Cela Conde C (2006) La Piedra que se volvió palabra: Las Claves Evolutivas de la Humanidad. Alianza, Madrid

Bakkaloglu B, O'Roak BJ, Louvi A, Gupta AR, Abelson JF, Morgan TM, Chawarska K, Klin A, Ercan-Sencicek AG, Stillman AA, Tanriover G, Abrahams BS, Duvall JA, Robbins EM, Geschwind DH, Biederer T, Gunel M, Lifton RP, State MW (2008) Molecular cytogenetic analysis and resequencing of contactin associated protein-like 2 in autism spectrum disorders. Am J Hum Genet 82:165–173

Balari S, Lorenzo G (2013) Computational phenotypes: towards an evolutionary developmental biolinguistics. Oxford University Press, Oxford

Balari S, Benítez-Burraco A, Camps M, Longa VM, Lorenzo G, Uriagereka J (2011) The archaeological record speaks: bridging anthropology and linguistics. Int J Evol Biol 2011:382679

Balari S, Benítez-Burraco A, Longa VM, Lorenzo G (2013) The fossils of language: what are they, who has them, how did they evolve? In: Boeckx C, Grohmann K (eds) The Cambridge handbook of biolinguistics. Cambridge University Press, Cambridge

Barbas H, García-Cabezas MA, Zikopoulos B (2012) Frontal-thalamic circuits associated with language. Brain Lang. doi:10.1016/j.bandl.2012.10.001

Benítez-Burraco A, Longa VM (2012) Right-handedness, lateralization and language in Neanderthals: a comment on Frayer et al (2010). J Anthropol Sci 90:187–192

Berwick RC, Beckers GJL, Okanoya K, Bolhuis JJ (2012) A bird's eye view of human language evolution. Front Evol Neurosci 4:5

Berwick RC (1998) Language evolution and the minimalist program: the origins of syntax. In: Hurford JR, Studdert-Kennedy M, Knight C (eds) Approaches to the evolution of language. Cambridge University Press, Cambridge

Berwick RC, Okanoya K, Beckers GJL, Bolhius JJ (2011) Songs to syntax: the linguistics of birdsong. Trends Cogn Sci 15:115–121

Bickerton D (1990) Language and species. University of Chicago Press, Chicago

Bickerton D (2009) Adam's tongue: how humans made language, how language made humans. Macmillan, New York

Boeckx C (2012) Homo combinans. Paper presented at the evolution of language conference 9 (EVOLANG9), Campus Plaza, Kyoto, 13–16 Mar 2012

Bolhuis JJ, Okanoya K, Scharff C (2010) Twitter evolution: converging mechanisms in birdsong and human speech. Nat Rev Neurosci 11:747–759

Brandon RN, Hornstein N (1986) From icons to symbols: some speculations on the origin of language. Biol Philos 1:169–189

Broca P (1861) Remarques sur le siège de la faculté du langage articulé: Suivics d'une observation d'aphemie. Bull Soc Anat (Paris) 6:330–357

Calvin WH, Bickerton D (2000) Lingua ex Machina. Reconciling Darwin and Chomsky with the human brain. MIT Press, Cambridge

Camps M, Uriagereka J (2006) The Gordian knot of linguistic fossils. In: Rosselló J, Martín J (eds) The biolinguistic turn. Issues on language and biology. Universitat de Barcelona, Barcelona

Cantalupo C, Hopkins WD (2001) Asymmetric Broca's area in great apes. Nature 414:505

Castro-Caldas A, Confraria A, Poppe P (1987) Non-verbal disturbances in crossed aphasia. Aphasiology 1:403–413

Cheney DL, Seyfarth RM (1990) How monkeys see the world. Inside the mind of another species. University of Chicago Press, Chicago

Chomsky N (1956) Three models for the description of language. IEEE T Inform Theory 2:113–124

Chomsky N (1959) On certain formal properties of grammars. Inform Control 2:137–167

Chomsky N (1982) The Generative Enterprise. a discussion with Riny Huybregts and Henk van Riemsdijk. Foris Publications, Dordrecht

Chomsky N (1988) Language and problems of knowledge. MIT Press, Cambridge

Chomsky N (2005) Three factors in language design. Linguistic Inq 36:1–22

Chomsky N (2010) Some simple evo devo theses: how true might they be for language? In: Larson RK, Déprez V, Yamakido H (eds) The evolution of human language. Cambridge University Press, Cambridge

Cochet H, Byrne RW (2013) Evolutionary origins of human handedness: evaluating contrasting hypotheses. Anim Cogn 16:531–542

Coolidge FL, Wynn T (2005) Working memory, its executive functions, and the emergence of modern thinking. Camb Archaeol J 15:5–26

Corballis MC (2002) From hand to mouth. Princeton University Press, Princeton

Corina DP, Bellugi U, Reilly K (1999) Neuropsychological studies of linguistic and affective facial expressions in deaf signers. Lang Speech 2:307–331

Corina DP, Poizner H, Bellugi U, Feinberg T, Dowd D, O'Grady-Batch L (1992) Dissociation between linguistic and nonlinguistic gestural systems: a case for compositionality. Brain Lang 43:414–447

Corina DP, San José L, Ackerman D, Guillemin A, Braun A (2000) A comparison of neural systems underlying human action and American sign language processing. J Cogn Neurosci Suppl, pp 414–447

Croft W (1995) Autonomy and functionalist linguistics. Language 71:490–532

Cziko G (1995) Universal selection theory and the second Darwinian revolution. MIT Press, Cambridge

d'Errico F, Henshilwood C, Vanhaeren M, van Niekerk K (2005) *Nassarius kraussianus* shell beads from Blombos cave: evidence for symbolic behaviour in the middle stone age. J Hum Evol 48:3–24

Deacon TW (1990a) Fallacies of progression in theories of brain-size evolution. Int J Primatol 11:193–236

Deacon TW (1990b) Problems of ontogeny and phylogeny in brain-size evolution. Int J Primatol 11:237–282

Del Vecchio D, Murray RM, Perona P (2003) Decomposition of human motion into dynamics-based primitives with application to drawing tasks. Automatica 39:2085–2098

Dennett DC (1995) Darwin's dangerous idea. Simon & Schuster, New York

Dennett DC (1996) Kinds of minds. Toward an understanding of consciousness. Basic Books, New York

Dessalles J-L (2000) Language and hominid politics. In: Knight C, Studdert-Kennedy M, Hurford JR (eds) The evolutionary emergence of language. Cambridge University Press, Cambridge

Dipietro L, Krebs HI, Fasoli SE, Volpe T, Hogan N (2009) Submovement changes characterize generalization of motor recovery after stroke. Cortex 45:318–324

Donald M (1999) Preconditions for the evolution of protolanguages. In: Corballis MC, Lea SEG (eds) The descend of mind. Psychological perspectives on hominid evolution. Oxford University Press, Oxford

Dunbar RI (1993) Coevolution of neocortical size, group size and language in humans. Behav Brain Sci 16:681–735

Dunbar RI (1996) Grooming, gossip and the evolution of language. Faber and Faber, London

Dunbar RI (1998) The social brain hypothesis. Evol Anthropol 7:178–190

Dunbar RI, Knight C, Power C (eds) (1999) The evolution of culture. An interdisciplinary view. Edinburgh University Press, Edinburgh

Ebbesson SOE (1980) The parcellation theory and its relation to interspecific variability in brain organization, evolutionary and ontogenetic development and neuronal plasticity. Cell Tissue Res 213:179–212

Eco U (1976) A Theory of Semiotics. Indiana University Press, Bloomington

Eden GF, Wood FB, Stein JF (2003) Clock drawing in developmental dyslexia. J Learn Disabil 36:216–228

Emmorey K (2002) Language, cognition, and the brain: insights from sign language research. Lawrence Erlbaum and Associates, Mahwah

Enard W, Przeworski M, Fisher SE, Lai CSL, Wiebe V, Kitano T, Monaco AP, Pääbo S (2002) Molecular evolution of FOXP2, a gene involved in speech and language. Nature 418:869–872

Evans PD, Anderson JR, Vallender EJ, Choi SS, Lahn BT (2004) Reconstructing the evolutionary history of microcephalin, a gene controlling human brain size. Hum Mol Genet 13:1139–1145

Falchook AD, Burtis DB, Acosta LM, Salazar L, Hedna VS, Khanna AY, Heilman KM (2013) Praxis and writing in a right-hander with crossed aphasia. Neurocase. doi:10.1080/13554794.2013.770883

Falk D (1990) The radiator hypothesis. Behav Brain Sci 13:333–381

Finlay B, Darlington R (1995) Linked regularities in the development and evolution of mammalian brains. Science 268:1578–1584

Fisher SE, Scharff C (2009) *FOXP2* as a molecular window into speech and language. Trends Genet 25:166–177

Fitch WT (2002) Comparative vocal production and the evolution of speech: reinterpreting the descent of the larynx. In: Wray A (ed) The transition to language. Oxford University Press, New York

Fitch WT, Reby D (2001) The descended larynx is not uniquely human. Roy Soc Lond B 268:1669–1675

Flash T, Hochner B (2005) Motor primitives in vertebrates and invertebrates. Curr Opin Neurobiol 15:660–666

Foundas AL, Leonard CM, Gilmore R, Fennell E, Heilman KM (1994) Planum temporale asymmetry and language dominance. Neuropsychologia 32:1225–1231

Frayer DW, Fiore I, Lalueza-Fox C, Radovčić J, Bondioli L (2010) Right handed Neandertals: Vindija and beyond. J Anthropol Sci 88:113–127

Gardner RA, Gardner BT, van Cantfort TE (1989) Teaching sign language to chimpanzees. State University of New York Press, New York

Geissmann T (2000) Gibbon songs and human music from an evolutionary perspective. In: Wallin NL, Merker B, Brown S (eds) The origins of music. MIT Press, Cambridge

Gentner TQ, Fenn KM, Margoliash D, Nusbaum HC (2006) Recursive syntactic pattern learning by songbirds. Nature 440:1204–1207

Gibson KR (1990) New perspectives on instincts and intelligence: brain size and the emergence of hierarchical mental construction skills. In: Parker ST, Gibson KR (eds) 'Language' and intelligence in monkeys and apes. Cambridge University Press, New York

Gould SJ (1991) Exaptation: a crucial tool for evolutionary psychology. J Soc Issues 47:43–65

Green RE, Krause J, Briggs AW, Maricic T, Stenzel U, Kircher M, Patterson N, Li H, Zhai W, Fritz MH, Hansen NF, Durand EY, Malaspinas AS, Jensen JD, Marques-Bonet T, Alkan C, Prüfer K, Meyer M, Burbano HA, Good JM, Schultz R, Aximu-Petri A, Butthof A, Höber B, Höffner B, Siegemund M, Weihmann A, Nusbaum C, Lander ES, Russ C, Novod N, Affourtit J, Egholm M, Verna C, Rudan P, Brajkovic D, Kucan Z, Gusic I, Doronichev VB, Golovanova LV, Lalueza-Fox C, de la Rasilla M, Fortea J, Rosas A, Schmitz RW, Johnson PL, Eichler EE, Falush D, Birney E, Mullikin JC, Slatkin M, Nielsen R, Kelso J, Lachmann M, Reich D, Pääbo S (2010) A draft sequence of the neandertal genome. Science 328:710–722

Gunz P, Neubauer S, Golovanova L, Doronichev V, Maureille B, Hublin J-J (2012) A uniquely modern human pattern of endocranial development. Insights from a new cranial reconstruction of the Neanderthal newborn from Mezmaiskaya. J Hum Evol 62:300–313

Hauser MD, Chomsky N, Fitch WT (2002) The faculty of language: what is it, who has it, and how did it evolve? Science 298:1569–1579

Henshilwood CS, Dubreuil B (2009) Reading the artifacts: gleaning language skills from the middle stone age in southern Africa. In: Botha R, Knight C (eds) The cradle of language. Oxford University Press, New York

Hickok G, Klima ES, Bellugi U (1996) The neurobiology of signed language and its implications for the neural basis of language. Nature 381:699–702

Holloway RL (1981) Volumetric and asymmetry determinations on recent hominid endocasts: spy I and spy II, Djebel Ihroud I, and the Salé *Homo erectus* specimen. With some notes on Neandertal brain size. Am J Phys Anthropol 55:385–393

Holloway RL (1996) Evolution of the human brain. In: Lock A, Peters CR (eds) Handbook of human symbolic evolution. Clarendon Press, Oxford

Holloway RL, De La Costelareymondie MC (1982) Brain endocast asymmetry in pongids and hominids: some preliminary findings on the paleontology of cerebral dominance. Am J Phys Anthropol 58:101–110

Hurford JR (1992) An approach to the phylogeny of the language faculty. In: Hawkins JA, Gell-Mann M (eds) The evolution of human languages. Addison-Wesley Publishing Company, Redwood City

Ifthikharuddin SF, Shrier DA, Numaguchi Y, Tang X, Ning R, Shibata DK, Kurlan R (2000) MR volumetric analysis of the human basal ganglia: normative data. Acad Radiol 7:627–634

Jerison H (1985) Animal intelligence as encephalization. In: Weiskrantz L (ed) Animal intelligence. Claredon Press, Oxford

Just MA, Carpenter PA, Keller TA, Eddy WF, Thulborn B (1996) Brain activation modulated by sentence comprehension. Science 274:114–116

King BJ (1996) Syntax and language origins. Lang Commun 16:193–203

Krause J, Lalueza-Fox C, Orlando L, Enard W, Green RE, Burbano HA, Hublin J-J, Hänni C, Fortea J, de la Rasilla M, Bertranpetit J, Rosas A, Pääbo S (2007) The derived *FOXP2* variant of modern humans was shared with Neandertals. Current Biol 17:1908–1912

Kurt S, Fisher SE, Ehret G (2012) *Foxp2* mutations impair auditory-motor association learning. PLoS ONE 7:e33130

Kyriacou A, Bruner E (2011) Brain evolution, innovation, and endocranial variations in fossil hominids. PaleoAnthropology 2011:130–143

Levinson SC, Jaisson P (eds) (2006) Evolution and culture. MIT Press, Cambridge

Lieberman P (1984) The biology and evolution of language. Harvard University Press, Cambridge

Lieberman P (2002) On the nature and evolution of the neural bases of human language. Am J Phys Anthropol 45:36–62

Liégeois F, Connelly A, Baldeweg T, Vargha-Khadem F (2008) Speaking with a single cerebral hemisphere: fMRI language organization after hemispherectomy in childhood. Brain Lang 106:195–203

Lightfoot D (1999) The development of language. Acquisition, change, and evolution. Blackwell, Oxford & Malden

Lipowska M, Czaplewska E, Wysocka A (2011) Visuospatial deficits of dyslexic children. Med Sci Monit 17:CR216–CR221

Lorenzo G (2012) The evolution of the faculty of language. In: Boeckx C, Horno MC, Mendívil JL (eds) Language, from a biological point of view: current issues in biolinguistics. Cambridge Scholars Publishing, Cambridge

Love AC (2007) Functional homology and homology of function: biological concepts and philosophical consequences. Biol Philos 22:691–708

Makuuchi M (2010) fMRI studies on drawing revealed two new neural correlates that coincide with the language network. Cortex 46:268–269

Makuuchi M, Kaminaga T, Sugishita M (2003) Both parietal lobes are involved in drawing: a functional MRI study and implications for constructional apraxia. Brain Res Cogn Brain Res 16:338–347

Maricic T, Günther V, Georgiev O, Gehre S, Curlin M, Schreiweis C, Naumann R, Burbano HA, Meyer M, Lalueza-Fox C, de la Rasilla M, Rosas A, Gajovic S, Kelso J, Enard W, Schaffner W, Pääbo S (2012) A recent evolutionary change affects a regulatory element in the human *FOXP2* gene. Mol Biol Evol 30:844–852

Marler P (1970) Birdsong and speech development: could there be parallels? Am Sci 58:669–673

Martínez I, Arsuaga JL (2009) El origen del lenguaje: la evidencia paleontológica. Munibe Antropologia-Arkeologia 60:5–16

Martínez I, Rosa M, Arsuaga JL, Jarabo P, Quam R, Lorenzo C, Gracia A, Carretero JM, Bermúdez de Castro JM, Carbonell E (2004) Auditory capacities in middle pleistocene humans from the Sierra de Atapuerca in Spain. PNAS 101:9976–9981

Mayberry RI, Squires B (2006) Sign language: acquisition. In: Brown K (ed) Encyclopedia of language and linguistics, vol 11. Elsevier, Oxford

McBrearty S, Brooks AS (2000) The revolution that wasn't: a new interpretation of the origin of modern human behavior. J Hum Evol 39:453–563

Mellars P (1996a) Symbolism, language, and the Neanderthal mind. In: Mellars P, Gibson KR (eds) Modelling the early human mind. McDonald Institute for Archaeological Research, Cambridge

Mellars P (1996b) The Neanderthal legacy: an archaeological perspective from Western Europe. Princeton University Press, Princeton

Mellars P (2002) Archaeology and the origins of modern humans: European and African perspectives. In: Crow TJ (ed) *The speciation of modern* homo sapiens. Oxford University Press, Oxford & New York

Mellars P (2005) The impossible coincidence. A single-species model for the origins of modern human behavior. Evol Anthropol 14:12–27

Meyer M, Kircher M, Gansauge MT, Li H, Racimo F, Mallick S, Schraiber JG, Jay F, Prüfer K, de Filippo C, Sudmant PH, Alkan C, Fu Q, Do R, Rohland N, Tandon A, Siebauer M, Green RE, Bryc K, Briggs AW, Stenzel U, Dabney J, Shendure J, Kitzman J, Hammer MF, Shunkov MV, Derevianko AP, Patterson N, Andrés AM, Eichler EE, Slatkin M, Reich D, Kelso J, Pääbo S (2012) A high-coverage genome sequence from an archaic Denisovan individual. Science 338:222–226

Miller GF (1999) Sexual selection for cultural displays. In: Dunbar RI, Knight C, Power C (eds) The evolution of culture. Edinburgh University Press, Edinburgh

Mithen S (1996) The prehistory of the mind. A search for the origins of art, religion, and science. Thames & Hudson, London

Mithen S (2006) The singing Neanderthals. The origins of music, language, mind and body. Weidenfeld & Nicolson, London

Morris C (1946) Sings, language, and behavior. Prentice-Hall, New York

Müller GB, Newman SA (eds) (2005) Evolutionary innovation and morphological novelty. J Exp Zool B Mol Dev Evol 304:485–486

Murdoch BE (2010) The cerebellum and language: historical perspective and review. Cortex 46:858–868

Natsopoulos D, Koutselini M, Kiosseoglou G, Koundouris F (2002) Differences in language performance in variations of lateralization. Brain Lang 82:223–240

Nettle D (2003) Hand laterality and cognitive ability: a multiple regression approach. Brain Cogn 52:390–398

Neville HJ, Coffey SA, Lawson DS, Fischer A, Emmorey K, Bellugi U (1997) Neural systems mediating American sign language: effects of sensory experience and age of acquisition. Brain Lang 57:285–308

Neville HJ, Bavelier D, Corina D, Rauschecker J, Karni A, Lalwani A, Braun A, Clark V, Jezzard P, Turner R (1998) Cerebral organization for language in deaf and hearing subjects: biological constraints and effects of experience. PNAS 95:922–929

Newmeyer FJ (1998) On the supposed 'counterfunctionality' of universal grammar: some evolutionary implications. In: Hurford JR, Studdert-Kennedy M, Knight C (eds) Approaches to the evolution of language. Cambridge University Press, Cambridge

Newport EL, Meier RP (1985) The acquisition of American Sign Language. In: Slobin D (ed) The cross-linguistic study of language acquisition. Erlbaum, Hillsdale, NJ

Okanoya K (2002) Sexual display as a syntactical vehicle: the evolution of syntax in birdsong and human language through sexual selection. In: Wray A (ed) The transition to language. Oxford University Press, New York

Ouattara K, Zuberbühler K, N'goran EK, Gobert J-E, Lemasson A (2009) The alarm call system of female Campbell's monkeys. Anim Behav 78:35–44

Pavlidou EV, Kelly ML, Williams JM (2010) Do children with developmental dyslexia have impairments in implicit learning? Dyslexia 16:143–161

Piatelli-Palmarini M (1989) Evolution, selection, and cognition: from learning to parameter setting in biology and the study of language. Cognition 31:1–44

Piatelli-Palmarini M (1990) An ideological battle over modals and quantifiers. Behav Brain Sci 13:752–754

Pinker S, Bloom P (1990) Natural language and natural selection. Behav Brain Sci 13:707–727

Poeppel D, Embick D (2005) Defining the relation between linguistics and neuroscience. In: Cutler A (ed) Twenty-first century psycholinguistics: four cornerstones. Lawrence Erlbaum, Hillsdale

Premack D (1971) Language in chimpanzee? Science 172:808–822

Raff RA (2000) Evo-devo: the evolution of a new discipline. Nat Rev Genet 1:74–79

Rilling JK, Glasser MF, Preuss TM, Ma X, Zhao T, Hu X, Behrens TEJ (2008) The evolution of the arcuate fasciculus revealed with comparative. Nature Neurosci 11:426–428

Rizzolatti G, Arbib MA (1998) Language within our grasp. Trends Neurosci 21:188–194

Robins Wahlin TB, Larsson MU, Luszcz MA, Byrne GJ (2010) WAIS-R features of preclinical Huntington's disease: implications for early detection. Dement Geriatr Cogn Disord 29:342–350

Rogers J, Pullum GK (2011) Aural pattern recognition experiments and the subregular hierarchy. J Logic Lang Inf 20:329–342

Sandler W, Lillo-Martin D (2006) Sign language and linguistic universals. Cambridge University Press, Cambridge

Sandler W (2006) An overview of sign language linguistics. In: Brown K (ed) Encyclopedia of language and linguistics, vol 11. Elsevier, Oxford

Savage-Rumbaugh S (1986) Ape language: from conditioned response to symbol. Columbia University Press, New York

Savage-Rumbaugh S, Lewin R (1994) Kanzi: the ape at the brink of the human mind. Wiley, New York

Savage-Rumbaugh S, Shanker SG, Taylor TJ (1998) Apes, language, and the human mind. Oxford University Press, New York

Selnes O, Whitaker HA (2006) Anatomical asymmetries versus variability of language areas of the brain. In: Brown K (ed) Encyclopedia of language and linguistics, vol 1. Elsevier, Oxford

Soffer O, Odovasio JM, Hyland DC (2000) The 'Venus' figurines: Textiles, basketry, gender, and status in the Upper Paleolithic. Curr Anthropol. 41:511–525

Strausfeld NJ, Homberg U, Kloppenburg P (2000) Parallel organization in honey bee mushroom bodies by peptidergic Kenyon cells. J Comp Neurol 424:179–195

Szaflarski JP, Binder JR, Possing ET, McKiernan KA, Ward BD, Hammeke TA (2002) Language lateralization in left-handed and ambidextrous people: fMRI data. Neurology 59:238–244

Tattersall I (1998) Becoming human: evolution and human uniqueness. Harcourt Brace, New York

Teichmann M, Dupoux E, Kouider S, Brugières P, Boissé MF, Baudic S, Cesaro P, Peschanski M, Bachoud-Lévi AC (2005) The role of the striatum in rule application: the model of Huntington's disease at early stage. Brain 128:1155–1167

Teichmann M, Dupoux E, Cesaro P, Bachoud-Lévi AC (2008) The role of the striatum in sentence processing: evidence from a priming study in early stages of Huntington's disease. Neuropsychologia 46:174–185

Teramitsu I, Kudo LC, London SE, Geschwind DH, White SA (2004) Parallel *FoxP1* and *FoxP2* expression in songbird and human brain predicts functional interaction. J Neurosci 24:3152–3163

Thompson RKR, Oden DL (2000) Categorical perception and conceptual judgments by nonhuman primates: the paleological monkey and the analogical ape. Cognitive Sci 24:363–396

Todt D, Hultsch H (1998) How songbirds deal with large amount of serial information: retrieval rules suggest a hierarchical song memory. Biol Cybern 79:487–500

Tomasello M, Carpenter M, Call J, Behne T, Moll H (2005) Understanding and sharing intentions: the origins of cultural cognition. Behav Brain Sci 28:675–691

Trimborn M, Bell SM, Felix C, Rashid Y, Jafri H, Griffiths PD, Neumann LM, Krebs A, Reis A, Sperling K, Neitzel H, Jackson AP (2004) Mutations in microcephalin cause aberrant regulation of chromosome condensation. Am J Hum Genet 75:261–266

Ulbaek I (1998) The origin of language and cognition. In: Hurford JR, Studdert-Kennedy M, Knight C (eds) Approaches to the evolution of language. Cambridge University Press, Cambridge

Ullman MT (2001) The declarative/procedural model of lexicon and grammar. J Psycholinguist Res 30:37–69

Uomini NT (2009) The prehistory of handedness: archaeological data and comparative ethology. J Hum Evol 57:411–419

van Heijningen CAA, de Visser J, Zuidema W, ten Cate C (2009) Simple rules can explain discrimination of putative recursive syntactic structures by a songbird species. PNAS 106:20538–20543

Vargha-Khadem F, Gadian DG, Copp A, Mishkin M (2005) *FOXP2* and the neuroanatomy of speech and language. Nat Rev Neurosci 6:131–138

Vernes SC, Newbury DF, Abrahams BS, Winchester L, Nicod J, Groszer M, Alarcón M, Oliver PL, Davies KE, Geschwind DH, Monaco AP, Fisher SE (2008) A functional genetic link between distinct developmental language disorders. N Engl J Med 359:2337–2345

Vicari S, Finzi A, Menghini D, Marotta L, Baldi S, Petrosini L (2005) Do children with developmental dyslexia have an implicit learning deficit? J Neurol Neurosurg Psychiatry 76:1392–1397

Walsh D (2007) Development: three grades of ontogenetic involvement. In: Matthen M, Stephens C (eds) Handbook of the philosophy of science, vol 3, Philosophy of Biology. North-Holland, Amsterdam

Wang YQ, Su B (2004) Molecular evolution of microcephalin, a gene determining human brain size. Hum Mol Genet 13:1131–1137

Watkins KE, Paus T, Lerch JP, Zijdenbos A, Collins DL, Neelin P, Taylor J, Worsley KJ, Evans AC (2001) Structural asymmetries in the human brain: a voxel-based statistical analysis of 142 MRI scans. Cereb Cortex 11:868–877

West-Eberhard MJ (2005) Developmental plasticity and the origin of species differences. PNAS 102:6543–6549

Wildgen W (2004) The evolution of human language: scenarios, principles, and cultural dynamics. Benjamins, Amsterdam

Wilkins WK, Wakefield J (1995) Brain evolution and neurolinguistic preconditions. Behav Brain Sci 18(161–182):205–226

Woods CG (2004) Human microcephaly. Curr Opin Neurobiol 14:1–6

Wynn T, Coolidge FL (2004) The expert Neandertal mind. J Hum Evol 46:467–487

Wynn T, Coolidge FL (2007) Did a small but significant change in working memory capacity empower modern thinking? In: Mellars P, Boyle K, Bar-Yosef O, Stringer S (eds) Rethinking the human evolution: new behavioural and biological perspectives on the origin and dispersal of modern humans. Cambridge University McDonald Institute Monographs, Cambridge

Xu X, Lee J, Stern DF (2004) Microcephalin is a DNA damage response protein involved in regulation of CHK1 and BRCA1. J Biol Chem 279:34091–34094

Zayan R, Vauclair J (1998) Categories as paradigms for comparative cognition. Behav Proces 42:87–99

What Can an Extended Synthesis do for Biolinguistics: On the Needs and Benefits of Eco-Evo-Devo Program

Cedric Boeckx

Abstract Recent publications exploring the links between linguistics and biology suggest that in sharp contrast to the overly adaptationist and genocentric framework provided by the modern synthesis and at the heart of evolutionary psychology, the conceptual pluralism made available by an evo-devo-inspired extended synthesis could lead to more productive investigations in the domains of language evolution and development. But such promises have yet to be articulated in detail, and the challenges ahead should also be made explicit. This chapter focuses on a range of historical, conceptual, and empirical issues surrounding language and seeks to address what evo-devo could do for biolinguistics.

Keywords Development • Evolution • Genetics • Linguistics • Neuroscience • Biolinguistics

1 Introduction

Biolinguistics, as used here, refers to a branch of the cognitive biosciences that seek to uncover the biological underpinnings of the human capacity to acquire at least one natural language. As such, and despite its name, it departs sharply from the many subdisciplines of linguistics, which focus on how human languages are put to use in various sociocultural contexts. That such uses require a (possibly complex and multifaceted) biological foundation cannot be seriously put into

C. Boeckx (✉)
ICREA—Catalan Institute for Advanced Studies, Barcelona, Spain
e-mail: cedric.boeckx@ub.edu

C. Boeckx
Department of General Linguistics, Universitat de Barcelona, Barcelona, Spain

M. Pina and N. Gontier (eds.), *The Evolution of Social Communication in Primates*, Interdisciplinary Evolution Research 1, DOI: 10.1007/978-3-319-02669-5_16,

doubt, and biolinguistics takes that fundamental aspect of human biology as its subject matter. Thus understood, biolinguistics asks questions concerning ontogeny, phylogeny, and neural implementation, exploiting results from theoretical linguistics that characterize linguistic knowledge in typical adult populations. In what follows, I will be concerned mainly with issues touching on phylogeny.

It is customary to allude to Theodor Dobzhansky's well-known dictum that "nothing makes sense in biology except in the light of evolution" (Dobzhansky 1973) whenever questions of origin are raised. The exquisite complexity of organisms can only be accounted for, so it seems, by means of natural selection. As Dawkins (1996, 202) puts it, "[w]henever in nature there is a sufficiently powerful illusion of good design for some purpose, natural selection is the only known mechanism that can account for it." Questions of origin pertaining to the mind, the "Citadel itself," as Darwin called it, are no exception. Indeed, the assumption that natural selection is the "universal acid" (Dennett 1995) is perhaps nowhere as strong as in the study of mental faculties, being the motto of evolutionary psychology [witness Pinker (1997)]. But the simplicity of Dobzhansky's assertion conceals layers of necessary refinements that cannot be ignored. Its meaning very much depends on what it means to make sense of life (including mental life) and what we understand by (Darwinian) evolution.

As Fox-Keller has made clear in her book *Making sense of life* (Keller 2002), the notion of explanation, of "making sense of life," cannot be uniformly defined across the life sciences. As for Darwinian evolution, Gould, more than anyone else, has stressed the richness and complexity of evolutionary theory [see Gould (2002)] and stressed the limitations of ultra-Darwinism and its narrowly adaptationist vision.

There are signs that the tide is changing. The promises of genome sequencing, and of the selfish gene, have not been met, and a growing number of biologists side with Lynch's 2007 opinion that "many (and probably most) aspects of genomic biology that superficially appear to have adapative roots...are almost certainly also products of non-adaptive processes." Speaking for all evo-devo adherents, Carroll (2005) points out that the modern synthesis has not given us a theory of form. A theory of form is at the heart of what Kirschner and Gerhart call "Darwin's Dilemma."

When Darwin proposed his theory of evolution, he crucially relied on two ingredients: variation and selection. Although he could explain the selection of variate, he could not explain the origin of this variation. The forms on which selection operated were taken for granted. Since *The Origin of Species,* at repeated intervals, and with accelerated pace in recent years, it has been suggested that several factors giving direction to evolution (facilitating variation, biasing selection, etc.) must be taken into account.

As Gould (2002, 347) clearly states,

> simple descent does not solve all problems of "clumping" in phenotypic space; we still want to know why certain forms "attract" such big clumps of diversity, and why such large empty spaces exist in conceivable, and not obviously malfunctional, regions of potential morphospace. The functionalist and adaptationist perspective

> ties this clumping to available environments, and to shaping by natural selection. Structuralists and formalists wonder if some clumping might not record broader principles, at least partly separate from a simple history of descent with adaptation principles of genetics, of development, or of physical laws transcending biological organization.

In this respect, Gould (2002, 21) calls for a renewed appreciation for "the enormous importance of structural, historical, and developmental constraints in channeling the pathways of evolution, often in highly positive ways", adding that "the pure functionalism of a strictly Darwinian (and externalist) approach to adaptation no longer suffices to explain the channeling of phyletic directions, and the clumping and inhomogeneous population of organic morphospace."

Echoing Gould Pigliucci (2007) writes that biology is in need of a new research program, one that stresses the fact that natural selection may not be the only organizing principle available to explain the complexity of biological systems. Pigliucci reviews numerous works that provide empirical evidence for non-trivial expansions of the modern synthesis, with such concepts as modularity, evolvability, robustness, epigenetic inheritance, and phenotypic plasticity as key components [see Pigliucci and Müller (2010) for a collection of important essays focusing on these concepts].

Once the richness of evolutionary biology is taken into consideration, it seems to me that one can begin to approach the central questions of biolinguistics in a new light. In this chapter, I will focus on two issues: the question of novelty and the question of complexity. I will show that in both cases, traditional conceptual dichotomies are harmful and propose that a pluralistic, integrative, "eco-evo-devo" approach may be far more promising.

2 Novelty

Perhaps no paper symbolizes the return of biolinguistics as well as Hauseret et al. (2002). The paper is famous for introducing the faculty of language in the broad [FLB]/narrow [FLN] sense distinction and notorious for suggesting that the content of FLN is "recursion." In an attempt to reconcile what makes the language faculty unique (human specific) and properly Darwinian descent-with-modification scenarios (a program aimed "at uncovering both shared (homologous or analogous) and unique components of the faculty of language"), Hauser et al. (2002) distinguished between FLB and FLN, with the latter being defined essentially as what makes language human specific. As Fitch et al. (2005) characterize it, "[i]t seems likely that some subset of the mechanisms of FLB is both unique to humans, and to language itself. We dub this subset of mechanisms the faculty of language in the narrow sense (FLN)."

Hauser, Chomsky, and Fitch went on to implicate "recursion" in trying to add content to FLN. I use words like "implicate" and put "recursion" in quotes for, as readers of Hauser et al. (2002) will immediately notice that the exact position defended in the paper is not clear. Consider the following passages, all taken from the same paper:

1. "We hypothesize that FLN only includes recursion and is the only uniquely human component of the faculty of language" (p. 1569, abstract).
2. "We assume …that a key component of FLN is a computational system that generates internal representations and maps them into the sensory-motor interface by the phonological system, and into the conceptual-intentional interface by the (formal) semantic system.… All approaches agree that the core property of FLN is recursion" (p. 1571, column 1).
3. "In fact, we propose in this hypothesis that FLN comprises only the core computational mechanisms of recursion as they appear in narrow syntax and the mappings to the interfaces" (p. 1573, column 2–3).
4. "At minimum, then, FLN includes the capacity of recursion" (p. 1571, column 3).

Unfortunately, this infelicity attracted most of the attention in the subsequent literature. This, in my opinion, was doomed to be unproductive because "the core computational mechanisms of recursion as they appear in narrow syntax and the mappings to the interfaces" alluded to by Hauser, Chomsky, and Fitch were never explicitly stated in that paper (nor in the 2005 sequel), making the claim about "recursion" too vague to be falsified. Because of this focus on "recursion," most of the literature has tried to determine "what is in FLN," but I think that this was the wrong strategy: The emphasis should have been on FLB, for the recognition that a significant amount of the language faculty could be neither specific to language nor unique to humans marked a rather sharp departure from the standard position in the dominant biolinguistic paradigm in its early days. As Chomsky (2007a) has observed, "[t]hroughout the modern history of generative grammar, the problem of determining the character of [the] F[aculty of] L[anguage] has been approached 'from top down': How much must be attributed to U(niversal) G(rammar) to account for language acquisition? The M(inimalist) P(rogram) seeks to approach the problem 'from bottom up': How little can be attributed to UG while still accounting for the variety of I-languages attained?" This significant shift of perspective in linguistics is, in fact, part of a sea change within the cognitive sciences that make what I have called "comparative biolinguistics" possible. This is well captured in the following passage from De Waal and Ferrari (2010):

> Over the last few decades, comparative cognitive research has focused on the pinnacles of mental evolution, asking all-or-nothing questions such as which animals (if any) possess a theory of mind, culture, linguistic abilities, future planning, and so on. Research programs adopting this top-down perspective have often pitted one taxon against another, resulting in sharp dividing lines. Insight into the underlying mechanisms has lagged behind…
>
> A dramatic change in focus now seems to be under way, however, with increased appreciation that the basic building blocks of cognition might be shared across a wide range of species. We argue that this bottom-up perspective, which focuses on the constituent capacities underlying larger cognitive phenomena, is more in line with both neuroscience and evolutionary biology.

Put succinctly, the more we attribute to FLB, the more biologically feasible FL becomes.

Having said this, I should hasten to add that the whole FLB/FLN distinction may not have been the most felicitous way of redirecting attention to the need for a comparative, as opposed to a contrastive, biolinguistics. Although Hauser, Chomsky, and Fitch wrote that "[l]inguists and biologists, along with researchers in the relevant branches of psychology and anthropology, can move beyond unproductive theoretical debate to a more collaborative, empirically focused and comparative research program" by adopting the FLN/FLB distinction ["the FLB/FLN distinction is critical for productive discussion of language evolution," according to Fitch et al. (2005)], I find the very B/N distinction dubious, for reasons that deserve to be elaborated on, because they bear on the potential benefits of adapting an evo-devo approach.

Hauser, Chomsky, and Fitch are certainly right in saying that "[s]omething about the faculty of language must be unique in order to explain the differences between humans and the other animals," but as Jackendoff and Pinker (2005) correctly point out, "the Narrow/Broad dichotomy [...] makes space only for completely novel capacities and for capacities taken intact from nonlinguistic and nonhuman capacities, omitting capacities that may have been substantially modified in the course of human evolution." Indeed, there are capacities that may have been substantially modified in the course of human evolution to be included in FLB or in FLN. Take FOXP2. While highly conserved, the gene has undergone two mutations in our lineage. Say, for the sake of the argument, that we succeed in establishing that these mutations led to specifically (and uniquely) linguistic changes. Would the role of FOXP2 be a component of FLB or of FLN? FLN, it seems to me, but not according to Fitch et al. (2005), who take the highly conserved nature of the gene to automatically make it part of FLB.

At the heart of the FLN/FLB distinction is the attempt on the part of Hauser et al. (2002) to grapple with the problem of innovation and novelty in biology. This problem is currently attracting a lot of attention in the evo-devo literature, after having been badly neglected in the context of the modern synthesis (as Mayr (1960) readily acknowledged). As is well known, while biologists have made great progress over the past century and a half in understanding how existing traits diversify, relatively little progress has been made in understanding how novel traits come into being in the first place. To remedy this explanatory deficit, evo-devo practitioners have first attempted to define what counts as a novelty [here the work of Gerd Müller has been especially illuminating; see especially Müller (2010) and Müller and Newman (2005)]. In this context, it is worth noting the similarity between how (Fitch et al. 2005) define FLN ("that which is specific to language and unique to humans") with the definition put forth in Müller and Wagner (1991): "a structure that is neither homologous to any structure in the ancestral species nor homologous to any other structure of the same organism."

Say that FLN counts as such a novel structure [I think this is the unstated intention of Hauser et al. (2002)]. If one turns to the relevant evo-devo literature, one finds a consensus regarding how such novel structures arise. The consensus position is what Müller (2010) characterizes as "type II" or "emergent" novelty. For Müller, and for many other biologists who worry about such novelties [see Wagner and Müller (2002), Moczek (2008), Moczek and Rose (2009), Prud'homme et al. (2011), who provide several concrete examples of novelties], "phenotypic novelty is largely reorganizational rather than a product of, say, innovative genes (West-Eberhard 2003) [cf. the notion of "deep homology" in Shubin et al. (2009), see also Balari and Lorenzo (2013); Fitch (2011a), Scharff and Petri (2011)]. In other words, novelty arises from the combination of generic mechanisms, whose collective effects give rise to what appears to be de novo characters (phase transitions).

Interestingly, the possibility of emergent novelty is alluded to in Fitch et al. (2005, 182): "Something about the faculty of language must be unique in order to explain the differences between humans and the other animals if only the particular combination of mechanisms in FLB." But they fail to note that this puts in jeopardy the very FLN/FLB distinction [with FLN taken to be a subset of FLB, as made clear in Hauser et al. (2002)]. Incidentally, the possibility of emergent novelty is present in some of Chomsky's own works Chomsky (1980, 2000), but is always put in doubt. ("Now a question that could be asked is whether whatever is innate about language is specific to the language faculty or whether it is just some combination of the other aspects of the mind. That is an empirical question and there is no reason to be dogmatic about it; you look and you see. What we seem to find is that it is specific.") This is unfortunate, for Chomsky's stance (which boils down to the FLN/FLB distinction) indeed "mak[e] some hypotheses—in our view the most plausible ones—impossible to state" (Jackendoff and Pinker 2005). Not surprisingly, one finds the following statement in Fitch's recent writings (in direct contradiction from the affirmation in Fitch et al. (2005) and already quoted above that the FLN/FLB distinction "critical for productive discussion of language evolution"):

> What all of these examples make clear is that the distinction between general and linguistically specialized mechanisms is hard to draw, even in those cases where the mechanisms themselves seem fairly clearly defined. Most areas of language are not, and will not soon be, so clearly defined, and thus the distinction itself is of little use in furthering our understanding of the mechanisms (Fitch 2011b, 384).

On the basis of this, I am led to conclude, with Bloomfield et al. (2011), that "[p]erhaps this is a good time to reconsider whether attempting to distinguish between qualitative and quantitative differences is helpful if the quantitative advantage is vast." It is indeed puzzling that so many researchers still cling to the FLN/FLB distinction when the rationale behind the distinction given at the outset of Hauser et al. (2002) is to reject the notion of the faculty of language as a monolithic object: How can we identify whether some mechanism is "specific to language" if "language" itself is not a well-defined, unique object. Perhaps it is for this reason that the phrase "unique/specific to language" does not appear in the original 2002 paper (Joana Rossello, personal communication), but only appears as such in Fitch et al. (2005), who take it from Pinker and Jackendoff (2005), where FLN is characterized in those terms for the first time [contrary to the literature that routinely attributes it to Hauser et al. (2002)]. The evo-devo stance on novelty seems to me to demand the abandonment of distinctions like FLN/FLB, where what is unique all too often takes center stage, over and above the logic of descent that Darwin was right to emphasize.

3 Complexity

The most promising aspect of Hauser et al. (2002) is the idea that much of what linguists have attributed to the language faculty falls in the broad characterization of it; that is, many properties of the human language faculty are neither specific to language nor specific to humans. This conclusion converges with the "minimalist" project that many formal linguists embarked on the 1990s [see Chomsky (1995); Boeckx (2006, 2010, 2011b), among many others]. Linguistic minimalism is an attempt to minimize the role of system-specific assumptions to account for properties of natural language grammars.

Over the years, it has become clear that the success of this minimalist enterprise depends on several factors that mesh well with biolinguistic concerns. First, it must rely on—and therefore assume the existence of—a rich cognitive apparatus with which the (minimal) specifically linguistic apparatus interfaces to yield the substantive universals that previous linguistic research had (somewhat blindly) attributed to a highly structured and specifically linguistic "Universal Grammar." (In other words, the success of the minimalist project depends on the success of what one may call "comparative biolinguistics.")

The rich cognitive apparatus in question covers both systems of interpretation ("meaning") and systems of externalization ("sound/sign"). This has effectively led to a model of the language faculty where many semantic and phonological aspects of language find their roots in capacities independently attested in other species. For phonology, I recommend Yip (2006), Samuels (2011), and Samuels et al. (in press); for semantics, Hurford (2007) is outstanding, but the following are also excellent sources of information: Burge (2010), Carey (2009), Carruthers (2006), Hauser (2001), Cheney and Seyfarth (1990, 2007), and Jackendoff (1990, 2010).

This is not to say, of course, that there is nothing specific about human language semantics or phonology. Rather, the emerging consensus is that specificity arises when ancient mechanisms are placed in a new context (see previous section). In particular, the dominant view in biolinguistics ought to be that the emergence of a capacity to form potentially unbounded hierarchical structures of a particular sort, characteristic of human language (technically known as "headed" or "endocentric" phrases, i.e., groupings organized around a prominent element, such as a verb in a verb phrase), significantly reshaped the semantic and phonological components inherited by descent. Such a possibility is defended in Samuels (2011), Boeckx and Samuels (2009) for phonology and in Uriagereka (2008), Hinzen (2007, 2011), Pietroski (2011), Boeckx (2009b), and Fujita (2013) for semantics.

It is also becoming extremely clear that the success of the biolinguistic program will depend on recognizing that the emergence of many grammatical properties of natural languages is the product of social transmission and cultural evolution. This effectively means that the success of this "Chomskyan" enterprise depends on the correctness of approaches that have (erroneously, in my opinion) traditionally been put in opposition with "Chomskyan" linguistics [e.g., Kirby (2001), Kirby and Hurford (2002), Kirby et al. (2008)].

Deacon (2006) usefully distinguishes between various notions or levels of emergence (and attendant complexity), and we think that they could be useful in the context of biolinguistics. Specifically, Deacon's notion of second- and third-order emergence has worth distinguishing. Deacon argues that many thermodynamic effects correspond to first-order emergent relationships. These arise when relational properties of systems amplify intrinsic material properties, eventually resulting in a reduction in complexity. Deacon's second level involves the self-organization of systems; what he calls "autopoietic" sets. Self-organization gives rise to what one might call spontaneous complexity. Deacon's third level encloses the additional factor of "recursive causality" of self-organized systems, arising from interaction among agents. As he notes, this type of emergence inevitably entails an evolutionary, historical character. Call this cumulative complexity. Traditionally, this latter type of complexity has not been given much attention in the Chomskyan literature: Social phenomena are often relegated to "E(xternal)-language," a notion distinct from what Chomskyan linguists focus on (I-language).

As discussed in Boeckx et al. (2013) on the basis of examples such as emerging sign languages, recognizing the influence of environmental factors on the range of properties grammatical systems manifest may be of interest in the context of complexity issues (specifically, Deacon's levels 2 vs. 3) and may enrich the biolinguistic enterprise. After all, in biology, it is standardly recognized that there is a mutual relation between what counts the genetic makeup of an organism and the environmental influences it undergoes. Genes determine the capacities of organisms, yet the limits of these capacities are affected by the environment and eventually may never be explored, depending on how adequate the environmental factor proves to be. In fax, approaching certain properties such as development of complex (grammatical) markers as environmentally driven adaptations of an innate capacity is nothing more than extending to human language what biologists do for the systems of communication of other animals (see, e.g., Okanoya (2012) on song complexity in domesticated song birds vs. those in the wild).

This perspective is typically resisted in "Chomskyan" circles, but it is, I think, inescapable. (If I am right, this illustrates how biolinguistics forces one to transcend the traditional "ideological" boundaries that populate linguistic circles.) As Hall (2012) correctly observes, "pursu[ing] a biolinguistic approach to … language,… [i]n one sense means that the field of inquiry becomes broader." A successful biolinguist must know enough about the cognitive systems of other species and about the properties of non-linguistic cognitive domains in humans to be able to make reasonable inferences about what each of them contributes toward the shape of the modern language faculty. But "[i]n another sense, the central object of study becomes much smaller" for Chomskyan linguistics, for many of the grammatical details that were often attributed to some rich innate component specifically dedicated to language ("Universal Grammar") are to be understand in terms of cultural evolution. The empirical and explanatory success of approaches like Blevins (2004) reinforces this conclusion, and I anticipate that this trend will become more prominent in the field in years to come, especially in the context of cross-linguistic variation (Newmeyer 2005; Boeckx In press).

In a certain sense, the off-loading of grammatical properties onto external factors converges with the claim advanced by Deacon (2010) that aspects of language should be understood not as the result of selection pressures (that often lead to the internalization or genetic assimilation of certain traits), but in terms of relaxed selection. The net effect of relaxed selection is for speakers to have to rely on social transmission to reliably develop complex grammatical systems and to possess a plastic, underspecified (i.e., minimally specified) "innate" initial state of the language faculty ("Universal Grammar").

As I have discussed at length elsewhere Boeckx (2012), the picture of the language faculty that emerges is one where the aspects of language geared toward externalization are much more complex and varied than those aspects dedicated to meaning. For the latter, the minimally specified structural properties of the language (what linguists often call "narrow syntax") appear to be sufficient to exploit the rich conceptual resources inherited by descent to yield the range of rich interpretations that characterize human thought. Put succinctly, the contribution of the internal structural component of the language faculty is asymmetric; it appears to be better designed for meaning than for articulation [on this point, see also Berwick et al. (2013)].

As should be obvious, if something like what I have touched on in this section is on the right track, the modern human language faculty is a mosaic, with old, inherited parts put together in a new context (which boils down to the presence of a mechanism for generating unbounded hierarchical expressions) and further modified (complexified) by external influences (the external context of cultural, imperfect transmission) [on the non-uniform character of the language faculty, see also Heinz and Idsardi (2013)]. It turns out, then, that debates about saltation versus gradualism surrounding the language faculty as a whole, though extremely common [see Pinker and Bloom (1990), Jackendoff (2002)], are misplaced and misleading [see also Clark (2013)], for the modern language faculty is not a uniform, monolithic object: Some parts have fairly direct homologs, while other parts are likely the result of abrupt phase transitions.

4 Evolutionary Pluralism

In its 60 years of history, the biolinguistic program has developed under the influence or inspiration of several biological paradigms. For instance, the role of ethology in the early days of biolinguistics has been well documented Chomsky (1972), Boeckx (2009a). Later on, in the course of the first biolinguistic meetings of the 1970s, Chomsky familiarized himself with the work of Francois Jacob and Jacques Monod on genetic regulation Jacob and Monod (1961), which provided the conceptual model for his parametric approach to cross-linguistic variation [as acknowledged in Chomsky (1981, 2007b, 2009)]. More recently, several biolinguists have acknowledged the role of recent developments in evolutionary developmental biology ("evo-devo") in their writings (Balari and Lorenzo 2013; Chomsky 2010; Hauser 2009).

The rise of evo-devo is, indeed, an important factor in the characterization of the revival of biolinguistics, for the field of evo-devo offers a morphospace of conceptual options to understand the origin, evolution, and development of the language faculty that is at once broader than and more congenial to long-standing themes in (bio)linguistics, than the neo-darwinian modern synthesis (see Boeckx (2011a) on this point; remember also the discussion above concerning evolutionary novelties, and also concerning the relevance of fossil, highly conserved genes). With its emphasis on gradualism and selectionism, the latter was the perfect incubator for evolutionary psychology [see, e.g., Pinker and Bloom (1990)]. But the difficulties of strict or exclusively selectionist approaches to language evolution (Piattelli-Palmarini 1989; Uriagereka 1998) made it look like linguistic investigations were out of tune with mainstream biology.

Not surprisingly, when a more extended synthesis is pursued in biology [see Pigliucci and Müller (2010)], it becomes much easier to relate biolinguistic proposals to the biology literature. As a matter of fact, several authors have pointed out that Lenneberg's research program was very much evo-devo *avant la lettre* (Boeckx and Longa 2011; Balari 2012) indeed, Lenneberg's writings display an awareness of the need to adopt a pluralist attitude toward the complex nature of the language faculty that is more in line with evo-devo calls in favor of an extended synthesis in biology than recent biolinguists' appeal to evo-devo, since these retain a more limited, molecular understanding of evo-devo [Benitez-Burraco and Longa (2010) on this point; see also Müller (2008) on why molecular evo-devo, of the sort popularized by Carroll (2005), is not the most distinctive feature of the evo-devo paradigm].

This is all to say that a complete assimilation of everything that evo-devo has to offer to biolinguistics is still very much a task for the future, but one that promises a lot, not only in the context of language evolution studies [and the evo-devo emphasis of the centrality of the problem of phenotypic innovations, already discussed in the previous subsection in the context of Hauser et al. (2002)], but also in the context of language development (Longa and Lorenzo 2008, 2012; Lorenzo and Longa 2009).

Acknowledgments The present piece is an overview of work in progress, much of it is carried out in collaboration with Evelina Leivada, Pedro Tiago Martins, Lluis Barceló i Coblijn, and Antonío Benitez-Burraco. I thank all of them for enriching my understanding of biolinguistics. Materials from Boeckx (2011a, 2013), Boeckx et al. (2013) have been reworked into the present piece. The present work was made possible through a Marie Curie International Reintegration Grant from the European Union (PIRG-GA-2009-256413), research funds from the Fundació Bosch i Gimpera, and a grant from the Spanish Ministry of Economy and Competitiveness (FFI-2010-20634).

References

Balari S (2012) Up the cudgels for lenneberg. In: Boeckx C, Horno M, Mendivĺl Giró JL (eds) Language from a biological point of view: current issues in Biolinguistics. Cambridge Scholars Publishing, Cambridge, pp 208–214

Balari S, Lorenzo G (2013) Computational phenotypes: towards an evolutionary developmental biolinguistics. Oxford University Press, Oxford

Benitez-Burraco A, Longa V (2010) Evo-devo—of course, but which one? Biolinguistics 4:308–323

Berwick RC, Friederici AD, Chomsky N, Bolhuis JJ (2013) Evolution, brain, and the nature of language. Trends in Cogn Sci 17:89–98

Blevins J (2004) Evolutionary phonology: the emergence of sound patterns. Cambridge University Press, Cambridge

Bloomfield TC, Gentner TQ, Margoliash D (2011) What birds have to say about language. Nat Neurosci 14:947

Boeckx C (2006) Linguistic minimalism: origins, concepts, methods, and aims. Oxford University Press, Oxford

Boeckx C (2009a) Language in cognition: uncovering mental structures and the rules behind them. Wiley-Blackwell, Malden

Boeckx C (2009b) Some notes on the syntax-thought interface. In: Proceedings of the Sophia University Linguistic Society vol 24. Sophia University Linguistic Society, pp 92–103

Boeckx C (2010) Linguistic minimalism. In: Heine B, Narrog H (eds) Oxford handbook of linguistic analysis. Oxford University Press, Oxford, pp 485–505

Boeckx C (2011a) Some reflections on Darwin's Problem in the context of Cartesian Biolinguistics. In: Di Sciullo A-M, Boeckx C (eds) The biolinguistic enterprise: new perspectives on the evolution and nature of the human language faculty. Oxford University Press, Oxford, pp 42–64

Boeckx C (ed) (2011b) The oxford handbook of linguistic minimalism. Oxford University Press, Oxford

Boeckx C (2012) The I-language mosaic. In: Boeckx C, Horno M, Mendivil J-L (eds) Language from a biological point of view: current issues in biolinguistics. Cambridge Scholars Publishing, Newcastle-upon-Tyne, pp 23–51

Boeckx C (2013) Biolinguistics: forays into human cognitive biology. J Anthropol Sci 91:63–89

Boeckx C (In press) What principles & parameters got wrong. In: Picallo C (ed) Linguistic variation and the minimalist program. Oxford University Press, Oxford

Boeckx C (To appear) Elementary syntactic structures. Cambridge University Press, Cambridge

Boeckx C, Leivada E, Martins PT (2013) Language and complexity considerations: a biolinguistics perspective. Llengua, Societat i Comunicacio 11:20–26

Boeckx C, Longa VM (2011) Lenneberg's views on language development and evolution and their relevance for modern biolinguistics. Biolinguistics 5:254–273

Boeckx C, Samuels B (2009) What emerges from merge in phonology. Presented at the 6th old world conference on phonology, Edinburgh

Burge T (2010) Origins of objectivity. Oxford University Press, Oxford

Carey S (2009) The origin of concepts. Oxford University Press, Oxford

Carroll SB (2005) Endless forms most beautiful: the new science of evo-devo and the making of the animal kingdom. Norton, New York

Carruthers P (2006) The architecture of the mind. Oxford University Press, Oxford

Cheney DL, Seyfarth RM (1990) How monkeys see the world. University of Chicago Press, Chicago

Cheney DL, Seyfarth RM (2007) Baboon metaphysics: the evolution of a social mind. University of Chicago Press, Chicago

Chomsky N (1972) Language and mind. Harcourt Brace Jovanovich, New York

Chomsky N (1980) Rules and representations. Columbia University Press, New York

Chomsky N (1981) Lectures on government and binding. Foris, Dordrecht

Chomsky N (1995) The minimalist program. MIT Press, Cambridge

Chomsky N (2000) New horizons in the study of language and mind. Cambridge University Press, Cambridge

Chomsky N (2007a) Approaching UG from below. In Interfaces+recursion=language? Chomsky's minimalism and the view from semantics, ed. U. Sauerland and H.-M. Gartner, 1–30. Mouton de Gruyter

Chomsky N (2007b) Of minds and language. Biolinguistics 1:009–027

Chomsky N (2009) Opening remarks and conclusion. In: Piattelli-Palmarini M, Salaburu P, Uriagereka J (eds) Of minds and language: a Basque encounter with Noam Chomsky. Oxford University Press, Oxford
Chomsky N (2010) Some simple evo devo theses: how true might they be for language. The evolution of human language: biolinguistic perspectives, p 62
Clark B (2013) Syntactic theory and the evolution of syntax. Biolinguistics 7:169–197
Dawkins R (1996) Climbing mount improbable. Norton, New York
De Waal F, Ferrari PF (2010) Towards a bottom-up perspective on animal and human cognition. Trends Cogn Sci 14:201–207
Deacon TW (2006) Emergence: the hole at the wheel's hub. In: Clayton P, Davies P (eds) The re-emergence of emergence: the emergentist hypothesis from science to religion. Oxford University Press, Oxford, pp 111–150
Deacon TW (2010) A role for relaxed selection in the evolution of the language capacity. Proc Natl Acad Sci 107:9000–9006
Dennett D (1995) Darwin's dangerous idea: evolution and the meanings of life. Simon & Schuster, New York
Dobzhansky T (1973) Nothing in biology makes sense except in the light of evolution. Am Biol Teach 35:125–129
Fitch WT, Hauser MD, Chomsky N (2005) The evolution of the language faculty: clarifications and implications. Cognition 97:179–210
Fitch WT (2011a) 'Deep homology' in the biology and evolution of language. In: Di Sciullo AM, Boeckx C (ed) The biolinguistic enterprise: new perspectives on the evolution and nature of the human language faculty. Oxford University Press, Oxford, pp 135–166
Fitch WT (2011b) Unity and diversity in human language. Philos Trans R Soc B Biol Sci 366:376–388
Fujita K (2013) Review of anna maria di sciullo and cedric boeckx eds. the biolinguistic enterprise: new perspectives on the evolution and nature of the human language faculty. Studies in English Literature 54:175–184
Gould SJ (2002) The structure of evolutionary theory. Harvard University Press, Cambridge
Hall DC (2012) Bridget d. samuels, phonological architecture: a biolinguistic perspective (oxford studies in biolinguistics 2). J Linguist 48:736–741 (Oxford: Oxford university press, 2011, pp. xii+ 252)
Hauser MD (2001) Wild minds: what animals really Think. Owl Books, New York
Hauser MD (2009) Evolingo: the nature of the language faculty. In: Piattelli-Palmarini M, Salaburu P, Uriagereka J (eds) Of minds and language: a Basque encounter with Noam Chomsky. Oxford University Press, Oxford, pp 74–84
Hauser MD, Chomsky N, Fitch WT (2002) The faculty of language: what is it, who has it, and how did it evolve? Science 298:1569–1579
Heinz J, Idsardi W (2013) What complexity differences reveal about domains in language*. Top Cogn Sci 5:111–131
Hinzen W (2007) An essay on naming and truth. Oxford University Press, Oxford
Hinzen W (2011) Language and thought. In: Boeckx C (ed) The oxford handbook of linguistic minimalism, Chapter. Oxford University Press, Oxford, pp 499–522
Hurford JR (2007) The origins of meaning. Oxford University Press, Oxford
Jackendoff R (1990) Semantic structures. MIT Press, Cambridge
Jackendoff R (2002) Foundations of language. Oxford University Press, New York, Oxford
Jackendoff R (2010) Meaning and the Lexicon: the parallel architecture 1975–2010. Oxford University Press, Oxford
Jackendoff R, Pinker S (2005) The nature of the language faculty and its implications for evolution of language (reply to Fitch, Hauser, and Chomsky). Cognition 97:211–225
Jacob F, Monod J (1961) On the regulation of gene activity. In: Cold spring harbor symposia on quantitative biology, vol 26. Cold Spring Harbor Laboratory Press, pp 193–211
Keller EF (2002) Making sense of life: explaining biological development with models, metaphors, and machines. Harvard University Press, Cambridge

Kirby S (2001) Spontaneous evolution of linguistic structure-an iterated learning model of the emergence of regularity and irregularity. IEEE Trans Evol Comput 5:102–110

Kirby S, Cornish H, Smith K (2008) Cumulative cultural evolution in the laboratory: an experimental approach to the origins of structure in human language. Proc Natl Acad Sci 105:10681–10686

Kirby S, Hurford J (2002) The emergence of linguistic structure: an overview of the iterated learning model. Simulating the evolution of language, Springer, London, pp 121–148

Longa VM, Lorenzo G (2008) What about a (really) minimalist theory of language acquisition? Linguistics 46:541–570

Longa VM, Lorenzo G (2012) Theoretical linguistics meets development: explaining FL from an epigeniticist point of view. In: Boeckx C, Horno M, Mendivíl Giró JL (eds) Language from a biological point of view: current issues in Biolinguistics. Cambridge Scholars Publishing, pp 52–84

Lorenzo G, Longa VM (2009) Beyond generative geneticism: rethinking language acquisition from a developmentalist point of view. Lingua 119:1300–1315

Lynch M (2007) The origins of genome architecture. Sinauer Associates, Sunderland

Mayr E (1960) The emergence of evolutionary novelties. Evol Darwin 1:349–380

Moczek AP (2008) On the origins of novelty in development and evolution. BioEssays 30:432–447

Moczek AP, Rose DJ (2009) Differential recruitment of limb patterning genes during development and diversification of beetle horns. Proc Natl Acad Sci 106:8992–8997

Müller GB (2008) EvoDevo as a discipline. In: Minelli A, Fusco G (eds) Evolving Pathways: Key themes in evolutionary developmental biology. Cambridge University Press, Cambridge, pp 3–29

Müller GB (2010) Epigenetic innovation. Evolution-the extended synthesis. MIT Press, Cambridge, pp 307–333

Müller GB, Newman SA (2005) The innovation triad: an Evodevo agenda. J Exp Zool Part B: Mol Dev Evol 304:487–503

Müller GB, Wagner GP (1991) Novelty in evolution: restructuring the concept. Annu Rev Ecol Syst 22:229–256

Newmeyer FJ (2005) Possible and probable languages: a generative perspective on linguistic typology. Oxford University Press, Oxford

Okanoya K (2012) Behavioural factors governing song complexity in bengalese finches. Int J Comp Psychol 25:44–59

Piattelli-Palmarini M (1989) Evolution, selection and cognition: from 'learning' to parameter setting in biology and in the study of language. Cognition 31:1–44

Pietroski PM (2011) Minimal semantic instructions. In: Boeckx C (ed) Oxford handbook of linguistic minimalism. Oxford University Press, Oxford, pp 472–498

Pietroski PM (To appear) Conjoining meanings: semantics without truth values. Oxford University Press, Oxford

Pigliucci M (2007) Do we need an extended evolutionary synthesis? Int J Org Evol 61:2743–2749

Pigliucci M, Müller G (eds) (2010) Evolution—The extended synthesis. MIT Press, Cambridge

Pinker S (1997) How the mind works. Norton, New York

Pinker S, Bloom P (1990) Natural selection and natural language. Behav Brain Sci 13:707–784

Pinker S, Jackendoff R (2005) The faculty of language: what's special about it? Cognition 95:201–236

Prud'homme B, Minervino C, Hocine M, Cande JD, Aouane A, Dufour HD, Kassner VA, Gompel N (2011) Body plan innovation in treehoppers through the evolution of an extra wing-like appendage. Nature 473:83–86

Samuels B (2011) Phonological architecture: a biolinguistic perspective. Oxford University Press, Oxford

Samuels B, Hauser MD, Boeckx C (In press) Do animals have universal grammar? A case study in phonology. In: Roberts IG (eds) The oxford handbook of universal grammar. Oxford University Press, Oxford

Scharff C, Petri J (2011) Evo-devo, deep homology and foxp2: implications for the evolution of speech and language. Philos Trans R Soc B Biol Sci 366:2124–2140

Shubin N, Tabin C, Carroll S (2009) Deep homology and the origins of evolutionary novelty. Nature 457:818–823

Uriagereka J (1998) Rhyme and reason: an introduction to minimalist syntax. MIT Press, Cambridge

Uriagereka J (2008) Syntactic anchors: on semantic structuring. Cambridge University Press, Cambridge

Wagner GP, Müller GB (2002) Evolutionary innovations overcome ancestral constraints: a re-examination of character evolution in male sepsid flies. Evol Dev 4:1–6

West-Eberhard MJ (2003) Developmental plasticity and evolution. Oxford University Press, Oxford

Yip MJ (2006) The search for phonology in other species. Trends Cogn Sci 10:442–446

Printed by Publishers' Graphics LLC
CAMZ140526.23.36.143